ASTRONOMY AND ASTROPHYSICS LIBRARY

Series Editors: G. Börner, Garching, Germany
A. Burkert, München, Germany
W. B. Burton, Charlottesville, VA, USA and
   Leiden, The Netherlands
A. Coustenis, Meudon, France
M. A. Dopita, Canberra, Australia
B. Leibundgut, Garching, Germany
A. Maeder, Sauverny, Switzerland
P. Schneider, Bonn, Germany
V. Trimble, College Park, MD, and Irvine, CA, USA

For further volumes:
http://www.springer.com/series/848

Peter H. Bodenheimer

Principles of Star Formation

Peter H. Bodenheimer
University of California, Santa Cruz
UCO/Lick Observatory
Santa Cruz, CA 95064
USA
peter@ucolick.org

ISSN 0941-7834
ISBN 978-3-642-15062-3   e-ISBN 978-3-642-15063-0
DOI 10.1007/978-3-642-15063-0
Springer Heidelberg Dordrecht London New York

Library of Congress Control Number: 2011933131

© Springer-Verlag Berlin Heidelberg 2011

This work is subject to copyright. All rights are reserved, whether the whole or part of the material is concerned, specifically the rights of translation, reprinting, reuse of illustrations, recitation, broadcasting, reproduction on microfilm or in any other way, and storage in data banks. Duplication of this publication or parts thereof is permitted only under the provisions of the German Copyright Law of September 9, 1965, in its current version, and permission for use must always be obtained from Springer. Violations are liable to prosecution under the German Copyright Law.

The use of general descriptive names, registered names, trademarks, etc. in this publication does not imply, even in the absence of a specific statement, that such names are exempt from the relevant protective laws and regulations and therefore free for general use.

Cover design: eStudio Calamar S.L.

Cover figure: "Orion's Dreamy Stars". Spitzer Space Telescope image of the star forming region Orion Nebula (M42). A colony of hot, young stars is stirring up the cosmic scene. This image is a combination of data from Spitzer and the Two Micron All Sky Survey (2MASS).

Credit: NASA/JPL-Caltech/J. Stauffer (SSC/Caltech).

Printed on acid-free paper

Springer is part of Springer Science+Business Media (www.springer.com)

Preface

In its simplest terms, star formation is the combination of processes that, in the context of very low-density interstellar gas, brings a certain amount of material to the point where the force of gravity exceeds all other opposing forces, resulting in collapse. On the theoretical side, one of the early significant accomplishments in the development of this concept was the determination, by Sir James Jeans in 1928, of what is now called the "Jeans length", one of the important conditions needed for the collapse in the interstellar medium to begin. Observationally, an early convincing piece of evidence in favor of the above hypothesis was the discovery, for example, by Walter Baade, that hot massive stars, that must be young, were associated with dust clouds and relatively dense interstellar gas in the spiral arms of galaxies outside the Milky Way, such as M31. In the last few decades, progress in understanding the details of this basic picture has been extraordinarily rapid. Observationally, the development of large ground-based telescopes and space observatories, and especially the opening up of the infrared and millimeter regions of the spectrum, have revealed the intricate structures of star-forming regions. Theoretical advances, based in part on the availability of high-powered computers, have begun to link together the various important physical and chemical processes that must be involved in star formation. Nevertheless, many basic questions, such as why the stars have the range in mass that is observed, and what decides whether a star is going to end up as a single object or a binary, are just beginning to be understood.

An impressive range of time scales and length scales are involved in the star formation process. Star formation began in the early Universe, only a few hundred million years after the Big Bang. Considerable research effort has been devoted to the chronicling of the history of the star formation rate throughout the entire history of the Universe. Star formation can occur at a relatively modest rate, such as in isolated regions of molecular clouds in our Galaxy, or at a moderate rate, such as in local regions of star cluster formation, or at quite a rapid rate, as in starburst galaxies. The length scales involved in star formation are measured in parsecs, the typical size of molecular clouds, while the sizes of the end products are measured in the range 0.1–10 solar radii, up to 9 orders of magnitude smaller. As a result a wide variety of physical processes must be considered. Clearly radiative transfer is of paramount importance, as the radiation from the star-forming regions forms the main link between the actual physical events and the observer. Hydrodynamical flows, including turbulent flows, and orbital mechanics are also key elements of the physical picture, but they must be supplemented by additional effects, including chemistry, atomic physics, nuclear physics, magnetic fields, and the properties of matter, for example, the equation of state of a gas. Thus the study of star formation brings together a rich diversity of physical processes, with the result that a rather complex theoretical model must be generated to interpret the observational data, which themselves are continuously increasing in resolution and detail.

The main theoretical methods applied in astrophysics range from the very approximate "back of the envelope" calculations, used to derive order-of-magnitude estimates from the fundamental laws of physics, to rigorous analytic theory, based in part on linear perturbation theory, to large-scale numerical simulations, based on approximate representation of the differential equations involved in the rigorous theory. There is usually an interplay between all three of these theoretical methods, and in this book the reader will find examples of each. However, the more recent theoretical results rely heavily on the numerical approach, as the complexity of the processes involved makes the purely analytical approach intractable. Analytical results play an important role as test cases against which to check the numerical codes, and the order-of-magnitude estimates serve as a "reality check" as to the physical reasonableness of the results. Even so, the reader is advised to maintain a critical attitude toward the details of particular numerical simulations, no matter how flashingly displayed. Many simulations involve magnetohydrodynamics in three space dimensions coupled with radiation transfer, a system of equations that can tax the computing power of even the most sophisticated modern machines. Is the numerical resolution sufficient to represent the important physical effects? Are the approximations, needed to weed out insignificant effects, justifiable? Has the simulation been run for a long enough time to provide a significant comparison with observations? Are there hidden numerical effects that significantly degrade the solution? Only by continual detailed testing of the numerical results can progress be assured.

As an introduction to the science of star formation, this book concentrates on the interpretation of observations in star-forming regions relatively nearby, in our own Galaxy, at the present time, where the most detailed and accurate results can be obtained. An excursion is made into the realm of the early Universe and the formation of the first stars, although no observations are as yet available to validate the theory. The range of observed phenomena includes turbulent clouds, magnetic fields, star-forming cores in molecular clouds, infrared protostellar sources, outflows and jets, Hertzsprung–Russell diagrams of young stellar clusters, disks around young stars, and multiple stellar systems. Observations over wavelengths ranging from the X-rays, at 1 nanometer or less, to the radio, at 1 centimeter or greater, are relevant. While the details of the observational and theoretical results are likely to change in the future as this developing subject evolves, many of the basic concepts should still remain appropriate. The on-going nature of research in this area is emphasized in the final chapter, which makes clear that there are numerous fundamental questions still to be resolved.

As an introductory text, this book does not, for the most part, go into "first-principles" derivations of the physical equations. Many other reliable sources are available to provide this information. The treatment may be regarded as somewhat simplified, in order to provide a general view of the subject rather than a rigorous discussion of the many important details. As such, the text should be useful for beginning-level graduate courses in astrophysics, as well as for the more advanced undergraduate courses for students who have had a few years of physics courses as well as an introduction to the basic concepts of astronomy. The student wishing to go into more depth on a specific topic is advised to first consult the *Annual Review of Astronomy and Astrophysics*, whose articles are written by experts in the field. And finally, the author acknowledges the many contributions to the substance of this book that were made by the graduate students in Astronomy and Astrophysics at the University of California, Santa Cruz, who took the course in Star Formation, the lecture notes for which formed the basis for this text.

Santa Cruz   *Peter Bodenheimer*
May 2011

Contents

1 **Overview** 1
   1.1   Basic Questions: Star Formation 3
   1.2   Observations of Objects in Star-Forming Regions 13
   1.3   Star Formation Phases 24
   1.4   Appendix to Chap. 1: Derivation of the Free-Fall Time 34

2 **Molecular Clouds and the Onset of Star Formation** 37
   2.1   Molecular Cloud Properties 37
   2.2   Initial Conditions for Star Formation 47
   2.3   Heating and Cooling 57
   2.4   Magnetic Braking 59
   2.5   Degree of Ionization 63
   2.6   Magnetic Diffusion 64
   2.7   How is Star Formation Initiated? 67
   2.8   Turbulence and Star Formation 68
   2.9   Induced Star Formation 77
   2.10 Summary 86
   2.11 Appendix to Chap. 2: Note on Numerical Methods 87
   2.12 Problems 89

3 **Protostar Collapse** 93
   3.1   Protostellar Initial Conditions 94
   3.2   Isothermal Collapse 99
   3.3   Adiabatic Collapse 102
   3.4   Accretion Phase 107
   3.5   Comparison with Observations 111
   3.6   Summary 122
   3.7   Problems 123

4	**Rotating Protostars and Accretion Disks**	127
	4.1 Disk Formation	128
	4.2 Observations of Disks	143
	4.3 Basic Theory of Disk Evolution	150
	4.4 General Results of Disk Evolution	160
	4.5 Disk Dispersal	167
	4.6 Winds and Outflows	170
	4.7 Summary of Disk Evolution	179
	4.8 Problems	179
5	**Massive Star Formation**	183
	5.1 Information from Observations	184
	5.1.1 The Present Massive Star Population	184
	5.1.2 Formation Sites	187
	5.2 The Problem of Radiation Pressure	193
	5.3 Full 3D with Radiation Transfer	200
	5.4 Massive Star Formation: Competitive Accretion or Monolithic Collapse?	207
	5.5 Summary	210
	5.6 Appendix to Chap. 5: Determination of the IMF	211
	5.7 Problems	219
6	**Formation of Binary Systems**	221
	6.1 Observational Data	222
	6.2 Basic Formation Mechanisms	230
	6.3 Capture	231
	6.4 Fission	232
	6.5 Fragmentation	233
	6.6 Summary	248
	6.7 Problems	249
7	**The Formation of the First Stars**	251
	7.1 Physics of the First Stars	252
	7.2 Sequence of Events During Formation	255
	7.2.1 Cosmological Phase	255
	7.2.2 Protostellar Collapse	256
	7.2.3 Accretion Phase	260
	7.2.4 Feedback Phase	264
	7.3 Dark Matter Annihilation in the First Stars	268
	7.4 Summary	275
	7.5 Problems	278

8	**Pre-Main-Sequence Evolution**		279
	8.1 Physical Relations		280
		8.1.1 Basic Equations of Structure and Evolution	280
		8.1.2 Equation of State	283
		8.1.3 Opacity	285
	8.2 Pre-Main-Sequence Evolutionary Tracks		290
	8.3 Comparison with Observations		295
	8.4 Summary		307
	8.5 Problems		308
9	**Summary: Issues in Galactic Star Formation**		311
	9.1 Molecular Clouds		311
	9.2 The Initial Mass Function		313
	9.3 Protostar Collapse		314
	9.4 Binary and Multiple Systems		316
	9.5 Star Formation in the Early Universe		317
	9.6 Massive Stars		319
	9.7 Young Stars and Disks		320
	9.8 Rate and Efficiency		322
References			325
Index			339

Chapter 1
Overview

The process of star formation occupies a critical position in astrophysics, because understanding of it is required for progress to be made on other fundamental problems, including stellar evolution, the energetics of the interstellar medium (ISM), galactic evolution, and the formation of planetary systems. However, in contrast to the theory of most phases of stellar evolution, the theory of star formation is not definitive, owing partly to the complexity of the physical processes involved, and partly to the fact that stars form in dusty regions of the Galaxy where they are obscured in the optical region of the spectrum. In recent years, however, observational progress has been rapid as a result of the opening up of substantial capabilities in the infrared and millimeter wavelengths, where the obscuration is much reduced. An extensive survey of the observed properties of star-forming regions within 2 kiloparsecs of the Sun has been published [430].

An example of the intersection of star formation with stellar evolution and with processes in the interstellar medium is given in Fig. 1.1. Although it just represents a hypothetical example, it illustrates the fact that cyclic processes do occur. The sequence of events starts with the formation of an association of massive stars in a molecular cloud and continues with the formation of HII regions from the ionizing radiation of the massive stars. The hot ionized regions provide a mechanism for the disruption of the molecular cloud [500, 547], and the strong winds from the massive stars as well as their sequential explosions as supernovae, when they reach the end of their life, contribute as well. New molecular clouds can form by a process that is not yet well understood. One possibility involves agglomeration of the shells of expanding material resulting from the disruption of the old clouds. Exactly how these shells are collected into giant molecular clouds is unknown, but the process could involve density waves, which are prominent features in the disks of spiral galaxies. A second possibility involves a large amount of material in the interstellar medium, which could become unstable to collapse under gravity. The cycle concludes with the formation of dense regions in the new molecular clouds and of stars within them. Alternatively, the supernova shocks could propagate through the interstellar medium and initiate star formation in a distant dense region. In Fig. 1.1, the upper line of text in the boxes refers to physical processes that are

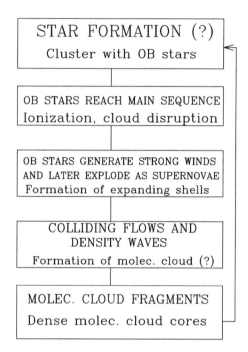

Fig. 1.1 A possible cycle connecting star formation, stellar evolution, and interstellar matter is illustrated. Note that two or more processes may be in operation at any given time. For example, the first supernova may occur before complete disruption of a cloud by ionization has occurred

occurring in stellar evolution or in the interstellar medium, and the lower line refers to the consequences of those physical processes.

This book concentrates on two aspects of the problem. First, the theory of star formation must be based on fundamental physical processes that interact over a rather wide range of conditions of temperature, density, and length scale. Second, observations, over a wide wavelength range, serve to test the theories, which is best done in relatively nearby regions in our galaxy; however more distant regions, also in external galaxies, are providing more and more information as observational sensitivity and resolution improve. Exciting progress has been made, both on the theoretical and on the observational side; however, the answers to many fundamental questions are not well understood. The question mark in the box labelled "Star Formation" means that although we do know the basic physical laws and processes that must be involved in star formation, the ways in which these processes interact in a complicated system is an unsolved problem. The range of length scales involved – from parsec scales in molecular clouds down to the order of a solar radius – adds to the difficulties. The following list gives, in no particular order, the major individual processes that are currently being investigated and which must be included in the solution of the overall problem.

- Hydrodynamics of collapse of material under gravity, including magnetic fields when they are significant
- The equation of state of a multi-phase gas, including heating, cooling, dissociation, and ionization
- Molecular chemistry and the determination of the abundances of the molecules whose line emission is observed
- Radiative transfer, as it affects the continuous and line spectra of clouds, protostars, disks, and young stars
- Turbulence and convection in the interstellar gas, in protoplanetary disks, and in protostars and young stars
- The physics of dust particles, including their formation, growth, destruction, interaction with the gas, and their properties for the absorption and emission of radiation
- Shock waves, for example resulting from accretion of gas onto a star or disk, or from a supernova explosion that later interacts with a relatively dense region of the interstellar medium and initiates star formation.

The complete solution to the star formation problem, including all of these effects and specification of relevant initial conditions, is so difficult that actual research work incorporates only a limited set of the important physical processes, along with some simplifying approximations. In the process of reducing the problem to the point where it is soluble, it is important to identify which processes are the most important in a particular region of density and temperature. Even reduced problems are not in general soluble by analytical techniques, and numerical simulations must be employed.

1.1 Basic Questions: Star Formation

The following questions represent some of the most important points among the wide range of problems that need to be investigated. Progress toward the solution of these problems is discussed in more detail later in the book.

1. **How does one understand the rate and efficiency of star formation as a function of time and position in our galaxy and in external galaxies, and how are these quantities measured?** The rate of recent star formation in the disk of the Galaxy is estimated to be about 1–4 $M_\odot$ yr^{-1}. A common procedure for the determination of the rate is to measure the intensities of emission lines, especially Hα, in the ionized regions surrounding hot O and B type stars. These measurements are converted into star formation rates through the use of the physics of the ionized regions, estimates of the masses and lifetimes of the stars involved, and an integration over the OB associations in the Galactic disk. There are various other ways of estimating this rate, in our galaxy and in external galaxies, based on luminosities in the ultraviolet (UV), optical, infrared, and radio portions of the spectrum [103,356]. The observations in almost all cases produce the star formation

rate of massive O and B stars, and this rate is extrapolated using a standard initial mass function (see below) to include the lower masses. The lifetime of the OB stars is short, so the rate is a measure of the current star formation rate; it does not necessarily reflect the rate of star formation in the distant past in the Galaxy. The problem is that the rate seems to be unexpectedly slow, based on the facts that (1) practically all star formation takes place in molecular clouds, (2) the thermal gas pressure in molecular clouds is not sufficient to keep them in hydrostatic equilibrium, and (3) thus the overall rate of star formation in the Galaxy might be expected to be roughly the total mass in molecular clouds (2×10^9 M$_\odot$) divided by their free-fall time (4×10^6 yr assuming a typical particle density of 100 H$_2$ molecules per cm^3), or 500 M$_\odot$ yr^{-1}. The free-fall time is the time required for a pressureless gas of initially uniform density ρ_0 to collapse to a point singularity of infinite density (see Appendix to this chapter for derivation):

$$t_{\rm ff} = \left(\frac{3\pi}{32G\rho_0}\right)^{1/2}. \tag{1.1}$$

Thus some other source of support, besides thermal pressure, is required to keep the molecular clouds from collapsing [576]. This conclusion is supported by the fact that the widths of spectral lines in molecular clouds are far broader than one would expect from thermal effects alone.

Another problem is the explanation of the density dependence of the local rate of star formation in galaxies. The original formulation by Schmidt [449], based on observations in our galaxy, stated that the rate of star formation is proportional to the square of the local gas density ρ. The modern version of Kennicutt [247], based on observations in many external galaxies, including starburst galaxies, gives

$$\dot{\Sigma}_* = {\rm constant} \times \Sigma_g^{1.4} \tag{1.2}$$

where $\dot{\Sigma}_*$ is the rate of conversion of gas into stars in, for example, solar masses per year per kpc^2 and Σ_g is the local surface density in solar masses per pc^2 (see Fig. 1.2). The power 1.4 is somewhat uncertain. An alternative form of this relation is that the star formation rate is proportional to the ratio of Σ_g to the orbital time scale $2\pi/\Omega$, where Ω is the average angular velocity of the orbit of the region being considered.

In general the surface density is defined to be

$$\Sigma = \int \rho dz \tag{1.3}$$

where z is the coordinate along the line of sight, and the integration proceeds through the entire thickness of the galaxy. In the case of disk galaxies, the quantity Σ_g is corrected for the inclination of the disk to our line of sight, so that the integration is normal to the plane of the galaxy. This quantity is more easily observationally determined in a distant galaxy than is the actual mass density ρ, and as a result the

1.1 Basic Questions: Star Formation

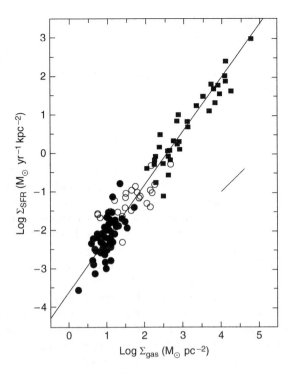

Fig. 1.2 The Kennicutt–Schmidt star-formation law [247], where the vertical axis gives $\dot{\Sigma}_*$. *Filled circles*: samples taken from normal disk galaxies; *open circles*: samples taken from the centers of normal disk galaxies; *squares*: samples of starburst galaxies. The *solid line* is a least-squares fit to the filled circles and squares; it has a slope of 1.4. The short diagonal gives an indication of the uncertainly in the slope. Its slope of 1.28 was derived by reducing the star-formation rate in the starburst galaxies by a factor 2, because systematic errors tend to overestimate the rate. Reproduced by permission of the AAS from [247]. © The American Astronomical Society

Kennicutt relation is an average star formation rate over some region of a galaxy. In the figure, the star formation rate in normal disk galaxies is measured by the strength of the Hα line of hydrogen, which again is an indicator of massive star formation through radiation from their HII (ionized hydrogen) regions, while the gas density is measured from HI (neutral hydrogen) plus CO (with a correction factor to convert it into the H$_2$ density) for the molecular gas. In high-luminosity starburst galaxies, the far-infrared luminosity is a good indicator of the star formation rate, since most of the radiation from the young massive stars is absorbed by the dust and is re-radiated in the infrared; in normal disk galaxies some of that radiation is not absorbed. CO is used to estimate the surface density. However the resulting relation must be true only in a rough overall sense. Local star formation in our Galaxy appears to occur only in the very densest regions of molecular clouds, while the Kennicutt relation is based on an average over a wide range of densities in external galaxies. A theory involving turbulence-induced star formation in molecular clouds [278, 282] in fact agrees very well with the data shown in Fig. 1.2 as well as with more modern data.

Correspondingly, the efficiency of star formation is found to be low, where we provisionally define "efficiency" to mean the fraction of the mass of a molecular cloud that is actually transformed into stars during the lifetime of the cloud. This quantity is crucial for models of galactic evolution. It is generally quoted to be only 2–5% in the galaxy as a whole, although it can be far higher in starburst galaxies. Roughly the efficiency can be estimated to be

$$\epsilon_{sf} \approx \dot{M}_* \tau / M, \qquad (1.4)$$

where $\dot{M}_*$ is the rate of star formation in the entire disk of the galaxy at the present time, M is the total mass in molecular gas, and τ is the lifetime of a molecular cloud. Thus this quantity is the ratio of the lifetime of a molecular cloud to the characteristic time for conversion of molecular cloud material into stars. The difficulty here is that τ is not easily measured. If one puts in approximate numbers, $\epsilon_{sf} \approx 4\ M_\odot\ yr^{-1} \times 10^7\ yr/(2 \times 10^9\ M_\odot) \approx 2\%$. Here, the lifetime of a typical molecular cloud, $\approx 10^7$ yr, is estimated from the ages of clusters as determined from stellar evolutionary tracks along with observations of the molecular gas that is associated with the clusters [313]. Young clusters (ages $< 5 \times 10^6$ yr) are associated with large amounts of molecular gas, while clusters with ages $>10^7$ yr are associated with much less such gas. Other methods of estimating the lifetime of clouds yield, however, different results, and there is considerable debate on this issue. For example, τ is found to be 27 ± 12 Myr for a sample of giant molecular clouds in our Galaxy with masses $\approx 10^6\ M_\odot$ [375]. In a different sample, in a lower mass range in the vicinity of the Sun, the estimate for τ is ≈ 5 Myr [28]. Clearly the term "efficiency" can have several different meanings, and it is further discussed in Sect. 2.8.

2. **How does a cloud fragment to form clusters and associations of stars?** We know that stars can form either in isolation or in small groups or in massive clusters and associations. Observational studies indicate that the typical star forms in a small cluster of a few hundred objects [289]. The fraction of stars formed in each mode still needs to be determined more precisely. The efficiency of star formation in the cluster mode must be locally much higher than in the molecular cloud as a whole. For example, measurements of the star formation efficiency in several very young clusters embedded in molecular clouds [289], simply from the ratio of star mass to total mass (stars plus gas), yield efficiencies ranging from 10% to 30%. However, theoretical results [3, 184, 229] show that if a cluster forms and reaches approximate virial equilibrium, and if its massive stars disperse the residual gas quickly (i. e. in less than a dynamical time, which is roughly the free-fall time of the cluster), through the effects of stellar winds, protostellar outflows, and ionizing radiation, the system becomes gravitationally unbound unless at least 50% of the original gas mass has been converted into stars [288]. Examples of this process in action are the so-called OB associations, which contain young massive stars; the associations are not gravitationally bound but are expanding. In general stars are probably formed predominantly in clusters, but in most cases the clusters disperse because the star

formation efficiency was relatively low. The typical dispersal time is estimated to be a few Myr. About 10% of the stars in the galaxy at present do exist in clusters, which suggests that at least in a few cases the star formation efficiency must have been around 50%. However it is also possible for at least some of the stars in a cluster to remain bound even for efficiencies considerably less than 50%, depending on the initial velocity distribution of the stars [3], or the spatial distribution of gas and stars, or if the gas is removed slowly, on time scales much longer than a dynamical time [184, 290]. Still, the basic question of how the clusters formed in the first place remains unresolved.

3. **Under what conditions is a "trigger", such as an external shock or a cloud–cloud collision, required to initiate star formation?** Is it required for high-mass star formation only? Or is it necessary only in an indirect sense, for example, to bring a region of a molecular cloud into a state where it can form stars by a more gradual, spontaneous process? There are several distinct scenarios for the initiation of star formation:

(a) Star formation by evolution of molecular cloud material on a relatively long time scale, controlled by the magnetic field, to the point where small regions reach high enough densities to become unstable to collapse [459];
(b) Star formation as a result of supersonic turbulence and random shock interactions in the interstellar medium, leading to transient dense regions which could, under the right conditions, collapse [158];
(c) Induced star formation by the action of supernova shocks, spiral density wave shocks, stellar winds, ionization fronts, cloud–cloud collisions, or galaxy mergers (e.g. [162]).

One of the main tasks of present star-formation research is to sort out the relative importance of these various possibilities. In the case of induced star formation there is at least some observational evidence that it takes place, but at least some of the plausible mechanisms require the prior presence of stars, formed by some other, spontaneous, process.

4. **Do high-mass stars and low-mass stars form by different processes?** By "high-mass" we mean the range 10 $M_\odot$ or above, while "low-mass" generally refers to the range around 1 $M_\odot$. Observations show regions of molecular clouds where both high-mass and low-mass stars are forming, and other regions where only low-mass stars appear to form. What is the relation between high-mass and low-mass star formation, both in time and in space? One suggestion [459] is that low-mass stars form by gradual diffusion of gas in magnetically controlled regions of molecular clouds, while high-mass stars form more quickly in regions where the gravitational energy dominates the magnetic energy, so that collapse occurs even in the presence of the field.

According to a different point of view, star formation is controlled primarily by turbulence in molecular clouds, rather than by magnetic fields [387]. In this case low-mass stars and high-mass stars form by the same process. The supersonic turbulence and the resulting shock fronts randomly compress cloud material, leading to collapse of a small fraction of that material. The statistical nature of the turbulence leads to a wide range in the masses of the elements of material that collapse; a relatively small number of these elements are of high mass.

Still another suggestion is based on a problem connected with high-mass star formation – that of radiation pressure. Once the high-mass core evolves to the point where it has built up a star of 10 or so $M_\odot$, the star's ultraviolet luminosity is high enough so that it can exert sufficient radiation pressure on the remaining infalling dust to reverse the collapse and prevent the star from collecting appreciable additional mass. Thus there must be additional physics connected with high-mass star formation that does not apply to low-mass star formation. Massive stars almost always form in clusters, leading to the suggestion [77] that massive stars form by collisional mergers of less-massive stars in regions of high stellar density. However this scenario is actually very unlikely. Instabilities in the collapse flow, as well as the formation of a disk, can help to solve the radiation pressure problem, as further discussed in Chap. 5.

5. **How is the mass spectrum determined? Why does the typical stellar mass fall in the range 0.1–1 $M_\odot$?** This question concerns the explanation of the observed distribution of stars according to mass. The maximum mass of a star is over 100 $M_\odot$, and the minimum mass is defined as that mass below which the object is never able, in its lifetime, to supply its entire radiated luminosity by nuclear burning of 4 protons to form ^{4}He. That limit falls around 0.08 $M_\odot$; objects below that point and above the planetary range (below about 0.01 $M_\odot$), are known as *substellar objects* or brown dwarfs. In its simplest form, the initial mass spectrum is given by

$$dN = f(m)dm \qquad (1.5)$$

where dN is the number of stars formed per unit volume over the history of (for example) the Galaxy with mass between m and $m + dm$. An often-used approximation is the Salpeter [443] mass spectrum $f(m) = \text{constant} \times m^{-2.35}$. It applies to the mass range from the highest masses down to about 0.5–1.0 $M_\odot$ but not for lower masses.

The alternative (and more usual) description of the mass distribution is the initial mass function (referred to as IMF)

$$\xi(\log m) = \frac{dN}{d\log m} = \text{constant} \times m^{-\Gamma} \qquad (1.6)$$

in which one considers the number of stars (per unit volume) per unit logarithmic mass interval that have formed over the history of the system being considered, where the last equality assumes a power-law relation which may not be accurate in general. In this formulation the Salpeter IMF is $\xi(\log m) = \text{constant} \times m^{-1.35}$. In fact the average IMF determined from observations in the Galaxy and in the Large Magellanic Cloud [270] does have almost the Salpeter value ($\Gamma = 1.3$) for

1.1 Basic Questions: Star Formation

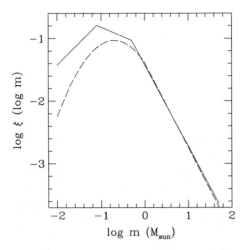

Fig. 1.3 Sketches of fits to the "standard" initial mass function, as determined from observations. The *solid line* is a power-law representation [271] while the smooth *dashed curve* is a fit using a lognormal function [108]. The function ξ is the number of stars per unit interval in log m per cubic parsec. In both cases the function is the single-star IMF, corrected for unresolved binaries. It is evident that the function is very uncertain in the substellar region ($m < 0.08$ M$_\odot$.)

masses above about 0.5 M$_\odot$, but Γ changes to 0.3 in the mass range 0.08–0.5 M$_\odot$ and to $\Gamma = -0.7$ for masses <0.08 M$_\odot$ (Fig. 1.3). Note that this relation is a rough average, and individual systems, such as young clusters, can exhibit substantial deviations from this IMF. The deviations are thought to result from small number statistical effects or the dynamical evolution of clusters, not from some systematic variation [270].

Another version of the observed IMF, for the solar neighborhood, is shown in Fig. 1.3. Here a lognormal function is used instead of a broken power law [108]. The IMF can vary, depending on what data are used and how the IMF is derived from the observations.

The Chabrier [108] fit to the IMF can be written

$$\xi(\log m) = 0.041 m^{-1.35 \pm 0.3} \qquad m \geq 1 M_\odot$$

$$\xi(\log m) = k_1 \exp\left[-\frac{(\log m - \log k_2)^2}{2 \times (0.55)^2}\right] \qquad \text{otherwise} \qquad (1.7)$$

where the normalization gives the function per unit logarithmic mass interval per cubic parsec. The single-object IMF has $k_1 = 0.093$ and $k_2 = 0.2$. The so-called system IMF has $k_1 = 0.076$ and $k_2 = 0.25$; it should be used to compare with observations where binary systems are not resolved. Note that k_2 gives the mass, in M$_\odot$, where the function peaks.

But there are further questions. What determines the upper limit to the mass of a star? In the 30 Doradus star formation region in the Large Magellanic Cloud it is suspected that stars well above 100 $M_\odot$ have formed. Current observations indicate that the IMF increases continuously as the mass decreases, down to near 0.1–0.2 $M_\odot$, beyond which it begins to fall. However it appears to remain continuous across the stellar/substellar boundary at about 0.08 $M_\odot$. Where is the peak? Does it vary in different kinds of stellar systems? What physically determines it? Do brown dwarfs form in the same way as stars?

6. **What parameters control the formation of multiple star systems, as opposed to single stars?** This question is also connected with that of the probability of forming planets, as certain types of binary systems do not favor planet formation. We define the *multiplicity fraction* as the percentage of observed stars which turn out to have one or more stellar companions. Among the nearby main-sequence stars in the range of spectral types from F to G, this fraction is at least 50% [2, 147]. Among other spectral types, the percentage is thought to be even higher for high mass stars, but for M stars the most accurate survey so far [135] gives a value of about 25%. Thus if one combines this information with the form of the IMF, which says that most stars, by number, are low mass stars, one finds [287] that about 2/3 of the observed main-sequence systems in the Galaxy are single stars. There also have been intensive searches for pre-main-sequence binaries, and it appears that in some regions, for example Taurus–Aurigae and Rho Ophiuchi [144, 342], the multiplicity fraction among them is significantly higher than that for main-sequence stars. In other regions, such as Orion, the fraction seems to be similar to that of main-sequence stars [264]. Are some binary systems disrupted soon after they form? How does one understand the possible difference in the multiplicity fraction in stars which form in the cluster mode as compared with those that form in a more distributed mode?

7. **How can it be determined observationally that an object is a protostar?** A protostar is a recently-formed object which still has a significant amount of material in hydrodynamic collapse onto a central core which is approximately in hydrostatic equilibrium, or onto a disk – that is, it is an accreting object. It obtains most of its radiated energy from accretion, either by conversion of infall kinetic energy to photons in an accretion shock at the edge of the stellar core or disk, or by accretion of disk material onto the central object. A related problem to the identification of protostars is the determination of the accretion rate as a function of time. Because the infalling material outside the core-disk structure is cool (30–100 K), dusty, and optically thick, this is the region that is observed at most wavelengths, and the peak intensity of the radiation falls in the mid infrared at 30–100 μm. Observing protostars is difficult because the collapse phase is relatively short (a few times 10^5 yr) and because the star-forming regions are largely obscured by dust. To prove that an object is a protostar one must show that first, its infrared continuous spectrum has the expected characteristics, and second, its line spectrum shows velocity shifts characteristic of infall at near free fall. Many objects have been claimed to be protostars on the basis of the first test (e.g. [295]), but it has proved to be more difficult to find objects which also satisfy the second test. The

relevant spectral lines fall in the millimeter region of the spectrum, and thus there is a problem of spatial resolution at these long wavelengths; furthermore, evidence for infall motions can be confused by other types of motion, such as expansion and rotation. To compare theory with observations, a detailed model of the density, temperature, and velocity structure of the protostar is needed, including rotation, along with radiative transfer calculations in the spectral lines. Nevertheless there is convincing evidence for infall in the spectrum of a number of low-mass protostars [381], and detailed comparisons with theoretical models are available in a few cases (e. g. [340, 535, 572]).

8. **What are the fundamental processes that govern the evolution of disks?** There is now considerable indirect evidence, for example spectra that do not represent a single black body, and some direct evidence (for example, adaptive optics direct imaging or HST imaging of disks in star-forming regions) that young stellar objects are associated with disks in the 10–1,000 AU size range. It is possible that all protostars evolve through a disk phase. There is evidence that matter is accreting from the disks onto the central stars and that there are associated outflows away from the stars. It has been suggested that in their earlier phases disks can evolve through the mechanism of gravitational instability. Here the term "gravitational instability" is taken to mean the development of non-axially symmetric structure in the density distribution of the disk, as a consequence of the disk's self gravity, resulting in transport of mass and angular momentum, and possible fragmentation. At later stages, disks may evolve under the action of turbulent viscosity, or magnetic transport of angular momentum; however many aspects of these theories still remain unclear. How do the gas and dust in the disk and the central object interact to determine an evolutionary time scale? How much of a star's mass is processed through a disk? Can binary companions or planetary companions form in disks? What is the lifetime of the gas component of a disk?

9. **What is the origin of bipolar flows from young stellar objects?** In this text the term *outflow* is used in connection with flows from young stars that generally are collimated and bipolar, and are always associated with the presence of disks. The origin of the flow is probably outside the star, near the inner edge of the disk. In contrast, the term *wind* or *stellar wind* applies in general to stars in any phase of evolution where the flow is quasi-spherical, and its origin is probably in the outer layers of the star itself. The terms *disk wind* and *X-wind* actually refer to particular models for the outflow phenomenon. All low mass protostars probably go through an outflow phase lasting about 10^5 yr with mass loss rates up to 10^{-5} $M_\odot$ yr^{-1} [22, 434]. Most of the mass is in a molecular component, observed in CO at velocities of 10–20 km s^{-1} on large scales, up to roughly a parsec. Jet-like structures with higher velocities are observed in the optical on smaller scales ($\sim$100 AU). Herbig–Haro objects – small knots of optical emission and reflection – are moving away from several young embedded protostellar objects at velocities of 100–300 km s^{-1}. Thus the outflowing material has significant interactions with the surrounding molecular cloud material. The source of energy for these flows is clearly close to the star, but the explanation of the driving mechanism for the jets is still a major

problem. However there appears to be a clear connection between the presence of an accretion disk and a bipolar flow. Does the magnetic-rotational interaction of the disk and star drive the outflow [461]? What produces the collimation?

10. **How does the star formation process produce favorable environments, particularly in protostellar disks, where planets could form?** What is the evolutionary history of giant planets? Such objects during their formation stage are considerably brighter, by up to a factor of 10^5, than Jupiter is today. Could their presence be detected in a protostellar disk? Our own planetary system can provide numerous clues regarding star formation. For example, the orbital distances of the planets give a hint as to the angular momentum of the protostellar cloud. From the compositions of the planets, satellites, and comets one can infer the conditions of temperature and density in the primitive solar nebula and obtain information on the planetary formation process. Now that extrasolar planets have been discovered, first around a pulsar [559] and then around a star similar to the Sun [347], what can we deduce about the interaction between disks and giant planets, and how can we determine the probability of formation of giant planets and terrestrial planets?

11. **What are the effects of rotation and magnetic fields in star formation?** What are the mechanisms for angular momentum redistribution at the various stages of star formation? What is the interaction between rotation and magnetic fields? During normal phases of stellar evolution one can generally make calculations of adequate accuracy by assuming spherical symmetry and hydrostatic equilibrium – that is, rotation and magnetic fields are not significant physical effects. But during star formation these effects can be of dominant importance at various stages, so that two- and three-dimensional hydrodynamical or magnetohydrodynamical calculations are required to follow the evolution. An overall definitive theoretical calculation of the phases of star formation and protostellar collapse is not yet available. Apart from the fact that the calculation is very complicated, the details of the initial conditions are not known. Qualitatively, it is clear that molecular clouds and cloud cores have much more angular momentum than the stars that form from them. But processes such as magnetic braking, formation of disks, and fragmentation into binary systems can go a long way towards resolving the difficulty.

Magnetic fields are observed to be significant in at least some molecular clouds and cores. The principal questions are, first, observationally, what is the strength and configuration of the fields over a wide range of densities in the interstellar medium? Second, to what extent are magnetic forces important, in all phases of star formation? The magnetic flux [defined in (2.23)] deduced from observations of clouds is far larger than that in stars. If strong enough, magnetic forces inhibit collapse and star formation. However the effects of magnetic fields can be reduced by at least two important effects. The first, known as *ambipolar diffusion*, involves the drift of matter across field lines when the degree of ionization becomes very low, which occurs in the dense cores of molecular clouds with densities in the range 10^4–10^6 particles per cm^3. As a result, in these regions, the ratio of mass to magnetic flux builds up to the point where collapse can start. The second effect occurs at somewhat higher densities where Ohmic dissipation (2.24) starts to occur, resulting in loss of magnetic flux [357].

12. **How did the first stars form?** And, after the first generation, how did star formation continue, under conditions of very low metallicity, such as in globular clusters? Conditions in the early history of the Universe were very different from what they are now. Two of the major ingredients which favor present-day star formation – low temperature and the presence of about 1% by mass of material in the form of dust grains – simply were not present. Also, magnetic fields were minimal, and no triggers were present. How did star formation differ during these early phases when the only elements in significant amounts were hydrogen and helium? Three-dimensional numerical simulations have been carried out (e. g. [1]) which start from small density fluctuations in the early Universe and lead to the formation of the first protostar, suspected to end up as a star around $100\,M_\odot$. Is the first actual star necessarily a single star?

1.2 Observations of Objects in Star-Forming Regions

We take a quick look here at some key observed objects, which have provided some clues to the answers to the questions posed above. These observed objects are ordered more or less in an evolutionary sequence, which is defined more precisely in the following section.

It is clear from both observations and theory that star formation takes place in the coolest and densest regions of the interstellar gas, where the absolute value of the gravitational energy of a region approaches the sum of its thermal, turbulent, and magnetic energies. Molecular clouds are the site of practically all star formation in the Galaxy. The clouds, which in fact show structure on a wide variety of length scales, are somewhat arbitrarily subdivided into clumps, which are observed in CO, with characteristic masses 10^3–10^4 $M_\odot$, radii 2–5 parsec, temperature 10 K, mean number density of H_2 of 10^2–10^3 cm^{-3} and magnetic field 3×10^{-5} gauss. The random velocities in these clumps, which represent a combination of the thermal particle motion and larger-scale gas motion, are determined from observations of the width (at half-maximum intensity) of Doppler-broadened molecular lines. This width would correspond to $0.2\,\mathrm{km\,s^{-1}}$ for purely thermal broadening at 10 K. However in the clumps the actual line widths are 2–3 km s^{-1}, highly supersonic, indicating that larger-scale gas motions, probably turbulent motions and/or magnetic Alfvén waves, are indeed present.

Embedded in the clumps are the higher-density cloud cores, observed in NH_3, CS, and other molecules. The masses here are typically about 1 to a few $M_\odot$, although a few range up to $1{,}000\,M_\odot$. The sizes are 0.05–0.1 pc, the temperature 10 K, and the density $\approx 10^4$–10^5 cm^{-3}. The linewidths in the low-mass cores are approximately thermal, indicating that magnetic and turbulent effects are no longer important and that these regions are likely to undergo gravitational collapse to form stars. In fact some cores already have embedded infrared sources [46], which are probably protostars. Others (starless cores) do not have such sources, but some of them may already have developed infall motions [195, 310]. The suggestion

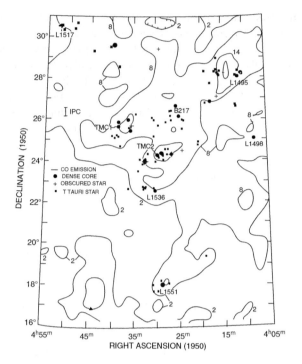

Fig. 1.4 Star formation in the Taurus–Aurigae region. *Solid curves:* contours of equal intensity of line emission in the CO molecule; *large filled circles:* dense cores in the molecular cloud, detected in the ammonia molecule; *small filled circles*: young stars, already formed; *crosses*: protostars, whose central regions, consisting of a young star and a disk, are highly obscured by infalling dust. The names refer to specific dark clouds. A 1–parsec scale is indicated. From *IAU Symposium No. 115: Star Forming Regions*, ed. M. Peimbert, J. Jugaku (D. Reidel Publ. Co., 1987), article by P. C. Myers: Dense Cores and Young Stars in Dark Clouds, p. 36, Fig. 2. Reproduced with kind permission of Springer Science and Business Media. © 1987 International Astronomical Union

that these regions represent the initial condition for low-mass star formation is strengthened by the close positional correlation (Fig. 1.4; from [377]) between the ammonia cores (large filled circles), recently-formed young stars known as T Tauri stars (small filled circles), highly obscured infrared sources (crosses) detected by IRAS (Infrared Astronomical Satellite), and sources with bipolar CO outflows [46, 49, 380]. In Fig. 1.4, which shows part of the Taurus region, outflow sources are not specifically plotted, but the infrared source (IRS5) in L1551 is a good example. Other, more isolated, dark regions in space which are likely sites of star formation are known as Bok globules. An example is shown in Fig. 1.5, which is a photograph by Bart Bok [72] of the globule Barnard 68, whose radius is about 0.05 pc. The argument that these cores are the likely sites of star formation is strengthened by observations that the distribution of masses of the cores is very similar to that of the IMF of stars.

1.2 Observations of Objects in Star-Forming Regions

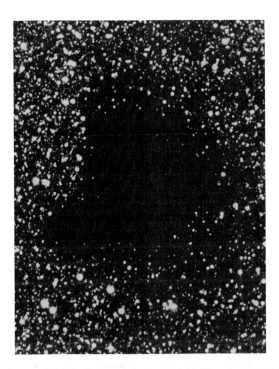

Fig. 1.5 Optical photograph of the Bok globule Barnard 68. The almost total lack of stars near the center of the globule suggests a visual extinction of more than 25 magnitudes. This figure originally appeared in the Publications of the Astronomical Society of the Pacific (B. Bok: *PASP* **89**, 597, 1977). © 1977 Astronomical Society of the Pacific; reproduced with permission of the Editors

Figure 1.6 shows the spectrum of a protostar, the object L1551 IRS5 [7]. There is practically no optical radiation, and the spectral energy distribution (SED) peaks at 60–100 μm. The data are given by the triangles and the lines are various model fits to the data. The models consist of a central (obscured) star and a disk around it, both embedded in a rotating collapsing dusty envelope, viewed at the same inclination to the rotation axis as that observed in the source.

A direct optical image of a disk around a protostar, with associated jets, is illustrated in Fig. 1.7, obtained by the Hubble Space Telescope (HST). The object is known as HH30, which shows an edge-on disk whose optically thick dusty material obscures the central star. Reflected light from this hidden central star is observed via scattering from the upper and lower surfaces of the disk. The disk itself, with a radius of about 200 AU, is the horizontal dark lane between the two reflecting surfaces. The vertical structure, observed in optical emission lines, is a bipolar high-velocity jet, originating near the central star. Typical jet velocities are a few hundred km s^{-1}.

The protostellar binary IRAS 16293-2422 [534] has two components separated by about 800 AU (Fig. 1.8). The objects are not observed in the optical, only in the radio at mm wavelengths and at 2 cm and 6 cm. The emission probably comes from

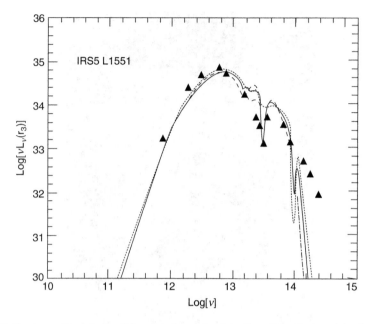

Fig. 1.6 Spectrum of a protostar. *Triangles*: observed points; *lines*: theoretical models of rotating, collapsing clouds. The spectrum is calculated at a radius $r_3 = 2 \times 10^{17}$ cm away from the protostar. The quantity L_ν is the observable luminosity of the protostar at frequency ν, per unit frequency interval; νL_ν is plotted as a function of ν, in *cgs* units. The total bolometric luminosity of the source is about $30\,L_\odot$. An absorption feature from silicate dust is observed at $\log \nu = 13.5$. Reproduced, by permission of the AAS, from [7]. © 1987 The American Astronomical Society

circumstellar disks. The lines of the molecule CS show a gradient in the line-of-sight component of the velocity, along the axis connecting the two components. The conclusion is that a rotating circumbinary disk is present, with a radius of several thousand AU. Each component is associated with a bipolar flow, whose directions indicate that the disks around the individual components are not coplanar. The high luminosity of the source ($30–40\,L_\odot$) suggests there still is accretion from the infalling cloud, and there is a possibility that the infall has been detected in the CS line profiles.

A classic protostar is L1551 IRS5 [491] which shows a bipolar outflow in CO at $10\,\mathrm{km\,s^{-1}}$, the first such flow discovered [472]. Other features include a disk observed in CS, an embedded infrared source, a reflection nebula which reveals some properties of the underlying star, and optical emission regions which correspond to collimated jets near the star and also to Herbig–Haro (HH) objects [127], luminous gas knots moving radially away from the star (Fig. 1.9).

Polarization measurements of the light from a few background stars indicate that the local magnetic field is roughly parallel to the outflow axis (although a couple of objects do not fit the pattern and may be in regions unrelated to IRS5.) It turned out later that the central object in L1551 IRS5 is a binary; each component is

1.2 Observations of Objects in Star-Forming Regions

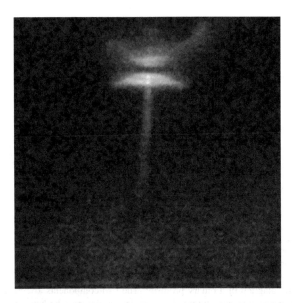

Fig. 1.7 The young object HH30: edge-on disk, bipolar jet, and scattered light (*bright regions*) from the obscured central star. Courtesy NASA, Space Telescope Science Institute, ESA

surrounded by a disk, whose radiation was detected by the Very Large Array (VLA) at a wavelength of 7 mm [440]. The separation is about 40 AU (Fig. 1.10).

Another example of HH objects is illustrated in Fig. 1.11 which shows the optical emission structure of the objects HH1 and HH2, which are moving in opposite directions from the deeply embedded source VLA1 [222]. Doppler measurements as well as proper motions give velocities of about 300 km s^{-1}.

Another unusual object associated with early stellar evolution is FU Orionis. It may be a transition object, on the borderline between protostar and young star. In 1936–1937 it flared up by 6 photographic magnitudes in 120 days [533], and it has remained at nearly constant brightness ever since. The star is associated with a dark cloud and young (T Tauri) stars; it has a high lithium abundance in the atmosphere and an infrared excess. A number of mechanisms have been suggested to explain the flareup. One possibility is that it represents a thermal instability in the inner part of the circumstellar disk, associated with the hydrogen ionization zone, that results in a temporary greatly enhanced rate of accretion from the disk onto the star [205, 322]. Other suggested mechanisms include perturbation of a disk by a passing star [176], gravitational instability triggered by infall onto a disk [531], or a combination of magnetohydrodynamic and gravitational effects in which gravitational instability in a disk triggers a magnetorotational instability (Chap. 4) that produces an episode of rapid accretion onto the star [23, 574]. In any case, observations of a small number of stars of the same class suggest that all stars, during their very early history, go through a process of brightening, followed by a return to the initial luminosity on a time scale of 100 yr. The process is thought to recur on a time scale of 10^3–10^4 yr

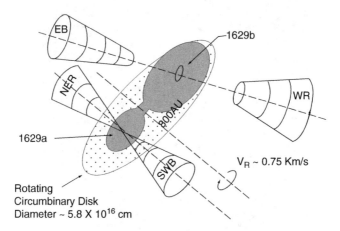

Fig. 1.8 Schematic diagram of a binary protostar. The two components 1629a and 1629b are in orbit, with the orbital angular momentum along the dashed line passing between the two objects. Each of the components has a small disk around it, and the angular momenta of the disks are not parallel. The component 1629a is driving a bipolar outflow, indicated by the cones NER and SWB. The component 1629b was driving an outflow in the past (EB and WR) but is not currently doing so. Both sources are embedded in a circumbinary disk which may not be in equilibrium, rather, still infalling. Darker/lighter shading indicates higher/lower dust column density between observer and source. The rotational velocity of the circumbinary disk, V_R, is estimated to be around 0.75 km s^{-1}. Reproduced, by permission of the AAS, from [534]. © The American Astronomical Society

so that each object goes through several events. The light curve of FU Ori, along with those for two other related objects V1057 Cyg and V1515 Cyg, is shown in Fig. 1.12. Note that V1057 Cyg has already faded by 3 magnitudes, and that the rise time for V1515 Cyg is considerably longer than that of the other two objects. The properties of these unusual objects are discussed in more detail by [47, 202, 206].

Young stars in general, which have just emerged from the collapse phase, show the following characteristics: they are associated with dark-cloud material, they are located well above the main sequence in the Hertzsprung–Russell diagram, they have a high abundance of the rare, light, easily-destroyed element lithium, and they show evidence of mass outflow and a large infrared excess, indicative of the presence of a disk. They also have rotation rates which are relatively slow, 5–30 km s^{-1}, emission lines, particularly Hα, which are a defining characteristic of T Tauri objects, and excess ultraviolet emission over a normal stellar photosphere. They are typically irregularly variable at all wavelengths. Two types of T Tauri stars are recognized, the classical or CTTS, which have an equivalent width (in emission) of H$\alpha > 10$ Å, and the weak-lined or WTTS, which have an equivalent width in H$\alpha <$ 10 Å. The WTTS have much less IR excess, much less UV excess, and weaker (or no) emission lines. Both types have X-ray emission. Magnetic fields, first measured

1.2 Observations of Objects in Star-Forming Regions

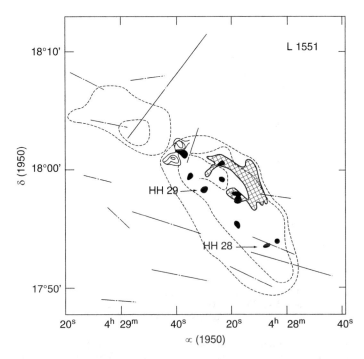

Fig. 1.9 Schematic diagram of the well-studied protostellar source L1551 IRS5 . The infrared source itself is indicated by the small triangle at $\alpha = 4^h29^m40^s$, $\delta = 18°016'$. *Solid lines*: contours of equal intensity in the CS molecule, interpreted as a rotating disk [246]; *dashed lines:* contours of CO emission, showing the bipolar outflow, with the approaching lobe extending toward the lower right. The *dark regions* are knots of optical emission, with a jet near IRS5 and several Herbig–Haro objects moving away from the source. *Cross-hatched region*: optical reflection nebula, which scatters the light from the (obscured) central star in the direction of the observer. *Dot-dashed lines* indicate the direction of the local magnetic field. Reproduced, by permission of the AAS, from [491]. © The American Astronomical Society

by [34], fall in the range 1–3 kiloGauss [242]. Figure 1.13 shows the continuous spectrum of the CTTS T Tauri, clearly indicating the IR excess. The spectrum in the infrared is practically flat, completely unlike a black body at 4,000 K, a temperature typical of this type of star. The infrared radiation is interpreted as dust emission from a circumstellar disk, and the observed spectrum is compared with theoretical disk models in the figure. The typical ages of CTTS are 2–3 million years. The lifetime of this phase provides an important constraint on the planet formation process.

Figure 1.14 shows how ages and masses of young stars are approximately derived, by comparing their observed positions in the H–R diagram with pre-main-sequence evolutionary tracks and isochrones. The particular objects shown [122] are the components of a pre-main-sequence double-lined eclipsing spectroscopic binary in the post-T Tauri phase, whose masses can be directly measured and compared with those obtained from the tracks, for example those calculated by [30]. The actual masses [123] obtained from the dynamical analysis of the system

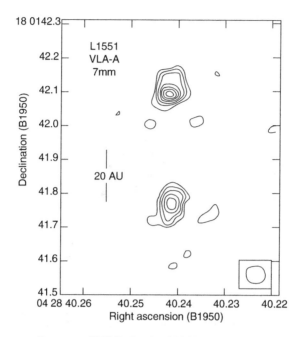

Fig. 1.10 The protostellar source IRS5 in the cloud L1551, imaged by the Very Large Array at a wavelength of 7 mm. The spatial resolution is 7 AU (beam in lower right). Contours of equal intensity are mapped. The source is found to be a binary (as previously suspected by [54]) with separation about 40 AU. The individual protostars are not seen at this wavelength; the radiation consists of emission from dust in the surrounding cool disks, each about 10 AU in radius and 0.05 $M_\odot$ in mass. Reprinted by permission from MacMillan Publishers Ltd., L. F. Rodriguez, P. D'Alessio, D. J. Wilner et al.: *Nature* **395**, 355 (1998). © 1998 Nature Publishing Group

are 1.27 and 0.92 $M_\odot$, which are in fairly close agreement with the "track" masses one reads off from the diagram.

The resolved optical image, taken with the Hubble Space Telescope, of a disk associated with the star AB Aurigae is shown in Fig. 1.15. A coronagraph occults the star and the inner disk out to 60 AU. The disk is about 1,300 AU in size and shows spiral features. This star is a so-called Herbig Ae star, and evidence for its youth (age 1–3 Myr) is provided by strong Hα in emission. It is closely related to the T Tauri stars, but its mass, of approximately 2 $M_\odot$, is higher than that of the typical T Tauri star, which falls in the range 0.5–1 $M_\odot$. The structure of this disk is being intensively studied through further ground-based observations [397] to determine the possible presence of a low-mass companion at a separation of about 100 AU, where a gap in the dust distribution has been observed.

The very late phase of a disk (and also the first disk to be directly imaged) is represented by that around the main-sequence star β Pictoris, about 1.75 solar masses. It was first detected by IRAS and later imaged in the optical [468]. The top part of the 1998 HST picture (Fig. 1.16) is a visible light image of the entire disk (diameter 1,500 AU), which is made up of microscopic dust grains of ices and

1.2 Observations of Objects in Star-Forming Regions

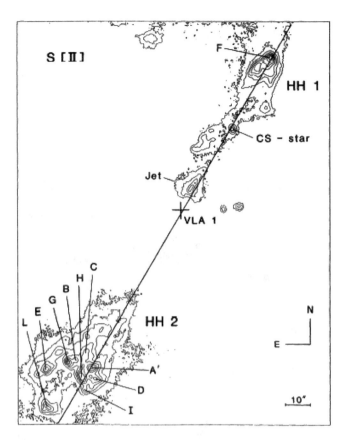

Fig. 1.11 Observation of the Herbig–Haro objects HH1 and HH2, from optical forbidden line emission of sulfur. Contours of equal intensity in this line are separated by a factor 2. HH2 is broken up into a number of emission knots (*capital letters*). The optically invisible radio source VLA 1 is thought to be the origin of the outflow that produces the HH knots. Reproduced, by permission of the AAS, from [474]. © The American Astronomical Society

silicates with a total mass of only about 10 lunar masses and practically no gas. The bright star has been blocked out in this image, so that the inner part of the very faint, nearly edge-on disk can be seen. There actually is a central clear region with a radius of about 15 AU. The presence of the dust, which is rapidly driven out of the system by radiation pressure from the central star, strongly suggests there is an underlying system of planetesimals (solid objects of radius roughly 1 km), which regenerate dust through their collisions. The dust is probably composed of silicates inside 120 AU, with additional ice and organic grains outside that point. The bottom frame shows the details of the inner part of the disk. Both pictures are in false color to accentuate details in the disk structure. The colors indicate the intensity of the starlight that is scattered by the dust particles to the observer. Such disks are important because of the clues they provide regarding the possible prior formation

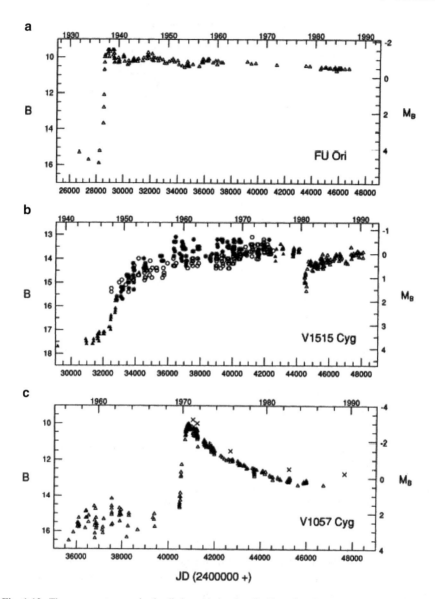

Fig. 1.12 The apparent magnitude (*left axis*) in the B filter (or the apparent photographic magnitude at the earliest times) as a function of Julian date for three FU Ori objects. The estimated absolute B magnitude is given on the right-hand scale. The time in years is given at the top of each diagram. Different symbols refer to different observers. Reproduced, by permission of the AAS, from [48]. © The American Astronomical Society

of planets in the system, and in fact a giant planet of about 10 Jupiter masses has been directly imaged in the region between 8 and 15 AU from the star [296].

Structure has also been observed in a nearly edge-on disk around AU Microscopii, a star of 0.5 solar masses about 9 parsecs away. The disk has been imaged and

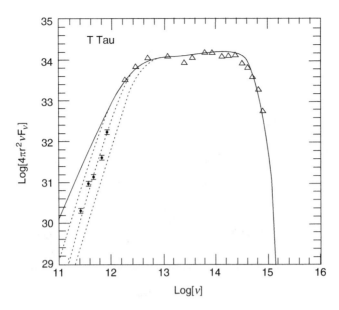

Fig. 1.13 The spectral energy distribution of the young star T Tauri. The quantity νL_ν, where L_ν is the energy per second per unit frequency interval, is plotted as a function of frequency. The quantity F_ν is the flux density [defined in (8.24)] received at Earth, and r is the distance of the object. *Open triangles:* observed data points in the optical and infrared. *Filled triangles with error bars:* submillimeter data points. *Solid curve:* theoretical spectrum under the assumption that the disk is optically thick at all frequencies. *Dotted curves:* theoretical spectra, from top to bottom, of disks with masses 1.0, 0.1, and 0.01 $M_\odot$, respectively, not assumed to be optically thick. The fact that the disks are optically thin at long wavelengths allows estimates of their masses to be made (Chap. 4). The derived disk mass depends on assumptions made regarding the dust opacity at long wavelengths. This particular source turns out to be a multiple system (Fig. 8.11); the plotted radiation represents the sum from all components. Reproduced, by permission of the AAS, from [5]. © The American Astronomical Society

resolved in the H band (1.65 μm) with the Keck II telescope using adaptive optics [324]. HST observations in the visual indicate a straight, narrow close-to-symmetric disk, viewed nearly edge-on, extending 150 AU away from the star, with an inner hole out to 12 AU [268]. These two examples of late-phase, low-mass disks, known as *debris disks*, have many counterparts among main sequence stars, as revealed by the Spitzer Space Telescope.

The cluster mode of star formation is illustrated in Fig. 1.17, which shows the Trapezium cluster in the Orion Nebula on a scale of 1/3 of a parsec. The bright, hot stars of the Trapezium appear in the center; they have masses in the 15–25 $M_\odot$ range. The faintest objects fall in the mass range of the brown dwarfs, between 0.01 and 0.08 $M_\odot$. The left-hand frame is in visible light; the right-hand frame in the near infrared (1 and 1.6 μm). At the age of the cluster, only 1 million years, the brown dwarfs are bright enough to be detectable. This cluster is a nearby example of

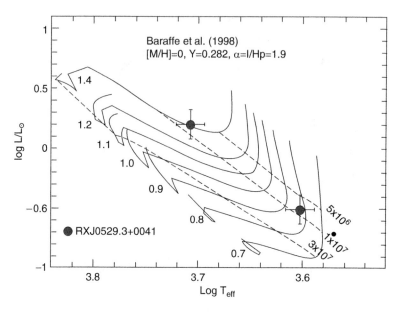

Fig. 1.14 Theoretical Hertzsprung–Russell diagram showing pre-main-sequence evolutionary tracks (*solid lines*), labelled by mass, in solar units, from 0.7 to 1.4 $M_\odot$. *Dashed lines:* lines of constant age, with age given in years. *Points with error bars*: the two components of an observed pre-main-sequence eclipsing binary. The evolutionary tracks are taken from [30], with solar metal abundance, helium mass fraction $Y = 0.282$, and a ratio of convective mixing length to pressure scale height of 1.9. Improved observational determinations of luminosity and $T_{\rm eff}$ [123] give better agreement between the track masses and the dynamical masses. Nevertheless, the degree of agreement depends on the particular set of tracks that is used. Credit: E. Covino et al., *Astron. Astrophys.* **361**, L49 (2000), reproduced with permission. © European Southern Observatory

a relatively large number of observed clusters which are embedded in interstellar gas and dust and are presumably still in the process of formation.

1.3 Star Formation Phases

The previous section shows observed examples of objects ranging from pre-stellar clouds to young objects which have the basic properties of stars. For convenience, the formation process can be divided into three phases, as indicated in Table 1.1, where the numbers given refer to the typical case of 1 $M_\odot$. The first, known as *star formation*, involves massive interstellar clouds or cloud fragments, which have cooled to the point where they are detectable in molecular lines (such as CO) but which are unable to collapse because of an excess of thermal, turbulent, rotational, and magnetic energy over gravitational. The phase, whose length is currently debated but which lies in the range 10^6–10^7 yr, involves dissipation of much of this energy and development of the dense cores. Much of the observational information

Fig. 1.15 Optical image from Hubble Space Telescope of the young star AB Aurigae. A coronagraph is used to occult the central star and reveal the surrounding disk, which is illuminated by scattered light. Courtesy NASA, Space Telescope Science Institute

about this phase is from the millimeter or submillimeter part of the spectrum. The second, *protostellar*, phase starts when cloud cores of $\sim 1\,M_\odot$ become unstable to collapse under gravity, with a resulting increase of 16 orders of magnitude in density and 5 orders of magnitude in temperature. The observable radiation produced during this phase is primarily in the midinfrared to submillimeter, and the time scale is approximately the free-fall time, about 2×10^5 yr at the typical density of a molecular cloud core (1.1).

At the end of the collapse phase, when the object has heated up to the point where gas pressure can support it in equilibrium against gravity, it has star-like properties, and it begins a slow *contraction phase*, in near hydrostatic equilibrium. For 1 $M_\odot$, the contraction time to the point where nuclear reactions are established at the zero-age main sequence is about 4×10^7 yr. This time scale, known as the *Kelvin–Helmholtz* time, is estimated simply from the gravitational energy at the final stage of the contraction divided by the average luminosity $\bar{L}$:

$$t_{\mathrm{KH}} \approx \frac{GM^2}{R\bar{L}} \qquad (1.8)$$

where R is the final radius. During this third phase, the objects can be observed in the optical and near infrared, although millimeter-wave observations are also relevant (Fig. 1.13) for detecting the surrounding disk. In this book we will cover the first

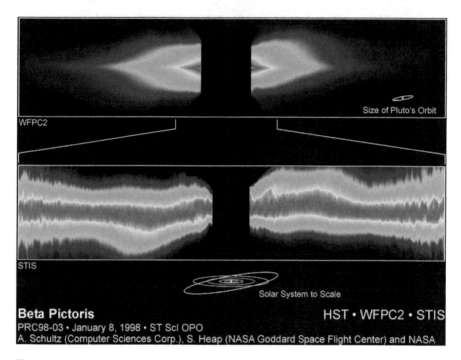

Fig. 1.16 Optical images of the debris disk around the star Beta Pictoris, taken with the Hubble Space Telescope. WFPC2 (*upper image*) is the Wide-Field and Planetary Camera 2, and STIS (*lower image*) is the Space Telescope Imaging Spectrograph. The color scale represents the intensity of the scattered light, with red corresponding to the highest intensity. Courtesy NASA, Space Telescope Science Institute

two phases, as well as the earlier part of the contraction phase, during which the star still has a disk, a remnant of the formation process.

A commonly used observational classification of young objects [285, 286, 291] involves use of the spectral energy distribution longward of $2\,\mu$m. Objects of Class I have a positive value of the spectral index $\alpha = -d\log(\nu F_\nu)/d\log\nu$ in the vicinity of $2.2\,\mu$m (1.36×10^{14} Hz). The flux density F_ν is the observed energy per unit area per second per unit frequency interval. The typical spectrum peaks at 60–$100\,\mu$m ($3 \times 10^{12} - 5 \times 10^{12}$ Hz). At $2.2\,\mu$m the flux is increasing toward longer wavelengths, and the spectrum is broader than a black body with little radiation in the near infrared, $\lambda < 2\,\mu$m. This type of object is also known as an "embedded" IR source, or as a candidate protostar; it is probably completely surrounded by dust. An object of Class II, on the other hand, has a zero or negative value of α; however it is also broader than a single black body, with a significant IR excess and also a UV excess, sometimes known as "veiling" (filling in) of the absorption lines in the blue region of the spectrum. Class II objects are often T Tauri stars; they have less circumstellar dust than Class I objects, and they are observable in the visible. Such objects are interpreted to be pre-main-sequence stars with disks. An object of

1.3 Star Formation Phases

Fig. 1.17 Hubble Space Telescope images in the optical (WFPC2; *left*) and in the infrared (NICMOS; *right*) of the Trapezium cluster, which consists of the inner few arc minutes of the Orion Nebula cluster, centered on the Trapezium stars themselves. [NICMOS = Near Infrared Camera and Multi-Object Spectrometer]. In the infrared the dust opacity is considerably reduced and the stars in the cluster are seen. The opacity as a function of frequency is usually expressed as $\kappa_\nu \propto \nu^\beta$, where usually $1 < \beta < 2$, but β is not well known and it depends on the composition of the dust. Courtesy NASA, Space Telescope Science Institute

Table 1.1 Major phases of early stellar evolution

Phase	Size (cm)	ρ (g cm^{-3})	T (K)	Time (yr)
Star formation	10^{20}–10^{17}	10^{-22}–10^{-19}	10	10^6–10^7
Protostar collapse	10^{17}–10^{12}	10^{-19}–10^{-3}	10–10^6	10^5–10^6
Pre-main-seq. contraction	10^{12}–10^{11}	10^{-3}–1	10^6–10^7	4×10^7

Class III has a negative value of α, its spectrum is close to that of a black body of a single temperature, and there is little or no evidence for excess IR radiation or dust. These objects are either pre-main-sequence stars beyond the T Tauri phase, that is without indications of significant disks, or young main-sequence stars. These three types of objects overlap the protostar collapse phase and the pre-main-sequence contraction listed in Table 1.1. For example, the object in Fig. 1.6 is clearly a Class I, while that in Fig. 1.13, with a flat spectrum, is technically a Class II, but it is really in the transition region between Class I and Class II.

A few "Class 0" sources have also been identified; they are thought to correspond to protostars at an even earlier evolutionary phase than that represented by Class I [18]. Some of the radiation is in the submillimeter, with hardly any radiation shortward of $10\,\mu$m. Their spectral energy distributions look like black bodies with temperatures of 15–30 K and a peak near $100\,\mu$m. One of the commonly used criteria for Class 0 is that the mass in the infalling region of such objects is greater than the mass in the central hydrostatic region, while the reverse is true in Class I. The four types of objects are illustrated in Fig. 1.18. If a "flat" spectral class is introduced along with the classes 0 through III [194], then the approximate slopes in the 2–10 μm spectral region can be summarized as follows:

- I: $\alpha \geq 0.3$
- Flat: $-0.3 \leq \alpha < 0.3$
- II: $-1.6 \leq \alpha < -0.3$
- III: $\alpha < -1.6$.

Note that the quantity α is not useful in connection with Class 0. In fact the boundaries between the classes of objects are somewhat fuzzy.

If one takes a census of the relative numbers of objects in Class 0, Class I, the "flat" class, and Class II in a given star formation region, one can make an estimate of the relative amounts of time spent in these four phases. Early studies of the Taurus region and the Rho Ophiuchi cloud indicate that the embedded protostellar phase (basically Class I) lasts $2-3 \times 10^5$ yr [249, 549]. A more extensive survey of star-forming regions with the Spitzer Space Telescope gives an average value of about 1×10^5, 4.4×10^5, and 3.5×10^5 years, respectively, for Class 0, Class I, and the "flat" class [164]. The embedded phase is then followed by a (classical) T Tauri phase (Class II) that lasts, on the average, $1 - 3 \times 10^6$ yr. However, these numbers are uncertain; sources of uncertainly include incompleteness, contamination by background objects, extinction corrections, and the assumption that star formation is occurring at a constant rate, allowing a continuous flow through the various classes.

Now consider an example of the phases of evolution (steps 1 to 7) that are followed by a low-mass star that forms as a single star in isolation by a spontaneous, rather than a triggered, process. This particular process is based on the idea of magnetically controlled star formation [459], about which there is considerable debate at the present time. Thus there are several different suggested processes by which a star can reach the onset of collapse (step 3). Only one of them is considered here; the others will be covered later. From that point onward, all cases join the same path.

1. Starting at densities characteristic of molecular clouds, the frozen-in magnetic field transfers much of the angular momentum out of the cloud on a time scale of 5×10^6 yr.
2. Matter becomes less and less tightly coupled to the field, and a molecular cloud core forms, through hydrostatic contraction, on a diffusion time $\leq 10^7$ yr. The gas, which is mainly neutral, drifts inward with respect to the magnetic field, which is tied to the ionized particles. Thus the gravitational energy increases relative to

1.3 Star Formation Phases

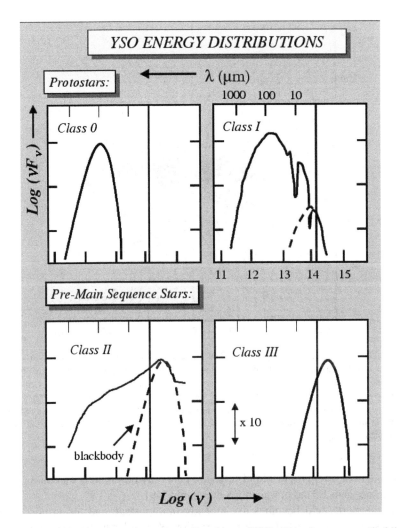

Fig. 1.18 Classification scheme for young stellar objects (YSO). For reference, a vertical line is plotted in each frame at a wavelength of 2.2 μm. Class 0 objects have distributions close to a black body at a single, but low, temperature. Class I objects have negative slopes in this diagram at wavelengths longer than 2 μm, and the spectral energy distribution is broader than that of a single black body. Class II objects have zero or positive slopes longward of 2 μm, and show an infrared excess over a black body at a single temperature. Class III objects are close to black bodies at a single temperature without any excess infrared. The absorption feature in the Class I object near 10 μm is caused by silicate dust. From *The Origins of Stars and Planetary Systems*, ed. C. J. Lada and N. D. Kylafis, article by C. J. Lada: The Formation of Low-Mass Stars, p. 172, Fig. 11. Reproduced with kind permission of Springer Science and Business Media. © 1999 Kluwer Academic Publishers

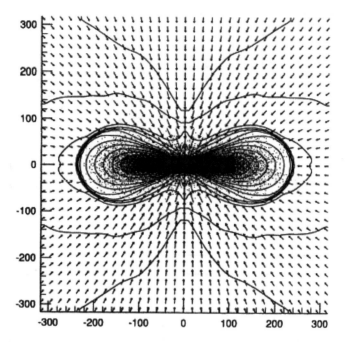

Fig. 1.19 Theoretical model of a protostar. The vertical axis corresponds to the rotation axis, the horizontal axis, to the equatorial plane. The scales are in AU. Arrows give gas velocity with length proportional to speed; the maximum velocity shown is about $5\,\mathrm{km\,s^{-1}}$. *Solid lines:* contours of equal density, separated by $\Delta \log \rho = 0.2$ and $\log \rho_{min} = -15.22$. The basic picture is that of infalling material passing through a shock and joining an equilibrium disk; an unresolved stellar core sits at the origin. Reproduced, by permission of the AAS, from [306]. © The American Astronomical Society

the magnetic energy. The cloud cores are generally observed in an emission line of ammonia at 1.3 cm or one in CS at 3.1 mm [49].

3. The magnetic field becomes dynamically unimportant, and the protostar collapses nearly in free-fall, with conservation of angular momentum, on a time scale of 10^5–10^6 yr.
4. The stellar core, material which has stopped collapsing and obeys the quasi-hydrostatic equations of stellar structure, forms with an accretion shock at its outer boundary. As material with higher angular momentum approaches equilibrium, a disk forms. The protostar becomes an infrared source with luminosity $\sim 1\,L_\odot$, because the optical photons released by the stellar core are absorbed by the dust in the cool infalling envelope and are re-radiated in the infrared. Figure 1.19 shows an example of a theoretical model in this phase [306], with about $0.6\,M_\odot$ unresolved in the central stellar core and about $0.4\,M_\odot$ in the surrounding disk. Densities in the disk range from 10^{-10} to $10^{-16}\,\mathrm{g\,cm^{-3}}$, and temperatures from $\log T = 2.9$ to 2.2. This object would be in Class I, but at

1.3 Star Formation Phases

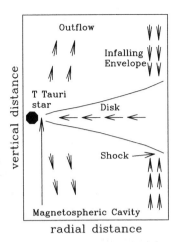

Fig. 1.20 Schematic picture of a disk around a classical T Tauri star. *Short arrows* indicate mass flows: the accretion flow through the disk onto the star, the outflow originating from the inner part of the disk, and the residual flow from the molecular cloud onto the outer part of the disk. Roughly 10% of the mass that is flowing through the disk toward the star is ejected in the outflow. The typical accretion rate through the disk for a young T Tauri star is 10^{-8} $M_\odot$ yr^{-1}, but it can increase to 10^{-4} $M_\odot$ yr^{-1} during an FU Orionis outburst. The diagram is not to scale. The radius of the magnetospheric cavity, inside of which the stellar field dominates, is only 0.1 AU, while the entire radial extent of the disk may be several hundred AU

earlier phases, when less mass has accumulated in the core, it would be in Class 0.

5. As the protostar accretes, the bipolar outflow phase starts. The outflow starts early; it is observed even in objects of Class 0. The ejected material first breaks through the infalling material at the rotational poles, where the density gradient is steepest and the infall ram pressure the smallest. The outflow velocities are 100–300 km s^{-1}, and near the star the jet is well collimated. Angular momentum from the star-disk system is transferred away in the outflow. Figure 1.20 shows a schematic diagram of a young object in the phase when it is still accreting from the surrounding infalling cloud and also producing an outflow. At this stage, near the end of protostar collapse, the object is optically visible at least when observed from the polar direction, so it would probably be placed in Class II. But if observed edge-on to the disk, it would be highly obscured and have the characteristics of Class I.

6. Infall stops because all the available material has either been accreted or swept away by the outflow. The stellar core emerges onto the H–R diagram as a visible T Tauri star (Class II), with infrared excess still remaining from the disk. Figure 1.21 shows the H–R diagram of young stars, including pre-main-sequence evolutionary tracks for various masses and showing the "birth line", the locus where stars of various masses first appear on the visible H–R diagram [407]. The

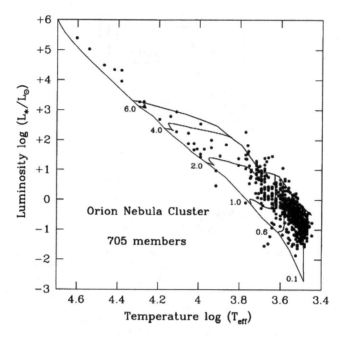

Fig. 1.21 Hertzsprung–Russell diagram: zero age main sequence labelled with the corresponding masses in $M_\odot$, pre-main-sequence evolutionary tracks for those masses, the birth line (upper envelope of the tracks), and observed positions of the stars in a young cluster. Reproduced, with permission from the AAS, from [407]. © 1999 The American Astronomical Society

filled circles are observed stars in the Orion star formation region, to which the birth line forms the upper envelope, as expected.

7. Disk evolution occurs, driven by the action of turbulent viscosity, gravitational instability, or magnetic fields. Mass is transferred inwards, angular momentum outwards. Figure 1.22 shows the evolution of the surface density of a standard thin accretion disk in which an arbitrarily prescribed viscosity produces the angular momentum transport [323]. The surface density decreases in the inner regions as material is accreted onto the star, and the outer regions expand as a result of angular momentum transport.

8. Planets form in the disk, starting with accretion of solid particles. Although planet formation will not be discussed in detail in this book, it should be pointed out that the process is closely coupled with the evolution of the disk. Firstly, the accretion of solid cores of 10–15 earth masses, which are indicated for the giant planets, must occur on a time scale short enough so that these planets can capture gas before the disk is dissipated. The dissipation occurs after an average disk lifetime of only of a few Myr, as a result of viscous evolution, photoevaporation of the disk gas, and stellar winds. Secondly, once the planetary core has reached a mass of order $1\,M_\oplus$, its gravity results in density perturbations in the nearby disk, resulting in a torque on the planet, causing its orbital radius to decrease. The

1.3 Star Formation Phases

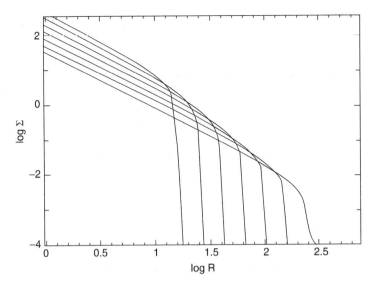

Fig. 1.22 Viscous evolution of a disk. The surface density of the disk Σ, mass per unit area integrated vertically through the disk, is plotted as a function of radius at various times. The initial state is given by the disk with the smallest outer radius, and the outer radius increases with time. Units are arbitrary, but the typical disk evolution time scale is 10^5–10^6 yr. Reproduced, with permission from the RAS, from D. N. C. Lin and J. E. Pringle: *MNRAS* **225**, 607 (1987). © 1987 Royal Astronomical Society

migration rate increases as the planet mass increases, and decreases as the disk mass decreases. Thirdly, once the planet reaches roughly 0.1–1 Jupiter masses (about 100 $M_\oplus$), its tidal effect on the disk can open up a gap in the nebula at the position of the planet and eventually suppress further accretion onto the planet. Inward orbital migration continues after the opening up of the gap, but at a slower rate than before. Observations of young-star disks with sufficient resolution could in principle detect the gaps and thereby infer the presence of giant planets.

The above sequence of events would lead, therefore, to a single star with a planetary system. However, surveys of star-forming regions [289, 294] indicate that most stars form in clusters. It is true that in the Taurus star formation region most of the stars seem to be forming as isolated objects, but in the Rho Ophiuchi cloud, 100 stars are seen forming in a cluster, in the L1641 molecular cloud in Orion about 85% of the stars have formed in clusters, and in the L1630 cloud in Orion the cluster mode also dominates, with most of the stars in 3 rich clusters[292]. Another example is the young cluster of stars in the center of the 30 Doradus giant HII region in the Large Magellanic Cloud (e.g. [543]). It is also true that many solar-type stars are found in binary or multiple systems rather than as single stars. Finally, primordial stars were probably not influenced significantly by the presence of dust or magnetic fields. The simple pathway outlined above will have to be modified significantly to take into account these other possibilities.

1.4 Appendix to Chap. 1: Derivation of the Free-Fall Time

The free-fall time (1.1) is derived for an assumed initial spherical cloud with zero temperature and therefore zero pressure, but with a finite mass and constant initial density ρ_0. If m is the mass enclosed within radius r, then r is a function of (m, t) where t is the time since beginning of collapse. Although r is a function of two variables, we use the ordinary derivative dr/dt for the velocity, to indicate that the derivative is taken in the Lagrangian sense, that is, following the motion of a mass element. For each such element, the equation to be solved is

$$\frac{d^2 r}{dt^2} = -\frac{Gm}{r^2}. \tag{1.9}$$

The initial condition is

$$r(m, 0) = r_0(m), \quad \frac{dr}{dt}(m, 0) = 0. \tag{1.10}$$

Multiply both sides of (1.9) by dr/dt and use

$$\frac{d}{dt}\left(\frac{dr}{dt}\right)^2 = 2\frac{dr}{dt}\frac{d^2 r}{dt^2}$$

to obtain

$$\frac{d}{dt}\left(\frac{dr}{dt}\right)^2 = -\frac{2Gm}{r^2}\frac{dr}{dt}. \tag{1.11}$$

Now integrate both sides over time to obtain

$$\left(\frac{dr}{dt}\right)^2 = 2Gm\left(\frac{1}{r} - \frac{1}{r_0}\right). \tag{1.12}$$

Each mass point is defined by its initial value: $m = 4\pi r_0^3 \rho_0 / 3$. Substituting into (1.12) one obtains

$$\frac{dr}{dt} = -\left[\frac{8\pi G r_0^2}{3}\left(\frac{r_0}{r} - 1\right)\right]^{1/2} \tag{1.13}$$

where the minus sign corresponds to collapse. Now introduce a new variable θ:

$$r(m, t) = r_0(m) \cos^2 \theta \tag{1.14}$$

which has the property $\theta = 0$ at $t = 0$, so that (1.13) becomes

$$\frac{d\theta}{dt} \cos^2 \theta = \frac{1}{2}\left(\frac{8\pi G \rho_0}{3}\right)^{1/2}. \tag{1.15}$$

1.4 Appendix to Chap. 1: Derivation of the Free-Fall Time

This equation may be integrated to obtain

$$\theta + \frac{1}{2}\sin 2\theta = \left(\frac{8\pi G\rho_0}{3}\right)^{1/2} t. \tag{1.16}$$

Then, from the definition of θ, all mass elements reach zero radius at the same time (as long as ρ_0 is a constant), when $\theta = \pi/2$, corresponding to a time

$$t_{\rm ff} = \left(\frac{3\pi}{32G\rho_0}\right)^{1/2}. \tag{1.17}$$

This time is identified as the free-fall time, and in this idealized problem the density $\rho(m, t_{\rm ff})$ becomes infinite then for all m.

Chapter 2
Molecular Clouds and the Onset of Star Formation

This chapter is concerned with the main physical processes that lead to star formation in gas that has already evolved to the molecular cloud state. Although this material is already relatively cold and dense, in general it has too much magnetic, rotational, thermal, and turbulent energy to allow collapse into low-mass stars. A number of physical processes, however, tend to dissipate these energies, or allow local collapse in spite of them, on time scales up to 10^7 yr. The main effects that need to be considered are heating, cooling, the role of shock waves, magnetic braking of rotation, diffusion of the magnetic field with respect to the gas, and the generation and decay of turbulence. After a review of the general properties of molecular clouds, the chapter discusses the physical processes relevant to the phase known as "star formation" (Table 1.1). It then discusses three important scenarios: magnetically controlled star formation, turbulence-controlled star formation, and induced star formation.

2.1 Molecular Cloud Properties

The molecular gas in the galaxy exhibits structure over a wide range of scales, from 20 pc or more in the case of giant molecular clouds down to 0.05 pc for molecular cloud cores. The clouds have been characterized as both clumpy and filamentary. The general properties on selected scales are listed in Table 2.1. In fact each type of structure exhibits a range of properties [335]. The clumpiness of a molecular cloud can be characterized by a so-called *volume filling factor* f_f. If a component of a molecular cloud has a particle density n (in cm^{-3}) which is greater than $<n>$, the average density of the molecular cloud as a whole, then $f_f = <n>/n$. Thus roughly f_f is the probability that the matter has density n. On the molecular core scale, where $n > 10^5$, and where star formation is believed to occur, this factor is only about 0.001. In addition to the listed properties, the structures on various scales exhibit rotation and magnetic fields, which are discussed in later sections of this chapter.

Table 2.1 Properties of molecular clouds

	Giant molec. cloud	Molecular cloud	Molecular clump	Cloud core
Mean radius (pc)	20	5	2	0.08
Density $n(H_2)$ (cm^{-3})	100	300	10^3	10^5
Mass ($M_\odot$)	10^5	10^4	10^3	10^1
Linewidth (km s^{-1})	7	4	2	0.3
Temperature (K)	15	10	10	10

On the giant molecular cloud scale, given an approximate mass one can calculate the gravitational potential energy and find that it is far greater in absolute value than the thermal energy. Thus the clouds are certainly highly gravitationally bound, but in general they are not collapsing. The linewidths in CO (several km/s) are always greater than the thermal width at 10 K (full width at half maximum of 0.13 km/s for CO), implying that some other mechanism, such as turbulence or magnetic fields, is helping to support the cloud against gravity. In fact, if these linewidths are caused by turbulence and/or magnetic fields, the deduced energies are close to sufficient to support the clouds. Inside molecular clouds, the gas pressures are higher than those in the surrounding interstellar medium, and almost all molecular clouds in our galaxy exhibit star formation. The deduced mean lifetime of molecular clouds is 10 Myr (subject to considerable argument). The suggested mechanism for breakup and destruction of the clouds is ionization effects from the most massive stars, particularly near the edge of the cloud, where the hot gas in the ionized region drives a flow of material away from the cloud. Evidence for this value of the age is provided by the fact that intermediate age clusters, with ages >10 Myr, do not have associated molecular material. By means of a similar method, the typical lifetime of a molecular cloud in the Large Magellanic Cloud is estimated to be 27 Myr ±50% [59].

Although the main constituents of molecular clouds are molecular hydrogen and atomic helium, it is very difficult to observe spectral lines of these species because the required excitation conditions do not exist, except possibly under special conditions, such as shocks. That is, the equivalent temperature needed to lift H_2 molecules to their lowest excited states, so they can emit a photon, is 512 K. Instead, trace species are used, in particular, emission lines of CO. The abundance of CO relative to H_2 is estimated to be about 10^{-4}, but with considerable uncertainty. The molecule can be observed in three different forms: $^{12}C^{16}O$, $^{13}C^{16}O$, and $^{12}C^{18}O$, the first of which is usually optically thick in molecular clouds, the second of which is marginally so, and the third of which is optically thin. At the higher densities in molecular clouds ($n \geq 10^5$ cm^{-3}) CO ceases to be a reliable estimator of the density of H_2, in part because the lines become optically thick, and in part because the CO tends to freeze out on grain surfaces at high densities, so that the gas phase is depleted [13].

To measure masses and densities, the main observed quantity is the particle column density N, that is, the number of molecules along the line of sight through,

2.1 Molecular Cloud Properties

say, a clump, divided by the projected area of the clump. To measure this quantity one needs an optically thin line, so that radiation from all molecules emitting that line is seen; on the larger scales $^{13}C^{16}O$ is often used for this purpose. The mechanism for production of the line radiation is collisional excitation of a low-lying level (for CO the equivalent excitation temperature is around 5 K) by the dominant molecular H_2, followed by radiative de-excitation; thus the strength of the line, integrated over the line profile (that is, over all velocities in the line) is proportional to the local emitting particle density. The details of the complicated and somewhat uncertain conversion from line emission to column density of $^{13}C^{16}O$, and then from column density of $^{13}C^{16}O$, to total column density N_{tot}, are given in [481]. Then the size of the clump is estimated from the linear scale of the region over which the line intensity drops by a factor e. Given the distance, the mass then follows from

$$M = \int m N_{tot} dA \qquad (2.1)$$

and the mean number density n (particles per unit volume) is obtained from

$$n = \frac{M}{mV} \approx \frac{3 N_{tot}}{4R} \qquad (2.2)$$

where m is the mean particle mass, V is the total volume, and R is the radius. The dimension of the cloud along the line of sight is not observed, but one assumes the cloud is spherical or cylindrical with length along the line of sight comparable to the linear dimension on the plane of the sky. Thus the typical column density of a molecular clump, with $n \approx 10^3$ cm^{-3} and a radius of 2 pc is 8×10^{21} particles per cm^2. An example of an observation in CO is given in Fig. 2.1.

There is a fundamental limitation to the use of a given molecule such as CO as a density probe, related to the concept of "critical density", defined as

$$n_{crit} = \frac{A_{ij}}{\gamma_{ij}} \qquad (2.3)$$

where A_{ij} is the Einstein probability for the radiative downward transition $i \to j$ per particle in level i per unit time, and $n\gamma_{ij}$ is the probability of a downward collisionally-induced transition per particle in level i per unit time. The critical density is that where the two probabilities are equal ($\gamma_{ij} n_{crit} = A_{ij}$). Thus at densities above n_{crit} the line intensity is no longer clearly related to the density, because the collisionally induced transition produces no photon. For the $J = 1 \to 0$ transition in CO, $n_{crit} = 3 \times 10^3$ particles cm^{-3}. In practice, a given transition is a good probe of regions somewhat below or near the critical density, but not above it.

Thus on the clump scale (Table 2.1) the $^{12}C^{16}O$ is near n_{crit} and it is useful for picking out density peaks on that scale (Fig. 2.1). But in general, no single observational tracer can represent a molecular cloud on all scales. For clumps, $^{13}C^{16}O$ or $^{12}C^{18}O$ can be used because of their much lower abundance relative to $^{12}C^{16}O$. On the even smaller scales of cores (0.05–0.1 pc), density can be traced

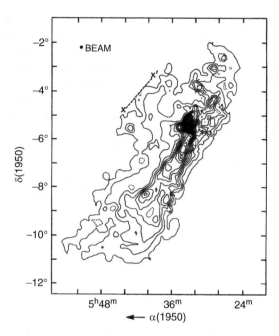

Fig. 2.1 Contours of equal intensity on the plane of the sky in the Orion A cloud. The transition is the $J = 1 \rightarrow 0$ in the $^{12}C^{16}O$ molecule at 2.6 mm. The intensity is integrated over the line width. The *dark region* is the star-forming dense gas associated with the Orion Nebula; it has a radius of roughly 2 pc. Reproduced, by permission of the AAS, from [337]. © The American Astronomical Society

by a transition in CS at 3.1 mm ($n_{crit} = 4.2 \times 10^5$) and one in H_2CO at 2.1 mm ($n_{crit} = 1.3 \times 10^6$).

Another way of getting the column density in a region is to use visual extinction of starlight by intervening dust. Dust, in the form of small particles with characteristic size 0.1–0.2 µm, accounts for about 1% of the mass of interstellar material. For example, there is an empirical relation

$$A_v = (N_H + 2N_{H_2})/2 \times 10^{21} \tag{2.4}$$

where the extinction A_v is given in magnitudes, N_H is the column density in particles cm^{-2} for atomic hydrogen, and N_{H_2} is the column density in molecular hydrogen [70]. To calibrate this relation, the quantity N_H is obtained, via satellite observations, from the strength of Lyα absorption of light from hot OB stars by material in intervening clouds, while N_{H_2} is obtained similarly from absorption in the Lyman molecular bands. Then A_v is obtained from the reddening of the light, also from the OB stars. The dust also provides thermal radiation at the dust temperature; this continuous spectrum in the 1 mm wavelength range, provides a nice complement to the molecular line radiation. The dust temperature, which is

2.1 Molecular Cloud Properties

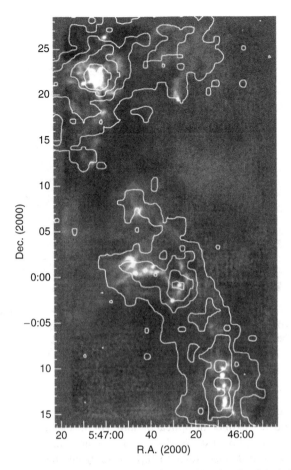

Fig. 2.2 Observed structure of a portion of the Orion B molecular cloud, both in the continuum at 850 μm (grey scale) and in a molecular transition of CS (contours). The bright spots in the continuum, an indicator of high dust density, correspond in general to the highest contours in CS emission, an indicator of high gas density. Those peaks have a size scale of ≈0.1 pc, and would be considered "cores", with masses in the range 0.2–12 $M_\odot$. Reproduced by permission of the AAS from [243]. © The American Astronomical Society

determined from analysis of the properties of the far infrared/millimeter emission as a function of wavelength, may be different from the gas temperature, which is determined from the relative strengths of molecular emission lines arising from energy levels of different excitation energy. Particularly at the lower densities, the gas and dust are not well coupled by collisions, and the physical processes that determine their equilibrium temperature are different. Figure 2.2 shows observations of cores both from continuum 850 μm observations of the dust and molecular line emission which traces the gas. The scale of the figure (top to bottom) is approximately 4 parsec. Cloud cores can also be observed by visual extinction of

Fig. 2.3 The Cone nebula, in the star-forming region NGC 2264, taken by the Advanced Camera for Surveys (ACS) on Hubble Space Telescope, April 2, 2002. The portion of the dark nebula that is shown measures about 0.8 parsec in length. Credit: NASA, H. Ford (JHU), G. Illingworth (UCSC), M. Clampin (STScI), G. Hartig (STScI), the ACS Science Team, and ESA

starlight (Fig. 3.2). Figure 2.3 shows a related observation of a clump-scale object that is being eroded by ionizing radiation from O and B stars. In this case the cold molecular gas and dust extinct the optical radiation of the background ionized region produced by the nearby hot O star S Monocerotis in the young cluster NGC 2264.

A number of well-defined relationships arise from the data, including the linewidth-size relation, the condition of near virial balance, and the mass spectrum (number of objects per unit mass interval as a function of mass). If the CO line widths are determined as a function of scale, a well-defined linewidth-size relation appears, as first determined by Larson [301]. Over a scale range (L) of 0.05–60 pc, he found $\Delta v \approx L^{0.38}$ where L is given in pc and Δv is the observed full width of a spectral line at half maximum, in km s^{-1} (see Fig. 2.4). This relation is similar to what one would expect for laboratory incompressible turbulence, which gives the so-called Kolmogorov spectrum with the exponent 1/3. Subsequent studies have confirmed the existence of the linewidth-size relation, but there is appreciable scatter in the data, and various determinations [224, 312, 475] give power law exponents ranging from 0.4 to 0.6, depending on the details of how velocities and sizes are measured. In view of the uncertainty, a value of 0.5 is commonly used. At the low-mass core scale (0.1 pc), the linewidth-size relation changes, so that there is no longer any significant correlation between linewidth and size [191]. On this scale, the line widths are usually dominated by the thermal component,

2.1 Molecular Cloud Properties

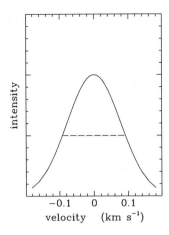

Fig. 2.4 Sketch of a molecular emission line, giving intensity in arbitrary units as a function of velocity in the line of sight. Zero velocity corresponds to the rest wavelength at the center of the line. The linewidth Δv is measured according to the dashed line. The line strength is the intensity integrated over all velocities. The profile shown is assumed to be a Gaussian. If it is assumed to arise only from thermal motions of the molecule, then $\Delta v_{\text{th}} = [8k_B T (\ln 2)/m]^{1/2}$, where m is the mass of the molecule in grams and k_B is the Boltzmann constant. In the figure, for the case of the ammonia molecule NH_3, the measured $\Delta v = 0.188$ km s^{-1} corresponds to a temperature of 13 K. Note that the velocity dispersion, namely the isothermal sound speed or equivalently the root mean square of the velocity of a particle in one dimension, $\sigma_{\text{th}} = (k_B T/m)^{1/2} = \Delta v/2.355$. If turbulence is also present, then the linewidth $\Delta v_{\text{tot}}^2 = \Delta v_{\text{th}}^2 + \Delta v_{\text{turb}}^2$ where $\Delta v_{\text{turb}} = 2.355\sigma_x$ and σ_x is the one-dimensional turbulent velocity dispersion, which is independent of the particle mass

$\Delta v / \Delta v_{therm} \approx 1.3$, so the non-thermal component, while still present, is now subsonic.

It is not clear exactly what this relation means. The standard interpretation is that some kind of compressible, highly supersonic, turbulence could be present, with properties somewhat different from laboratory turbulence (lab turbulence is subsonic). Under this basic assumption, one can approximately state Larson's findings (often referred to as "Larson's Laws") as follows. The first is the linewidth-size relation [475]:

$$\sigma_x \approx (0.7 \pm 0.07) R_{pc}^{0.5 \pm 0.1} \text{ km s}^{-1} \qquad (2.5)$$

where σ_x is the one-dimensional velocity dispersion, in this case the mean turbulent velocity measured in the line of sight, and R is approximately the radius of the cloud. Clearly on all scales larger than about 0.1 pc, the turbulence is supersonic, because the thermal velocity dispersion at 10 K is 0.19 km s^{-1} for a gas with mean molecular weight 2.37 (approximately 70% H and 28% He by mass).

The second finding is essentially that molecular clouds and clumps within them are gravitationally bound and are close to virial equilibrium. The Virial Theorem for a quasi-spherical cloud in force balance can be written

$$2E_{\text{kin}} + 2E_{\text{th}} + E_{\text{mag}} - 3P_{\text{surf}}V + E_{\text{grav}} = 0. \qquad (2.6)$$

where E_{kin} is the total macroscopic kinetic energy, including rotation and turbulence, E_{th} is the total thermal energy, E_{mag} is the total magnetic energy (magnetic surface terms are not considered here), E_{grav} is the total gravitational energy, P_{surf} is the external pressure, and V is the total volume. Assuming that the kinetic energy is dominated by turbulence, that the only other term of importance is E_{grav}, and that the cloud can be approximated by a uniform-density sphere with mass M, then the Virial Theorem gives

$$M\sigma^2 = \frac{3}{5}\frac{GM^2}{R} \qquad (2.7)$$

where the three-dimensional velocity dispersion σ is related to σ_x by $\sigma^2 = \sigma_x^2 + \sigma_y^2 + \sigma_z^2$. Assume isotropic velocities and solve for the mass, which is denoted by the virial mass M_{vir}:

$$M_{vir} = \frac{5\sigma_x^2 R}{G}. \qquad (2.8)$$

For an actual cloud one defines $\alpha_{vir} = M_{vir}/M$, so that a cloud with $\alpha_{vir} = 1$ is in virial balance but not necessarily in hydrostatic equilibrium, because of the assumption regarding uniform density. But if $\alpha_{vir} < 1$ then the cloud is definitely not in force balance and is unstable to collapse.

Note that the mean column density $\bar{N}_H = M/(\pi m R^2)$, where m is the mean mass (in grams) per particle and R is the radius of a clump, so that $\sigma^2 \propto \bar{N}_H R$, so $\bar{N}_H$ = const. (there is some observational evidence in support of this but the suggestion of virial equilibrium according to (2.8) is oversimplified). The conclusion that all molecular clouds have similar column densities, on all length scales where the linewidth-size relation applies, is Larson's third finding, even though it can be derived from the first two. One can also deduce from Larson's findings that $M \propto R^2$ and $\rho \propto R^{-1}$.

Another explanation of Larson's findings [459] is that they represent virial equilibrium of masses on various scales with the magnetic field supporting the region against gravity. The linewidth in that case would arise from Alfvén waves, and it can easily be shown that $\Delta v \propto R^{0.5} B^{0.5}$ (see below).

The mass spectrum of the molecular clouds, clumps and cores has been measured over a wide range of masses by several different methods. The results are usually expressed in terms of a power law

$$dN/dM = \text{const.} \times M^{-x} \qquad (2.9)$$

where dN is the number of clouds in the mass range M to $M + dM$. On the largest scales, corresponding to giant molecular clouds in the mass range above about 3×10^4 $M_\odot$, the preferred method is to use the total luminosity in a spectral line of ^{12}CO or ^{13}CO as an indicator of mass. The latter molecule is preferred because it is usually optically thin, but the former is often used simply because its intensity is greater. In either case corrections must be applied [352] to convert from molecular column density to total mass. The typical result [552] is $x \approx 1.6$ with an upper mass cutoff at about 6×10^6 $M_\odot$.

2.1 Molecular Cloud Properties

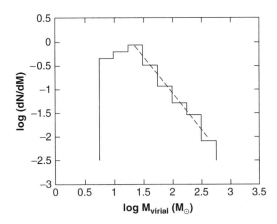

Fig. 2.5 Mass spectrum of clumps in the L1630 molecular cloud (Orion B). The spectrum is plotted as a function of the virial mass in solar masses. The *dashed line* represents a slope of −1.6. Below about 20 M$_\odot$ the observations are incomplete. Reproduced by permission of the AAS from [293]. © The American Astronomical Society

On smaller scales, one can use line widths and sizes of various regions to get $M_{\rm vir}$ from (2.8). An example of such a mass spectrum obtained by use of the CS molecule in the Orion region, in the mass range 20–500 M$_\odot$, is shown in Fig. 2.5, where the power-law index is also $x = 1.6$. This spectrum does not match that of stars, which if fitted to a power law in the mass range 2–10 M$_\odot$, has an exponent of about −2.3. The diagram shown represents clumps in the Orion region; however very similar relations are obtained in different molecular clouds and with different line tracers; the power law seems to hold in a wide range of masses. However there is somewhat of a range in the power laws deduced, so a better representation of the situation would be that x falls in the range 1.3–1.9 [335]. Still, the power law in the clump mass spectrum is consistent with that for molecular clouds as a whole.

At even lower masses, in the core-mass range, a somewhat different result is obtained by a different method [19, 370]. Continuum dust emission at 1.3 mm is used to estimate actual masses (rather than virial masses). The flux in the (optically thin) millimeter continuum can be converted, using a standard dust-to-gas ratio, to column density N_H and then to total mass (the method is described in more detail in Chap. 4). The observations show that pre-stellar condensations in the Rho Ophiuchi region follow approximately $dN/dM \propto M^{-1.5}$ below 0.5 M$_\odot$, but above that mass it steepens to $dN/dM \propto M^{-2.5}$. This observation (which has been confirmed by other studies) is not too different from the stellar mass spectrum, which gives $dN/dM \propto M^{-2.3}$ for masses above 0.5 M$_\odot$ but $dN/dM \propto M^{-1.3}$ for lower masses. The switch in slopes occurs at about 0.6 M$_\odot$, at about the same mass where a corresponding change occurs in the stellar mass spectrum. There is a factor 2 uncertainty in core masses, and it is not entirely clear that the slope change is not due to some selection effect, but this result suggests that the IMF for stars in clusters is determined at the pre-stellar stage.

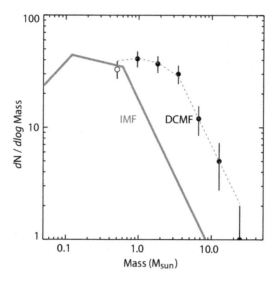

Fig. 2.6 Mass function of dense molecular cores in the Pipe Nebula (*filled circles*), compared with the initial mass function for stars in the Orion Nebula cluster (*grey line*). The *dashed line* refers to the stellar mass function shifted to higher masses by a factor of about 4. Credit: J. F. Alves, M. Lombardi, C. J. Lada: *Astron. Astrophys.* **462**, L17 (2007), reproduced with permission. © European Southern Observatory

An independent method for the core mass function [15] involves observations of the nearby (130 pc) molecular cloud known as the Pipe Nebula, whose background consists of large numbers of stars in the Galactic bulge. By measuring the near IR extinction of those background stars caused by the dust in low-mass dense cores in the Pipe Nebula, one can obtain the mass of each core, using the standard conversion (2.4) between extinction in magnitudes and hydrogen column density N_H. The results are shown in Fig. 2.6 (note that the mass function, not the mass spectrum, is plotted). The advantage of this method is that it avoids the uncertainties in the determination of masses through dust emission. The typical core measured here has density $\sim 10^4$ particles/cm^3. The broad grey line is the stellar IMF as measured in the Orion Nebula cluster. The circles with error bars are the mass function determined in the dense cores. The dotted grey line is the stellar IMF shifted over by a factor 4 in mass. Clearly the shape of the mass function of the dense cores is very similar to that of the stars and the change in slope at lower masses is evident; the change in slope occurs at 2–3 $M_\odot$, rather than about 0.6 $M_\odot$ for the IMF. If a correction for background extinction is made then the difference in mass between the two curves reduces to a factor 3. The conclusion is that the stellar IMF is set by the mass function of dense cores, and there is an efficiency factor of about 30% in the process of turning the mass of the core into a star. This conclusion is supported theoretically by a calculation [460] that includes the effects of magnetic fields and outflows and which also shows an efficiency of about 30%.

2.2 Initial Conditions for Star Formation

The requirement on a region in a molecular cloud that must be satisfied before it can collapse and form a star was first derived by Sir James Jeans [241]. His method was a linear stability analysis performed on the basic hydrodynamic equations, assuming an isothermal gas (Problem 2), including only thermal and gravitational effects. A simpler method of defining the initial conditions for collapse, which, however, includes all relevant physical effects, is to require that the absolute value of the gravitational energy must exceed the sum of the thermal, rotational, turbulent, and magnetic energies. This requirement defines a mass (M) of gas that is gravitationally bound. For this mass to be as small as a solar mass, the requirement can be satisfied only in the coolest, densest parts of the interstellar medium. Thus the requirement is

$$|E_{\text{grav}}| > E_{\text{th}} + E_{\text{rot}} + E_{\text{turb}} + E_{\text{mag}}. \qquad (2.10)$$

For an assumed spherical configuration,

$$E_{\text{grav}} = -C_{\text{grav}} \frac{GM^2}{R} \qquad (2.11)$$

where C_{grav} is a constant depending on the mass distribution and equals 3/5 for uniform density. The total thermal energy for an isothermal ideal gas with temperature T is

$$E_{\text{th}} = \frac{3}{2} \frac{R_g T M}{\mu} \qquad (2.12)$$

where $R_g = k_B/m_u$ is the gas constant, k_B is the Boltzmann constant, m_u is the atomic mass unit, and μ is the molecular weight of the gas in atomic mass units. The rotational energy is

$$E_{\text{rot}} = C_{\text{rot}} M R^2 \Omega^2 \qquad (2.13)$$

for an assumed uniform angular velocity Ω, where C_{rot} depends on the mass distribution, and equals 1/5 for uniform density. The turbulent kinetic energy is

$$E_{\text{turb}} = \frac{1}{2} M \sigma^2 \qquad (2.14)$$

where σ is the mean turbulent velocity, as in (2.7). The magnetic energy is given by the volume integral

$$E_{\text{mag}} = \frac{1}{8\pi} \int B^2 dV \approx \frac{1}{6} B^2 R^3 \qquad (2.15)$$

where B is the assumed uniform magnetic field.

Now consider thermal and gravitational effects alone, as did Jeans in his original analysis. Although his analysis contains a physical inconsistency, the result is still very similar to that obtained by energy considerations. The requirement that a uniform-density, uniform-temperature sphere be gravitationally bound $[(3/5)GM^2/R = (3/2)R_g TM/\mu]$ leads to the determination of the Jeans length:

$$R_J = \frac{0.4GM\mu}{R_g T}, \tag{2.16}$$

where $\mu \approx 2.37$ for solar composition with molecular hydrogen. For a cloud of a given mass and temperature, the radius must be smaller than R_J to be unstable to gravitational collapse. Alternatively, we can eliminate the radius from (2.16) in favor of the density ρ, assuming again that the gas is a sphere, to obtain an expression for the Jeans mass, which is the minimum mass that the cloud of given (ρ, T) must have to be unstable:

$$M_J = \left(\frac{5}{2}\frac{R_g T}{\mu G}\right)^{3/2} (\frac{4}{3}\pi\rho)^{-1/2} = 8.5 \times 10^{22} \left(\frac{T}{\mu}\right)^{3/2} \rho^{-1/2} \text{ g}. \tag{2.17}$$

Another commonly used version of the Jeans length is obtained by eliminating the mass in (2.16) in favor of density and radius:

$$R_J \approx \left(\frac{R_g T}{\mu}\right)^{1/2} \frac{1}{\sqrt{G\rho}} \approx c_s t_{\text{ff}} \tag{2.18}$$

where c_s is the isothermal sound speed and t_{ff} is defined by (1.1). Here, given T and ρ, the radius of the cloud must be larger than R_J for collapse to occur.

An alternate form of the thermal Jeans mass is known as the *Bonnor–Ebert mass*. The situation envisaged here is slightly different: an isothermal cloud exists in equilibrium, with the effects of the internal pressure gradient, plus an external confining pressure P_{surf}, balancing gravity. Only gravitational effects and thermal effects are considered in the cloud interior. This problem, involving a bounded isothermal sphere, is treated in more detail in Chap. 3, where it is shown that the corresponding critical length for instability to collapse is essentially identical to (2.16).

We now consider rotational effects in addition to thermal effects and gravity. We define $\alpha = E_{\text{th}}/|E_{\text{grav}}|$ and $\beta = E_{\text{rot}}/|E_{\text{grav}}|$, where the density of the sphere and its angular velocity Ω are assumed constant. The revised expression for the Jeans mass becomes

$$M_J = \left(\frac{\frac{3R_g T}{2\mu} + 0.2\Omega^2 R^2}{0.6G}\right)^{3/2} (\frac{4}{3}\pi\rho)^{-1/2}, \tag{2.19}$$

and $M > M_J$ is the condition for the cloud to collapse. Alternatively,

$$\alpha \leq 1 - \beta \tag{2.20}$$

for collapse, although numerical studies indicate that actually $\alpha \leq 1 - 1.43\beta$ is a more realistic criterion. Clearly rotation has a stabilizing influence, but in the typical observed cloud core β is relatively small (see below). However, even if criterion (2.20) is satisfied and the cloud starts to collapse (assuming conservation of angular momentum), the rotational energy $E_{\rm rot} = J^2/2I \approx J^2/(MR^2)$, where J is the total angular momentum and I is the moment of inertia, increases faster than the gravitational energy, and the collapse could be stopped at relatively low density.

Rotation is detected in molecular clouds and cloud cores through observations of a gradient in the radial velocity dv/ds, where s is a distance in the plane of the sky. This quantity is measured at various points across the cloud, for example, by the Doppler shifts of emission lines in NH_3 [188, 191]. A linear velocity variation with the spatial coordinate across the cloud is consistent with uniform rotation. From the velocity gradient, and from an assumed model of uniform rotation which is consistent with the observations within the errors, the angular velocity $\Omega \approx dv/ds$ and the specific angular momentum $j \approx 0.4\Omega R^2$ are derived. The inclination angle of the rotation axis to the line of sight is not known. About half of the clouds observed show measurable rotational velocities; the remainder are presumably rotating below the observational limit. On the cloud core scale, typical values of Ω are 10^{-13}–10^{-14} s^{-1}. Rotation has also been detected on larger molecular cloud scales. For example, CO measurements indicate that a gradient in the line-of-sight velocity exists in the Orion A cloud (Fig. 2.1), extending from its upper right end to its lower left end [337], implying rotation about an axis perpendicular to the galactic plane, but in the opposite sense to the Galactic rotation.

Rotation does not appear to be a major factor in the support of clouds against collapse. Nevertheless there is an angular momentum problem, as indicated by the data in Table 2.2. An example of the angular momentum problem was stated by Spitzer [476] as follows: Consider a gas cylinder 10 pc long and 0.2 pc in radius (a filament) with density 5×10^{-23} g cm^{-3}. Its mass is about a solar mass. Let it rotate about its long axis, say with the typical galactic rotation, $\Omega = 10^{-15}$ s^{-1}. Its contraction parallel to J is not opposed by rotation, but to reach stellar size the radius perpendicular to J must contract by 7 orders of magnitude. Since if angular momentum is conserved $\Omega R^2 = $ constant, Ω must increase by 14 orders of magnitude, to 10^{-1}. The corresponding rotational velocity of the star then would be 6×10^9 cm s^{-1}, or 0.2 c! The centripetal acceleration would be 10^4 times that of gravity.

Table 2.2 Characteristic values of specific angular momentum

Object	J/M (cm^2 s^{-1})
Binary orbit (10^4 yr period)	4×10^{20}–10^{21}
Binary orbit (10 yr period)	4×10^{19}–10^{20}
Binary orbit (3 day period)	4×10^{18}–10^{19}
100 AU disk (1 M$_\odot$ star)	4.5×10^{20}
T Tauri star (rotation)	5×10^{17}
Jupiter (orbit)	10^{20}
Present Sun (rotation)	10^{15}
Molecular clump (scale 1 pc)	10^{23}
Cloud core (scale 0.1 pc)	1.5×10^{21}

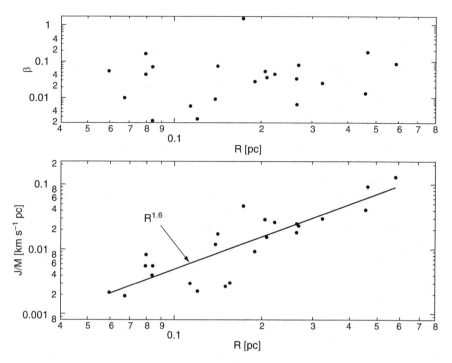

Fig. 2.7 Ratio β of rotational energy to absolute value of gravitational energy (*upper*), and specific angular momentum (*lower*), as a function of the size of a molecular cloud core, based on observations of the velocity gradient across the core. $1\,\text{km s}^{-1}\,\text{pc} = 3.08 \times 10^{23}\,\text{cm}^2\,\text{s}^{-1}$. Reproduced, by permission of the AAS, from [191]. © 1993 The American Astronomical Society

The results concerning rotation show that first, the angular velocity Ω is similar on all scales; it is only slightly higher in cloud cores than in molecular cloud clumps as a whole, suggesting magnetic coupling between the two regions. Second, $j = J/M$ decreases to smaller scales, also suggesting magnetic braking, but this conclusion is not entirely clear because different mass elements are being sampled. A survey [191] of rotational velocities in cloud cores on scales of 0.06–0.60 pc shows values of $j \sim 6 \times 10^{20}$ to $3 \times 10^{22}\,\text{cm}^2\,\text{s}^{-1}$. A fit to the data gives $j \propto R^{1.6}$, implying $\Omega \propto R^{-0.4}$, where R is the size of the core. The values of β and j derived from observations [191] are shown in Fig. 2.7. Note that β is small on average, only 2 or 3%. But, third, if the cores were to collapse with conservation of angular momentum into disks, they would reach equilibrium with an outer keplerian radius of 180–4,500 AU, comparable in size to many observed disks. The fact that cloud cores contain far too much angular momentum to be compatible with stellar rotational velocities of, for example, T Tauri stars, is demonstrated in Table 2.2.

2.2 Initial Conditions for Star Formation

Turbulent motions are important in molecular clouds, and one can define a turbulent Jeans mass in analogy to the thermal Jeans mass. Equating turbulent energy to gravitational energy

$$M_{\text{turb}} = \frac{5\sigma_x^2 R}{2G} \tag{2.21}$$

where σ_x is the turbulent velocity dispersion in the line of sight [see (2.5)]. This mass is clearly closely related to M_{vir}, as defined by (2.8), but in that equation the Virial Theorem is used, which accounts for the additional factor 2. For example, in a turbulent core of radius 0.1 pc with a velocity dispersion of 1 km s^{-1} the mass must be above 50 $M_\odot$ for collapse. This situation is appropriate for the formation of massive stars (Chap. 5), but, as mentioned earlier, in low-mass cores the thermal effect dominates.

Finally we consider the magnetic Jeans mass, under the assumption that thermal, turbulent, and rotational effects are unimportant. Using the magnetic energy as simply $E_{\text{mag}} = \frac{B^2}{8\pi}\frac{4}{3}\pi R^3$ one obtains instability to collapse in a cloud of density ρ, radius R, and uniform magnetic field B if its mass is greater than M_ϕ, where

$$M_\phi = \frac{BR^2}{(3.6G)^{1/2}} = \left(\frac{5}{18\pi^2 G}\right)^{1/2} \phi = \frac{B^3}{(3.6G)^{3/2}(\frac{4}{3}\pi\rho)^2}$$

$$\approx 10^3 M_\odot \left(\frac{B}{30\mu G}\right)\left(\frac{R}{2\text{pc}}\right)^2 \tag{2.22}$$

where ϕ is the magnetic flux. Thus on the clump scale (Table 2.1) where typical measured fields are around 30 μG, the magnetic Jeans mass is close to the clump mass.

If one defines a closed loop threaded by a uniform magnetic field B, the magnetic flux is defined by

$$\phi = \int_S \mathbf{B}\cdot\mathbf{n}\,dA \tag{2.23}$$

where S is the surface enclosed by the loop and $\mathbf{n}$ is the normal to that surface. The magnetic induction equation reads

$$\frac{\partial B}{\partial t} = \nabla\times(\mathbf{v}\times\mathbf{B}) - \nabla\times(\eta_e \nabla\times\mathbf{B}) \tag{2.24}$$

where $\mathbf{v}$ is the velocity vector and η_e is the electric resistivity (also referred to as the magnetic diffusivity)

$$\eta_e = \frac{c^2}{4\pi\sigma_e}. \tag{2.25}$$

Here c is the velocity of light and σ_e is the electric conductivity. The second term on the right-hand side of (2.24) represents the decrease in magnetic field as a result of

Ohmic dissipation. The ratio of the first and second terms is known as the *magnetic Reynolds number*, usually expressed as the dimensionless quantity

$$R_{em} = \frac{vL}{\eta_e} \qquad (2.26)$$

where L is a characteristic length and v is a characteristic velocity. If $R_{em} \gg 1$, then dissipative effects are negligible. In the interstellar gas, σ_e is high enough so that indeed R_{em} is very large, so that the second term in (2.24) can be neglected. Then, using the condition $\nabla \cdot \mathbf{B} = 0$ and the Gauss theorem, one can prove that, as the material associated with the loop evolves in time, the flux remains constant, a condition known as *flux-freezing*.

If the cloud is collapsing quasi-spherically, then the magnetic flux $\phi \approx \pi B R^2$ remains constant as the cloud collapses, and (2.22) shows that M_ϕ is constant, that is, it would not be possible for the cloud, for example the $1,000 \, M_\odot$ clump just mentioned, to fragment into smaller masses. Furthermore, under these conditions of quasi-spherical collapse $B \propto \rho^{2/3}$ during collapse. In general, however, if the collapse is not spherical or the field is not frozen in, the field will increase as $B \propto \rho^\kappa$, implying $M_\phi \propto \rho^{-3(2/3-\kappa)}$. As long as $\kappa < 2/3$, M_ϕ decreases on compression, and fragmentation eventually becomes possible. It has generally been believed, on both theoretical and observational grounds, that $\kappa \approx 0.5$. In this case, as a cloud contracts, magnetic energy becomes less important relative to gravitational energy. However an extensive set of measurements [126] suggests that κ is actually higher, closer to 2/3, at densities ranging up to those in molecular cloud cores (Fig. 2.9).

The criterion for collapse in the presence of a magnetic field is often stated as a critical mass-to-flux ratio:

$$\left(\frac{M}{\phi}\right)_{crit} = \frac{0.17}{\sqrt{G}} \qquad (2.27)$$

where the constant depends on the details of the geometry. If the ratio of the actual M/ϕ to the critical value is greater than unity, the cloud is *supercritical* and can contract. In the *subcritical* case (ratio < 1), the field dominates, preventing overall contraction; however the cloud can contract in the direction parallel to field lines and become quite flattened (neglecting other effects).

The magnetic field influences the evolution of molecular cloud material in various ways. It is important, first, for at least partial support of the magnetic cloud clumps against collapse. If they were collapsing, the star formation rate would be far higher than currently observed. The typical observed field at the mean clump density is $30 \, \mu G$. Consideration of the magnetic energy and turbulent energy in comparison with the gravitational energy indicates that the clumps are unlikely to collapse. The measured magnitude of the field at these early stages leads to the magnetic flux problem. For the same interstellar filament of about $1 \, M_\odot$ that we considered in connection with the angular momentum problem, suppose it has a magnetic field of $3 \, \mu G$ (appropriate for the low-density interstellar gas) parallel to the long axis. Then the magnetic flux $\pi B R^2 = \pi \times 3 \times 10^{-6} \times (6 \times 10^{17})^2 \approx \pi \times 10^{30}$. Material

2.2 Initial Conditions for Star Formation

can contract along the magnetic field, and we assume that the contraction across the magnetic field takes place with conservation of flux. The final stellar radius is about 6×10^{10} cm, about 7 orders of magnitude smaller than that of the cloud, so the field must increase by 14 orders of magnitude to 3×10^8 gauss, much higher than observed (the mean field on the solar surface is only 1 gauss) and giving a magnetic energy about the same as the gravitational energy, even if the field is uniform through the star. Despite extensive calculations [318], this difficult problem has not yet been resolved.

A second important effect is the braking of rotation. The field is the most likely mechanism to transport angular momentum out of the dense cores of molecular clouds, as discussed in more detail below. Third, if a region of the cloud is subcritical, then evolution of the region to the point of collapse requires that the field diffuse with respect to the matter. Estimated field diffusion times (3×10^6–10^7 yr) are consistent with the spread of ages of stars in at least some young clusters. This process, known as *ambipolar diffusion*, or plasma drift, is treated in more detail in Sect. 2.6. Finally the suprathermal line widths observed in molecular cloud clumps could be accounted for by the Alfvén waves associated with the field.

It is simple to derive an approximate relation between line width Δv, field strength, and linear scale. Assuming approximate balance between magnetic and gravitational forces:

$$\frac{4}{3}\pi R^3 \frac{B^2}{8\pi} = \frac{GM^2}{R} \tag{2.28}$$

and multiplying and dividing the left-hand side by ρ one obtains

$$V_A^2 \approx \frac{2GM}{R} \approx 2\pi G \Sigma R \tag{2.29}$$

where $\Sigma = M/(\pi R^2)$ is the surface density (in g cm^{-2}), and the Alfvén velocity

$$V_A = \frac{B}{\sqrt{4\pi\rho}}. \tag{2.30}$$

Then from the critical mass-to-flux ratio, and assuming that the Alfvén speed determines the line width

$$\frac{\Sigma}{B} = \frac{0.17}{\sqrt{G}} \tag{2.31}$$

and

$$\Delta v = 0.79(\sqrt{G}\,BR)^{1/2} \approx 1.23 \left(\frac{B}{30\mu G}\right)^{1/2} \left(\frac{R}{1\text{pc}}\right)^{1/2} \text{ km s}^{-1}. \tag{2.32}$$

Figure 2.8 shows a plot [373] of this relation (solid line) compared with the observations of regions where R, B, and Δv have been measured. This interpretation of the line widths is in contrast to the alternative view that they are primarily a result of turbulent broadening. Equation (2.32) can be thought of as the alternate form

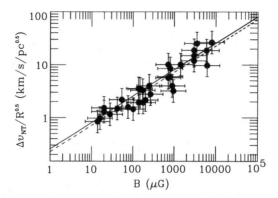

Fig. 2.8 Relationship between the magnetic field, the line width, and the cloud size, as observed (*points with error bars*) and as predicted by (2.32) (*solid line*). The *dashed line* is a fit to the data. The quantity Δv_{NT} is the non-thermal component of the velocity dispersion, in this case caused by magnetic effects. Reproduced by permission of the AAS from [373]. © The American Astronomical Society

of Larson's first finding, when magnetic effects are important. The result strongly suggests that at least in regions where B is strong enough to be measured, the clouds are close to the point where they are magnetically critical.

The evidence for existence of the field is given by measurements of Zeeman splitting of molecular lines. The Zeeman effect is strong enough to be measurable only in certain lines, for example in OH, CN, and the 21 cm line of neutral H. The measurements are extremely difficult, because the Zeeman splitting is only a small fraction of the line width. The goals of such observations are to determine to what extent the magnetic field can support a cloud or clump against collapse, and the relation between field strength and density. In addition, there are measurements of the polarization of starlight [190, 532] which give only the direction of the field but show that the field in a cloud is at least in part well-ordered, rather than random. The Zeeman measurements give only the line-of-sight component of the field, but the results can be deprojected under the assumption of random orientation of the field lines. Relatively few detections exist [125, 126, 166, 510], and in a number of clouds the field has been looked for and not found. Plots of magnetic field against gas density, while subject to large uncertainties in both coordinates, show a general trend that B increases with n with an approximate power law of 0.65 ± 0.05 (Fig. 2.9), as long as the density is above about $300\,\mathrm{cm}^{-3}$; below that value the field does not depend appreciably on density. These data support the following conclusions which apply in cases where there are actual detections of B.

1. The ratio of thermal to magnetic pressures is low, averaging about 0.04.
2. The mass to magnetic flux ratio for the average observed cloud is within a factor 2 of the critical value where gravitational energy and magnetic energy are equal, depending on the assumed geometry. Some observations show that the field is subcritical, others that it is supercritical. However the mean value is supercritical.

2.2 Initial Conditions for Star Formation

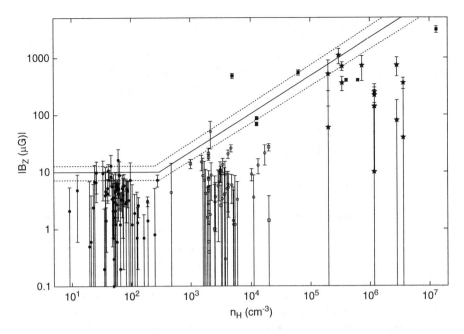

Fig. 2.9 Relationship, in the interstellar gas, between the observed component of the magnitude of the magnetic field in the line of sight $|B_z|$ and the hydrogen volume density in cm^{-3}. In molecular clouds, the horizontal axis is $2n(H_2)$. *Filled circles:* HI diffuse clouds; *open circles:* OH dark clouds; *filled squares and stars:* molecular clouds. The solid line, obtained from a statistical analysis, represents the maximum of the total field strength B as a function of density. In molecular clouds, the actual values of B are randomly distributed between a very small value and this maximum. The dotted lines give the uncertainty in the statistical model. Reproduced by permission of the AAS from [126]. © 2010 The American Astronomical Society

But the large number of non-detections suggests that on the average the magnetic field alone does not prevent material from collapsing.
3. Kinetic energy in macroscopic gas motions is roughly a factor 2 higher than magnetic energy in the average observed cloud. Clouds are in approximate virial balance, with turbulent kinetic plus magnetic energies comparable to gravitational energy (Larson's second finding).
4. At densities above $n \approx 300$ particles cm^{-3}, the maximum magnetic field B scales with density as $|B| \propto \rho^\kappa$ with $\kappa \approx 0.65$. This value agrees with the theoretical value of 0.67 for the case in which the cloud is relatively spherical and has a weak magnetic field [362]. However a number of numerical simulations in which the cloud was initially rapidly rotating, e. g. [98], show that the collapse to a disk results in a value for $\kappa \approx 0.5$.

As a specific example, 34 dark-cloud regions were extensively observed [510] with the Arecibo radio telescope in the 1665 and 1667 MHz lines of the molecule OH. Nine significant detections were made. The number density of H_2 in these regions ranges from 1,500 to 6,600 cm^{-3}. Taking a well-defined set of detections in the range of column densities $N(H_2) = 4 \pm 2 \times 10^{21}$ cm^{-2}, the mean magnetic field component in the line of sight is 17 µG, not including the non-detections. We can now estimate the ratio (R) of the observed mass-to-flux ratio to the critical value:

$$R = \frac{(M/\phi)_{\text{obs}}}{(M/\phi)_{\text{crit}}} = \frac{mN\sqrt{G}}{0.17B} = 6 \times 10^{-21} N/B \qquad (2.33)$$

where N is the total column density in particles cm^{-2}, m is the mean mass per particle, taken to be 2.4 atomic mass units, taking into account the helium, and B is the magnetic field in µG. The actual magnetic field strength is statistically obtained from the line-of-sight component by multiplying it by a factor 2, and the observed column density of H_2 is corrected by 20% to get N, taking into account the helium. With the very specific set of data chosen, R is very close to 1, perhaps even slightly subcritical. If the cloud region is disk-like rather than spherical, there is an additional small correction which makes R even more subcritical. In view of the uncertainties, the observations do not rule out the picture of magnetically controlled star formation, at least in some regions. However, the large number of non-significant detections or detections showing a very weak field indicate that there are many regions where the magnetic field is not a controlling effect.

Under typical interstellar conditions outside molecular clouds, where the hydrogen is neutral and T $\approx$ 100 K, $n \approx 10$ particles cm^{-3}, and the median field $B \approx 6 \times 10^{-6}$ gauss (Fig. 2.9), the thermal Jeans mass M_J from (2.17) is $10^4 M_\odot$, and the magnetic Jeans mass from (2.22) is of the same order of magnitude. The HI (so-called "diffuse") clouds have a wide range of properties; however the typical mass is only $10^3 M_\odot$. Thus it is clear that in order to get a region with mass of order 1 $M_\odot$ to collapse, we must consider regions, in molecular clouds, that are denser and cooler than average. In molecular clumps with temperature 10 K, mean number density of H_2 of 10^3 cm^{-3}, and magnetic field 3×10^{-5} gauss, the thermal Jeans mass (assuming a mass density of $\rho = 3.3 \times 10^{-21}$ g cm^{-3} and a mean molecular weight $\mu = 2$) under these conditions is only about 10 $M_\odot$, while the magnetic Jeans mass and the turbulent Jeans mass (2.22) and (2.21) are both comparable to the clump mass, about $1 \times 10^3 M_\odot$. Thus probably turbulence and the magnetic field combine to keep the typical clump from collapsing. In the higher-density cloud cores where masses are a few $M_\odot$, sizes 0.05 pc, temperature 10 K, and density about 10^5 cm^{-3}, the thermal Jeans mass is down to about 1 $M_\odot$, and the turbulence is subsonic, so $M_{\text{turb}} < M_J$. The magnetic Jeans mass, assuming that the field strength is roughly 100 µG (Fig. 2.9), is about 2 $M_\odot$. However here the assumptions of spherical contraction and flux-freezing that led to (2.22) are starting to break down; also not all cores have magnetic fields that strong.

2.3 Heating and Cooling

Star formation is clearly favored in cloud cores where the temperature is only 10 K. But why is the temperature so low? The rates of heating and cooling are important for determining the temperature as a function of density and for maintaining the low temperature of molecular cloud material once it starts to compress under the influence of gravity. The details of the heating and cooling processes are complicated, but the somewhat surprising result is that for densities ranging from the clump value of 10^3 cm^{-3} to as high as 10^{10} cm^{-3} the temperature stays near 10 K. The only exception is for molecular cloud material in the vicinity of newly formed stars, where heating to 30 K or even higher is possible.

Let Γ be the rate of energy gain and Λ the rate of energy loss, both per unit volume. Then, if $\Lambda > \Gamma$ the cooling time scale is given by

$$t_c = \frac{3 R_g T \rho}{2\mu(\Lambda - \Gamma)}. \qquad (2.34)$$

The basic assumption is that cooling times and heating times are short enough so that an equilibrium is reached, with cooling rate balancing heating rate in determining an equilibrium temperature. A wide range of physical processes is considered to determine the rates [73, 134, 187, 189, 476, 481]; we summarize the important ones here.

The dominant external heating processes are: (1) the photodissociation of molecular H by interstellar photons, (2) the photoionization of carbon atoms by interstellar radiation, (3) the ionization of H and of H_2 by low-energy cosmic rays, and (4) the production of photoelectrons liberated from grains by interstellar UV photons. In these heating processes, the extra energy delivered by the photon or energetic particle, above the ionization or dissociation energy, goes into heating the gas. The heating from these external sources is proportional to the first power of the local gas density. At relatively high densities, above 10^3 cm^{-3}, where most of the gas is in molecular form, cosmic ray heating dominates. The main constituent of cosmic rays is high-energy protons. The heating rate [481] is approximately

$$\Gamma_{CR} = 1.1 \times 10^{-11} \zeta \, n_{H2} \approx 3 \times 10^{-28} n_{H2} \, \text{erg cm}^{-3} \, \text{s}^{-1}, \qquad (2.35)$$

where ζ is the rate of ionization (molecules s^{-1}) of H_2 by cosmic rays, estimated to have a value of $\approx 3 \times 10^{-17}$. This rate actually must be corrected for absorption of cosmic rays in the outer parts of the molecular region.

At still higher densities, where the mass exceeds the local Jeans mass and collapse starts, a fifth process, compressional heating, becomes important. It can be calculated under the assumption that the gas is collapsing at half the free-fall rate:

$$\Gamma_f = -P\rho \frac{dV'}{dt} = \frac{P}{\rho} \frac{d\rho}{dt} = (8G\rho)^{1/2} \frac{R_g T \rho}{(3\pi)^{1/2}\mu} = 2 \times 10^4 \rho^{3/2} T/\mu \, \text{erg cm}^{-3} \, \text{s}^{-1} \qquad (2.36)$$

where Γ_f is the rate of work done by gravity per unit volume, and $V' = 1/\rho$. This approximate relation is obtained by setting dt equal to twice the free-fall time (1.1) and $d\rho/\rho$ to unity, and by using the ideal gas equation for P.

The dominant cooling processes are (1) collisional excitation of atoms, molecules, and ions, by electrons, H, or H_2, followed by radiative decay and escape of the photon, and (2) grain cooling. The cooling rates are all proportional to the square of the particle density. For collisional excitation, the important contributors are C^+ at low densities and C, O, and CO at higher densities; the details of the cooling rates are given in [481]. Grain cooling involves collision of molecules with grains, heating them somewhat but also cooling the gas. The grains then radiate their excess energy in the infrared; they are assumed to behave as miniature black bodies, radiating a Planck spectrum at the grain temperature T_g. As long as the cloud has $\rho < 10^{-13}$ g cm^{-3}, it is optically thin to this radiation, and it escapes. A molecule arrives on a grain with a temperature T and leaves with the (lower) T_g. The rate of transfer of energy from gas to dust grains is then given by the collision rate per unit volume times the energy loss per collision:

$$\Lambda_g = n_g n_{H_2} v_{H_2} \pi r_g^2 k_B (T - T_g) = 3.2 \times 10^{13} \rho_{H_2}^2 T^{1/2}(T - T_g) \text{ erg cm}^{-3}\text{s}^{-1}, \quad (2.37)$$

where the n's are number densities, v_{H_2} is the mean of the magnitude (in 3 dimensions) of the velocity of an H_2 molecule $[8k_B T/(\pi m)]^{1/2}$, and r_g, the grain radius, is assumed to be 2×10^{-5} cm. The sticking coefficient is assumed to be 1 and the number density of grains is assumed to be 2×10^{-13} that of H_2. The grain temperature is then determined by the condition that the heating rate of the grain by the above process is equal to the cooling rate by emission of infrared radiation. The emission coefficient is then assumed to be given by $j_\nu = \kappa_\nu B_\nu(T)$, so that $j = 2.3 \times 10^{-4} \kappa_p T_g^4$ erg g^{-1} s^{-1} where κ_p is the Planck mean opacity (5.13). The emission coefficient is the energy emitted per unit mass per unit time per unit solid angle per unit frequency interval, so $j\rho$, which is integrated over frequency and solid angle, is the energy emitted by the grains per unit volume per unit time.

To obtain the temperature one makes the assumption that at equilibrium the total rate of heating equals the total rate of cooling:

$$\Gamma_{CR} + \Gamma_f + \Gamma_{\text{photo}} = \Lambda_{ion} + \Lambda_{atomic} + \Lambda_g + \Lambda_{molec} \quad (2.38)$$

where Γ_{photo} is the heating rate from the sum of all processes involving photons, and Λ_{ion}, Λ_{atomic}, and Λ_{molec} refer to collisional excitation processes. Then one solves this expression together with $j\rho = \Lambda_g$ to find T and T_g. It turns out that at lower densities, typical of HI clouds, collisional excitation of C^+ and photoheating are, respectively, the dominant cooling and heating processes, while at molecular cloud densities cosmic ray ionization dominates the heating while CO as well as grains provide the cooling.

Results of heating–cooling balance, based on a particular model of a molecular cloud and under the assumption that $\zeta = 10^{-17}$ [134] show that $T = 70$ K at $n = 35$,

T = 21 K at $n = 4 \times 10^3$, and T = 10 K at density $n = 10^4$ cm^{-3}. Further calculations [300] show that at still higher densities ($\approx 10^6$ cm^{-3}) the gas cools further to 5 K. Nevertheless, the equilibrium temperature varies only slowly for $10^4 < n < 10^{10}$, or $4 \times 10^{-20} < \rho < 4 \times 10^{-14}$ g cm^{-3}. Thus the gas is often assumed to be isothermal at 10 K in that density range. The gas tends to cool on compression, because the heating rate per unit volume is proportional to particle density (also the extinction increases with density), while the cooling rate per unit volume, dominated by collisions, increases with the square of the density. The grain temperature does indeed turn out to be less than the gas temperature and nearly constant as a function of density at about 10 K. At the higher densities, above 10^6 cm^{-3}, the dust and gas temperatures are practically equal at about 5 K.

2.4 Magnetic Braking

The first stage of the solution of the angular momentum problem is to explain why the small-scale, high density regions of molecular clouds have specific angular momentum considerably less than that of the larger-scale regions. One way to solve the problem is to imagine that the turbulent properties of the cores determine their angular momentum [95]. A turbulent velocity distribution consistent with Larson's first finding can be shown to match the observations of angular momentum. The line-of-sight component (v) of the velocity field can be interpreted as rotation, as it shows a velocity gradient across the cloud. Roughly, the specific angular momentum $j \propto vR$ and $v \propto R^{1/2}$ so $j \propto R^{3/2}$, a relation that closely fits the observations (Fig. 2.7). The turbulent velocity field also gives $\beta \approx 0.03$, independent of the size of the cloud, also in agreement with the data in Fig. 2.7. It turns out that the turbulent motions do in fact give an overall net angular momentum to the cloud as a whole.

Another approach, for example in the situation where turbulent effects are not important, is to assume that the high-density regions are connected by magnetic field lines to the lower-density material. Suppose a high-density core has a uniform magnetic field passing through it, and it is rotating faster than the background medium of lower density. If the magnetic field is coupled to both regions, then the field lines will become twisted as a result of the rotation. The equations of magnetohydrodynamics (MHD) show that the twist generates a torque, which slows down the rotation of the high-density material and transfers its angular momentum to the low-density material. The braking of the rotation will continue as long as the time scale for transport of angular momentum is shorter than the contraction time of the high-density region (as the contraction would tend to spin it up).

The braking time was developed in a heuristic manner [151]; more detailed numerical solutions of the equations of MHD have shown that this (highly idealized) argument, summarized in the following paragraph, gives the correct time scale for braking of the cloud.

Consider a spherical cloud with radius R, density ρ, and mass M, threaded by a uniform magnetic field B; an ordered field is required. The cloud is surrounded by external material with lower density ρ_{ext}, which is also threaded by the field. The field is frozen into the gas in both media. The cloud is rotating with uniform angular velocity Ω, parallel to the direction of B, and the external medium is initially not rotating. The field lines become twisted because of the discontinuity in Ω, and the twist propagates outward along the field lines at the Alfvén velocity $V_A = B/(4\pi\rho_{ext})^{1/2}$ (in the external medium). Suppose the initial field is in the Z-direction in a cylindrical coordinate system. Then the twist in the field lines will generate components of the field in the R and ϕ-directions. The torque on an element of material in the cylinder that contains the cloud is then derived from the ϕ-component of the Lorentz force

$$\frac{1}{4\pi}[(\nabla \times \mathbf{B}) \times \mathbf{B}]_\phi.$$

But in the argument, a detailed calculation of the torque is bypassed, and it is simply assumed that the wave spins up the external material within a cylinder with the same radius as that of the cloud until it corotates with the cloud. The angular momentum of the cloud itself decreases correspondingly.

The amount of material accelerated per second in the external medium in a cylindrical shell of radius R and thickness dR as the Alfvén wave passes through it is $2\pi\rho_{ext} R V_A dR$ g s^{-1}. Multiply by the specific angular momentum ΩR^2 to get the angular momentum increase per second for the mass element: $2\pi\rho_{ext} R V_A \Omega R^2 dR$. We assume that the moment of inertia of the sphere $I = 0.4MR^2$ and that its angular momentum is J. We equate the angular momentum loss of the sphere to the angular momentum gain of the external medium, and we integrate over the cylindrical region of external matter through which the Alfvén waves propagate:

$$\frac{dJ}{dt} = 0.4MR^2\frac{d\Omega}{dt} = -\left(\frac{dJ}{dt}\right)_{ext} = -\pi\rho_{ext} V_A R^4 \Omega \qquad (2.39)$$

where the cloud radius is assumed to be constant and where a factor 2 has been introduced to account for wave propagation in both directions.

The braking time is then given by

$$t_b = \frac{\Omega}{\frac{d\Omega}{dt}} = \frac{0.4MR^2}{\pi\rho_{ext} V_A R^4} = \frac{0.4\Sigma}{\rho_{ext} V_A} = \frac{4M}{5(\pi\rho_{ext})^{1/2} BR^2} \approx \frac{0.5\rho}{\rho_{ext}} \frac{R}{V_A} \qquad (2.40)$$

where $\Sigma = M/(\pi R^2)$, the surface density of the cloud, and ρ is the density of the cloud. The braking time is thus closely related to the propagation time of the Alfvén

2.4 Magnetic Braking

wave into the external medium. The time can be re-written in terms of the critical magnetic Jeans mass: $M_\phi = (BR^2)/(3.6G)^{1/2}$,

$$\frac{M}{BR^2} = (3.6G)^{-1/2}\frac{M}{M_\phi}. \tag{2.41}$$

Then using the free-fall time of the cloud, $t_{\rm ff} = [3\pi/(32G\rho)]^{1/2}$ we obtain

$$t_b \approx 0.4\frac{M}{M_\phi}(\rho/\rho_{ext})^{0.5}t_{\rm ff}. \tag{2.42}$$

For $M/M_\phi \approx 1$, the marginally critical case, the braking time is comparable to the free fall time [357]. For example, if the cloud has $\rho = 10^{-21}$ g cm^{-3} and the density ratio is 10, then we have $t_{\rm ff} = 2 \times 10^6$ yr and t_b about the same for $M/M_\phi \approx 1$. If the cloud is magnetically subcritical the braking is very fast. Then

$$\Omega_{cloud}(t) = \Omega(0)\exp(-t/t_b). \tag{2.43}$$

But if the cloud is supercritical and is already collapsing, a detailed calculation is required to determine how much angular momentum loss occurs. In the limit where the collapse velocity exceeds the Alfvén velocity, magnetic braking is not effective and the cloud collapses with approximate conservation of angular momentum. Also, when the density of the collapsing cloud becomes quite high, the degree of ionization becomes so low that the assumption of complete coupling between matter and field breaks down (Sect. 2.5).

Figure 2.10 shows the results of an analytical solution of the MHD equations for the case of a uniform-density cylinder with radius R and half-height Z_1, linked to the external medium by a uniform field **B** along the Z direction [372]. The braking time can be understood as the time required for the Alfvén wave in the external medium to sweep across enough material so that its moment of inertia equals that of the cloud. The figure shows the results for the cloud's $\Omega(t)$ in units of its initial value, assuming it is always uniformly rotating, and that the external medium is initially at rest. The e-folding time for decrease in Ω is found to be

$$t_b = \frac{Z_1}{V_A} \tag{2.44}$$

where Z_1 is the initial half-height of the cylinder, when $\rho = \rho_{\rm ext}$. As the cylinder contracts in Z at constant R, the braking time is independent of the stage of contraction indicated by $\rho/\rho_{\rm ext}$. As an example take a typical molecular clump with $R = Z_1 = 2$ pc and a density of 10^3 cm^{-3}. If the mean particle mass is 2.3 atomic mass units, then $\rho = 3.8 \times 10^{-21}$ g cm^{-3} and the total mass is 2.8×10^3 M$_\odot$. Assume that initially $\rho = \rho_{\rm ext}$. Take a field of 30 μG, which gives $M_\phi = 1.2 \times 10^3$ M$_\odot$ and

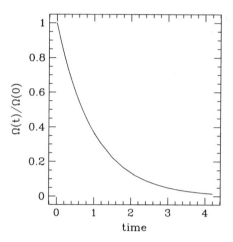

Fig. 2.10 Magnetic braking. The angular velocity of a rotating cloud, in units of its initial value, is plotted as a function of time. The configuration is a uniform-density, uniformly rotating cylinder. The unit of time is the Alfvén wave crossing time Z_1/V_A, where Z_1 is half of the height of the cylindrical cloud at its initial state, when $\rho = \rho_{\rm ext}$, and V_A is the Alfvén velocity in the external medium. Result from [372]

$t_b = 4.5 \times 10^{13}$ sec, only slightly longer than the free-fall time of 3.4×10^{13} s. This assumed cloud is unstable to collapse, and as the density increases, the free-fall time decreases but the braking time remains the same, so eventually the braking becomes ineffective.

In the case where the field and the angular momentum are perpendicular, a similar estimate [371] gives a braking time that is much shorter, up to a factor 10, than in the parallel case. Both of these times are confirmed by detailed complete numerical simulations of the MHD equations [35, 139]. However it is quite possible that the assumed geometry, a uniform parallel or perpendicular magnetic field, is not quite appropriate, particularly when the cloud has contracted to relatively high density. In the case where the magnetic field lines are assumed to diverge spherically from the boundary of the (contracted) cloud, the ratio of braking times between the cases where the angular momentum (**J**) is parallel to **B** to that where **J** is perpendicular to **B** is only about a factor 2 [361]. In this case again the times, assuming the mass-to-flux ratio is close to the critical value, are close to the free-fall time of the cloud. In summary, in the case where a rotating cloud is initially subcritical, its angular velocity will go through three stages. First, the angular velocity will decrease rapidly as a consequence of braking. Second, the angular velocity will remain constant at the background value, as there is no longer any torque on the cloud. Third, when the cloud starts to collapse, the angular velocity will increase again, with approximate conservation of angular momentum, as its collapse time becomes shorter than the braking time.

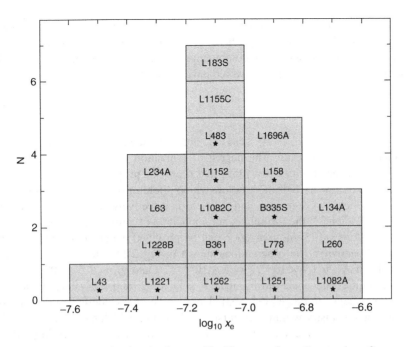

Fig. 2.11 Ionization in molecular cloud cores. The histogram shows the number of cores as a function of the ratio x_e of charged particles to neutral particles. The numbers in the boxes give identifications of the cores observed. A star in a box means that the core has at least one embedded infrared source. Reproduced by permission of the AAS from [551]. © 1998 The American Astronomical Society

2.5 Degree of Ionization

The discussion of magnetic braking assumed that the magnetic field was firmly coupled to the matter, or in other words, magnetic flux was conserved. Coupling requires the presence of charged particles, as well as a sufficient rate of collisions between charged particles and neutral particles such that the neutrals are coupled to the charged particles and therefore are coupled to the field. Define the degree of ionization $x = n_i/n_n$ to be the ratio of the number densities of ionized (positively charged) and neutral particles. In the outer parts of clouds this ratio is determined by photoionization by UV photons from external hot stars. At core densities ($n_n \approx 10^4 \, \text{cm}^{-3}$) the UV is attenuated and cosmic ray ionization dominates. It is difficult to measure x in molecular clouds, but a value of 10^{-7} is inferred from observations of the ion HCO^+, which is expected to be one of the dominant charged particles. A set of observations [551] of 20 low-mass cores (mean density $2.5 \times 10^4 \, \text{cm}^{-3}$) shows a peak at the above value (Fig. 2.11).

Theoretical work [156] assumes that, at a given density, the cosmic ray ionization is balanced by 2-body recombination of charged particles and recombinations on charged grains, so that a steady-state value of x can be determined. The cosmic ray ionization rate is ζ per particle per second. In the simplest case of balance, one would have $\zeta n_{H_2} = \alpha_R \times n_i^2$ where the left and right-hand sides, respectively, are the ionization rate and the recombination rate per unit volume, and α_R is the recombination coefficient. So one would have $n_i \propto n_{H_2}^{0.5}$ or

$$x = K_i n_{H_2}^{-0.5} \tag{2.45}$$

where $K_i = (\zeta/\alpha_R)^{1/2} \approx 10^{-5} \, \text{cm}^{-3/2}$, although actually it is a weak function of temperature and density. The degree of ionization is then 10^{-7} at a density of $10^4 \, \text{cm}^{-3}$. It has been shown [357] that the $n^{-0.5}$ proportionality is applicable for densities as low as $10^3 \, \text{cm}^{-3}$, below which photoionization effects rather than cosmic rays begin to dominate, and the field is effectively coupled. At very high densities, $>10^8 \, \text{cm}^{-3}$, the formula is not applicable because cosmic ray ionization effectively shuts off. Some ionization can be provided under these conditions by natural radioactivity. The observations shown in Fig. 2.11 actually are consistent with (2.45) if $K_i \approx 1.4 \times 10^{-5}$, $\zeta = 5 \times 10^{-17} \, \text{s}^{-1}$, and $\alpha_R = 2.5 \times 10^{-7} \, \text{cm}^3 \, \text{s}^{-1}$. These quantities are uncertain by at least a factor 2, and the values quoted are within the acceptable range [131, 503]. The recombination coefficient depends on complicated chemistry within molecular clouds and on which particular molecular ion dominates in the recombination process. Thus the relation shown in (2.45) is oversimplified, but nevertheless useful.

2.6 Magnetic Diffusion

In clouds which are not massive enough to contract across magnetic field lines, star formation can occur only if the ratio of mass to magnetic flux is increased. An important process that is considered in this regard is "plasma drift", or "ambipolar diffusion." This effect can be significant at relatively high densities where the degree of ionization is very low and ideal MHD begins to break down. Even if a cloud is magnetically supercritical, but the magnetic field is of some importance, this process will act to reduce the magnetic field effects as the cloud contracts and so contribute to the solution of the "magnetic flux" problem. The process in effect represents the onset of decoupling between the neutral particles and the magnetic field.

Consider a cloud which is in equilibrium with magnetic forces balancing gravitational forces. Neglect pressure forces and assume the cloud is a sphere with radius R, threaded by a uniform field B. Then a rough estimate of the diffusion time, that is, the time required for the neutral particles to drift significantly with respect to the ionized particles, can be derived as follows (from [476]).

The neutral particles are unaffected by the field and tend to drift inward. The ions feel the magnetic force, which is transmitted to the neutrals by collisions. For the neutrals, the force of gravity per unit volume is balanced by the frictional force on

2.6 Magnetic Diffusion

the neutrals by collisions with ions, which in turn is given by the collision rate times the momentum exchange per collision:

$$\frac{GM\rho}{R^2} = n_i \langle \sigma v \rangle n_{H_2} m_{H_2} u_D .\tag{2.46}$$

Here v is the actual velocity of the particles, assumed to be $\sim 10^5$ cm s^{-1}, $\sigma \approx 10^{-14}$ cm^2 is the cross section for collision, $\rho = m_{H_2} n_{H_2}$ is the total density of the molecular gas, u_D is the drift velocity of ions relative to neutrals, n_i is the number density of ions, and M is the mass of the cloud. The ions, whose motion is controlled by the magnetic force, satisfy a similar equation, except that the left hand side is replaced by the magnetic force per unit volume, $(\nabla \times B) \times B/(4\pi) \approx B^2/(8\pi R)$. Solving the above equation for u_D, one obtains

$$u_D = \frac{4\pi}{3} \frac{G\rho R}{n_i \langle \sigma v \rangle} \approx R(\frac{10^{-8}}{x}) \text{ km s}^{-1} \tag{2.47}$$

where R is given in parsecs and x is the degree of ionization, n_i/n_{H2}, which is assumed to be small. The time scale for drift out of the core of the cloud is then given by

$$t_{AD} = \frac{R}{u_D} = 5 \times 10^5 (\frac{x}{10^{-8}}) \text{ yr.} \tag{2.48}$$

Even if x is as small as 10^{-5}, the time scale is long, about 5×10^8 yr. At a number density of 10^6, the degree of ionization (2.45) is small enough (10^{-8}) for the time scale to come down to 5×10^5 yr. But this result shows that in most of the mass of molecular clouds, which has much lower densities, ambipolar diffusion is not important on reasonable time scales. However if turbulence is present and there are local strong compressions of the material, the ambipolar diffusion time is speeded up by a factor of a few [384].

Going back to (2.47), we find that the diffusion time is given by

$$t_{AD} = \frac{xn_n \langle \sigma v \rangle}{4/3\pi G\rho_n} = \frac{3K_i n_{H_2}^{0.5} \langle \sigma v \rangle}{4\pi Gm_{H_2} n_{H_2}} \tag{2.49}$$

and

$$\frac{t_{AD}}{t_{ff}} = \frac{3K_i \langle \sigma v \rangle}{4\pi G^{1/2} m_{H_2}^{0.5} (3\pi/32)^{0.5}} = 19. \tag{2.50}$$

The numerical value depends on the composition, the cloud geometry, and K_i but in any case it is of order 10. This relation is very important because it implies that subcritical non-turbulent molecular clumps, which are supported by the magnetic field, evolve quasistatically. Note that the above ratio depends both on the assumption of equilibrium and on the validity of the ionization law. Although the value of the field B does not appear in the above expression, it is implicitly present through the assumption of equilibrium. To see how the result depends on B, just replace the left-hand side of (2.46) by $B^2/(8\pi R)$ (follows from equating magnetic energy to gravitational energy GM^2/R), solve as before for u_D and t_{AD}, and set $K_i = 1.4 \times 10^{-5}$ to obtain

$$t_{AD} \approx 4 \times 10^6 \text{ yr} \left(\frac{n_{H2}}{10^4 \text{ cm}^{-3}}\right)^{3/2} \left(\frac{B}{30\mu G}\right)^{-2} \left(\frac{R}{0.1 \text{ pc}}\right)^2. \qquad (2.51)$$

The times given by (2.48) and (2.51) are in good agreement with more detailed numerical modelling of equilibrium cloud structures, including ambipolar diffusion, supported against gravity by magnetic fields and pressure effects [325, 386, 507], as well as with calculations that extend into the early part of the collapse phase [35, 171]. During the quasistatic phase the central value of the field evolves approximately as $B_c \propto n_c^{0.5}$. When the central densities increase to the range 10^5-10^6 cm^{-3}, sufficient diffusion has taken place so that the central regions become unstable to gravitational collapse. At this point the magnetic field is still present, and it is still coupled to some extent to the gas even though the degree of ionization is very low. However it is no longer dynamically important; a relatively small flux loss (factor 2) goes a long way in allowing the central regions to go supercritical, partly because of flow along field lines.

Some other conclusions from such calculations are as follows: (1) Ambipolar diffusion can produce cloud cores but the process requires 10 initial free-fall times, so if one starts from $n_{H2} = 10^3$ cm^{-3}, long time scales ($\approx 10^7$ yr) are required. Only moderate flux loss is required. Therefore the magnetic flux problem remains to be solved at higher densities. (2) This long time scale is definitely a problem, since practically all (90%) molecular clouds in the Galaxy show some evidence of star formation. Supersonic turbulence might help, compressing the magnetic field in shocked regions and reducing t_{AD} in some regions. Also, analytical and numerical calculations show [115] that if the cloud is approximated as a thin disk, t_{AD} can be considerably reduced from 10 t_{ff} if the initial mass-to-flux ratio is close to the critical value. (3) Once the cloud core has been formed, runaway collapse can occur at the center after $\sim 2 \times 10^5$ yr. The radial density distribution in the core as it begins collapse is close to $\rho \propto R^{-2}$. (4) Observed core shapes are not good indicators of physical conditions, that is, directions of angular momentum and magnetic field. A flattened cloud core does not necessarily mean that the angular momentum or magnetic field vectors lie along the short axis. (5) The diffusion times derived above assumed equilibrium between magnetic and gravitational forces, that is $M \approx M_\phi$. If, as observed, some regions are already supercritical, they could evolve quickly to the core stage, limited by the decay time of turbulence. Thus the diffusion model does allow star formation over a range of time scales (slow star formation) and could be consistent with the range of ages in a young cluster, such as NGC 2264.

Nevertheless some criticisms of the ambipolar-diffusion picture remain. (1) Observations of magnetic fields suggest that most clumps and cores, over a range of densities, are already supercritical, so diffusion isn't needed. Even though there are observational uncertainties in magnetic field observations, there are enough clumps and cores with unobserved or low fields to account for the rate of star formation. (2) If ambipolar diffusion were the dominant process controlling star formation, there would be many more high-density cores in molecular clouds without embedded infrared sources (starless cores) than cores with embedded protostellar objects, because the time scale for the protostellar phase is only about $2 - 3 \times 10^5$ years. A starless core is defined as one which is gravitationally bound and therefore likely to eventually become a protostar, but which has no detectable evidence for an embedded protostar. Surveys have been made of the numbers of starless cores versus cores containing protostars over various star-forming regions [442]. The ratio is found to be about 3 to 1, but different studies have given different results. Thus the lifetime of a starless core, at a density close to 10^5 cm^{-3} is $\approx 6 \times 10^5$ yr. The free-fall time in the density range of the observed cores is $1 - 2 \times 10^5$ yr. The lifetime is definitely shorter than one would expect from the ambipolar diffusion model; on the other hand it is longer than the lifetime of a core that one would expect in a supersonically turbulent model (next section), which is only 1–2 free-fall times. This kind of study has been done for various density ranges [537], with the same result. The lifetime of a core is always a few free-fall times, but definitely less than 10. If true, the implication is that cores, once formed, do not immediately go into dynamical collapse, but they do evolve on a time scale shorter than that of ambipolar diffusion.

2.7 How is Star Formation Initiated?

Three different scenarios are being discussed regarding the process by which a cloud core is brought to the onset of collapse:

1. Low-mass cores are assumed to be magnetically subcritical, that is magnetic effects prevent collapse and support molecular clouds. The densest regions evolve to the onset of collapse, controlled by ambipolar diffusion (slow drift of neutral particles across magnetic field lines). The time scale is a few million to 10^7 yr.
2. Star formation is controlled by turbulence. Supersonic turbulence generates a complicated shock pattern. Randomly produced shock-compressed regions of high density can occasionally reach the point of instability to collapse. The time scale is much shorter, more like 10^6 yr, and the efficiency of star formation is tied to the properties of the underlying turbulence.
3. Cores that are intrinsically stable to collapse are forced into collapse by a specific event, an external trigger such as an ionization front, a supernova shock wave, or a cloud–cloud collision. The time scale then is the shock crossing time, which could be as short as 10^5 yr.

The first of these possibilities was discussed in the previous section. Existing observations of magnetic fields in the precursors of cloud cores indicate that most of them, but not all, are supercritical. Even when they are supercritical, the magnetic field is still of some importance and cannot be left out of a theory of star formation. For example, in supercritical regions, ambipolar diffusion will still take place, but it is not necessary for the initiation of star formation. It is likely that all three of the possibilities just mentioned play some role in the theory. Nevertheless, more emphasis is now being placed on the second and third alternatives, which are discussed in the next two sections.

2.8 Turbulence and Star Formation

The study of turbulence in the laboratory involves relatively incompressible fluids and subsonic velocities. The general criterion for onset of turbulence is a Reynolds number ($R_e = vL/\nu$, where v is a typical velocity, L is a typical length, and ν is the molecular viscosity) greater than a critical value, which differs for different situations but is typically several thousand. Once turbulence develops, motions develop on a wide range of scales. The general picture is that energy is fed into the turbulence on the largest scale, and it is transmitted through the so-called "turbulent cascade" to smaller and smaller scales, until it reaches a very small scale on which it is dissipated into heat. The relation between typical velocity and length scale is known as Kolmogorov's law, in which $v \propto L^{1/3}$. Thus most of the kinetic energy is on the large scales. The dissipation scale is given roughly by

$$\frac{L_{\text{diss}}}{L_m} = R_e^{-3/4} \qquad (2.52)$$

where L_m is the scale of energy input. In molecular clouds the dissipation scale is estimated to be only $\approx 3 \times 10^{-5}$ pc [352]. Reynolds numbers on the molecular cloud scale (Table 2.1) can be estimated using $v \approx 3$ km s^{-1}, $L \approx 5$ parsecs, $\nu \approx c_s \lambda_m$ where $\lambda_m = 1/(n\sigma_m)$ is the particle mean free path (about 10^{13} cm), and σ_m is the particle scattering cross section, typically 10^{-15} cm^2 for neutral particles. The result is $R_e \sim 10^7$, well in excess of the critical value for the onset of turbulence.

The origin of interstellar turbulence is unclear [29]. Some theories of the origin include instabilities induced by colliding flows in the interstellar gas [213], magnetorotational instability in galactic disks [417], and instabilities that develop behind the spiral waves in the galaxy [80]. It is clear that the turbulence is compressible, and it is supersonic, on the larger scales, with respect to the cold molecular cloud material. Nevertheless it retains some properties of laboratory turbulence. We have seen, from Larson's first finding, that approximately $v \propto L^{1/2}$; thus most of the kinetic energy is still on large scales. There also is a turbulent cascade, or an approximation to it, but the energy input is not necessarily limited to the largest scales, although these scales for energy input may dominate. Also, the dissipation is not necessarily occurring on the smallest scales, because the random supersonic motions produce shock waves, which can dissipate kinetic energy into heat on various scales. On a given scale, the flow can be characterized by a turbulent Mach number, the ratio of the root-mean-square turbulent velocity (in 3 space

2.8 Turbulence and Star Formation

dimensions) to the sound speed. On the molecular clump scale (2 pc) this ratio is of order 10. On the other hand, the flow speed is comparable to the Alfvén speed. Measurements on various scales in molecular clouds indicate rough equipartition between gravitational, magnetic, and turbulent energy; thus the clouds are supported against gravity in the global sense. A detailed discussion of interstellar turbulence appears in [352].

The key to star formation by turbulent effects is the local phenomenon of transient compression of certain regions by shock waves. If the overdense regions behind the shocks are massive enough and long-lived enough, they can become Jeans unstable and begin to collapse. The random nature of the strength of these shock events suggests that star formation is possible but inefficient. Shock wave dissipation also implies that supersonic turbulence, without continuous energy input, must decay on a time scale comparable to that for a strong shock to propagate across the largest eddies, which for a 5 pc scale is $O(10^6)$ yr [186].

Numerical simulations of turbulence, which require 3-dimensional hydrodynamics, can only approximate the actual situation. However such simulations, both with and without magnetic fields [29, 335] verify the qualitative picture described above that the kinetic energy in turbulence, at least on the larger scales, decays. Some of the important points that emerge include: (1) A clumpy structure develops with the maximum fluctuation in density increasing with turbulent Mach number. (2) Turbulence decays, even with the presence of magnetic fields, on a time scale L/v_{rms}, where L is approximately the scale of the system and v_{rms} is the rms velocity of the turbulence. This time scale for a 1 parsec cloud and an rms velocity of 1 km/s again is about 10^6 yr.; it scales as $L^{1/2}$. If this occurred in a cloud of a few hundred solar masses, the cloud after decay would have several hundred thermal Jeans masses, and could form a small cluster. (3) Even if turbulence is maintained by some driving mechanism and it is able to support the cloud overall, locally high-density regions are generated randomly as a result of the turbulence. Some of them could exceed the local Jeans mass for long enough to allow collapse to stars. The resulting star formation efficiency turns out to depend on the driving scale of the turbulence. (4) Some mechanism to maintain the turbulence in the ISM is required, in order to explain the observations that all molecular clouds have line widths above thermal, and to prevent overall collapse of molecular clouds and a star formation rate that is much higher than that observed. If there is no continual energy input, one is driven to the conclusion [203] that the turbulence must have been generated as part of the process by which the molecular cloud was formed, for example by large-scale colliding gas flows. In this scenario star formation would have to occur rather quickly, in less than a crossing time, in a relatively small amount of material that becomes gravitationally bound. Then, on the same time scale as that of the decay of turbulence, the cloud is disrupted by stellar feedback. The conclusion then would be that lifetimes of even the largest clouds are only 2–3 Myr, a result that is being debated.

With regard to point number (2), numerical simulations both with and without magnetic fields show that without external energy input the turbulence does decay on short time scales [334, 494]. Simulations (Fig. 2.12) show that in 0.4 sound

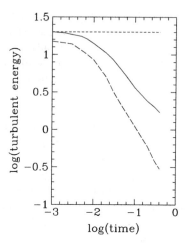

Fig. 2.12 Decay of turbulence in molecular clouds, according to three-dimensional numerical simulations, after [494]. The horizontal axis gives the log of the time in units of the sound crossing time. The vertical axis gives the log of the kinetic energy plus the magnetic energy in turbulent fluctuations, in non-dimensional units. *Short dashed line:* Turbulent energy is continuously supplied to the grid, at a fixed rate, leading to a steady state. *Solid line:* The turbulence is allowed to decay, with no energy input. The initial ratio of gas pressure to magnetic pressure is 0.01, corresponding to a magnetic field far stronger than observed. *Long dashed line:* decay of turbulence with no magnetic field

crossing times, which corresponds to about 2 turbulent crossing times for the calculation shown, the kinetic energy decreases by a factor 10 even if a very strong magnetic field is present. Without the field the decay in the energy is closer to a factor 50. Much of the dissipation occurs in shocks, but some energy is lost by energy cascading to smaller length scales and then being numerically dissipated at the grid scale. The upper curve shows how the energy in turbulence saturates if energy is supplied at a given rate by input of a random velocity field.

An independent calculation [334] yielded similar results. The turbulent energy decayed as approximately t^{-1}, fairly independent of whether or not magnetic fields were present, and the characteristic decay time is about 0.5 of the turbulent crossing time, based on a turbulent Mach number of 5. The calculations are performed with both a smoothed-particle hydrodynamics (SPH) numerical code and a standard fixed-grid code, with very similar results. These simulations support the suggestion [202, 387] that the time scale of star formation is that of turbulent decay rather than ambipolar diffusion, under the further assumption that there is no significant source of driving for the turbulence. However if star formation took place on a time scale of only a million years in molecular clouds, even if the efficiency were only 2%, the rate of star formation in the galaxy would be far higher than observed.

The alternative to pure turbulent decay in molecular clouds is continuous driving of the turbulence by some process that is not well understood (point 4). Some of the possibilities are supernova shocks, outflows from embedded young stars, HII

2.8 Turbulence and Star Formation

regions, or galactic rotational shear. Nor is it understood how the energy, if injected on large scales, cascades down to small scales to limit star formation. Conversely, if energy is injected on small scales, as it would be from bipolar flows from newly formed stars, it is difficult to explain the high turbulent velocities observed on large scales. Of course there could be more than one driving mechanism, and the expansion of HII regions is a promising way to do the driving on large scales. Nevertheless if it takes 10 cloud free-fall times to form a significant number of stars, which can occur for reasonable assumptions regarding the scale of the turbulence, and the molecular cloud is blown away on that time scale, it may be possible to explain a low efficiency of star formation. If the driving scale of the turbulence is long, or if it is not driven at all, local collapse will occur, star formation will be efficient, and cluster formation could occur. The alternative, mentioned above is that the turbulence is not driven, but is generated by initial conditions, and that the time scale for star formation is the same as that for turbulent decay, and once the star formation occurs, the molecular cloud is dissipated.

Extensive work has been done to simulate numerically the effects of turbulence on star formation. The typical simulation has a box size of 0.1–1 pc and a mass 100–1,000 $M_\odot$. Thus, assuming that the initial cloud fragments, these simulations represent cluster formation. The main question to be considered here is whether the simulations come up with the same approximate star formation efficiency as is observed. The different simulations include different physical effects, and the point is to determine which of these effects is the most important in determining the star formation efficiency. Such simulations address other questions as well, notably the form of the initial mass function and the properties of the resulting binary and multiple systems; these questions are discussed in other chapters of this book.

Suppose one considers pure hydrodynamic turbulence with no driving, no magnetic fields, and no feedback effects from the stars that have formed. The initial turbulence just decays and the cloud fragments into protostars. SPH calculations [258] were performed for an isothermal gas in a box which originally contained 222 thermal Jeans masses but had turbulent kinetic energy comparable to gravitational energy. Random density fluctuations are introduced to represent the initial turbulence. The simulation included 500,000 particles and is shown in Fig. 2.13. When the gas density in a local region exceeds a given minimum, and the particles in the region are gravitationally bound, the collection of particles is collected into one "sink" particle, that represents an unresolved collapsing protostar. The first box represents the initial condition, and the three remaining boxes represent times at which 10%, 30%, and 60% of the mass has collected into the collapsing protostellar cores (black dots). The resulting clump mass spectrum looks something like $dN/dM \propto M^{-1.5}$ (similar to what is observed in the interstellar medium). The core mass spectrum has a lognormal shape – that is, a Gaussian in log M – peaked at about twice the Jeans mass of the original cloud. The conclusion is that after only 2 free-fall times (final panel) 60% of the mass of the initial cloud has been converted into a small cluster of protostars, an efficiency much larger than observed, for the overall ISM, and larger even than what actually occurs locally in cluster-forming regions (<30 %).

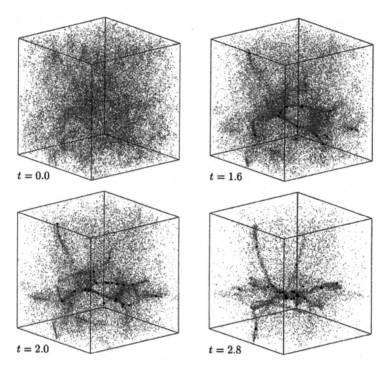

Fig. 2.13 Decay of turbulence in molecular clouds and formation of a protostellar cluster, according to three-dimensional numerical simulations with an SPH code. If the initial density in the cloud is taken to be 10^5 particles cm^{-3} then the size of the box is 0.32 parsec and the unit of time is 6.9×10^4 years. The total number of particles in the simulation is 500,000, not all of which are shown. Reproduced by permission of the AAS from [258]. © 1998 The American Astronomical Society

Further simulations investigate the question of whether driving the turbulence continuously has an effect on reducing the star formation efficiency. Here the overall picture would be that energy is injected at a sufficient rate to match the decay rate, so that the molecular cloud reaches a steady state, supported overall against collapse by the turbulence, and having a lifetime of say 3 crossing times before the onset of disruption by stellar feedback. In one example [259] the velocity fluctuations that are introduced to drive the turbulence have a scale of 1/4 to 1/3 the size of the box. Calculations have been performed with both grid-based and SPH numerical codes. The cloud is globally stable against collapse. The high-density compressed regions randomly produced by shocks in such flows are the sites of local collapse and star formation even if turbulent energy integrated over the cloud is sufficient to support it against gravity. At the end about 50% of the mass of the cloud has been converted to dense collapsing cores, but the time scale (seven initial free-fall times) is about three times longer than in the case of decaying turbulence. Thus even driven turbulence gives fairly efficient star formation, but further simulations show that the efficiency drops as the driving scale decreases. The only way, in this purely hydrodynamic picture, to prevent star formation altogether (or at least make it very inefficient) is to introduce high Mach number turbulence, so that even on the smaller scales the turbulent energy is greater than gravitational, and also to drive the turbulence on very small scales (say 30 or 40 wavelengths per box) so that the driving wavelength is smaller than the local Jeans length. However the results show that the Mach numbers needed to explain the observed low efficiency are unrealistically high, and the driving scales needed are unrealistically low, so this approach does not necessarily explain the low efficiency of star formation.

Another SPH calculation [79] with turbulent initial conditions but no driving and no feedback includes 500,000 particles and starts with a cloud of 1,000 solar masses and a diameter of 1 parsec. The cloud initially has about 1,000 thermal Jeans masses, but it is supported by a random turbulent velocity field with total turbulent energy equal to the gravitational energy; the corresponding mean line-of-sight turbulent velocity is about 1.3 km/s, approximately Mach 6. The gravity is softened on a scale of 160 AU, which determines the effective resolution, so many binaries are not resolved, nor are disks. The lowest mass that is resolved is about 0.1 $M_\odot$. The calculation proceeds through a dissipation phase which lasts about 1 free-fall time. During the decay of the turbulence it produces filamentary structures and locally high-density regions. These regions form the nucleus of about 5 subclusters, in each of which a few tens of stars fragment out. Figure 2.14 shows the column density through the calculated box of 1 pc on a side. The subclusters begin to merge (panel C) and at the termination of the calculation (2.6 free-fall times) a single centrally condensed cluster of more than 400 stars has formed, with a maximum stellar mass of 27 $M_\odot$. At this time 42% of the initial mass remains in gas, giving a high star formation efficiency. The mass function at the end of the simulation has a slope of about -1.0 at the high-mass end, while the Salpeter slope is -1.3. The turnover at masses below solar is in qualitative agreement with observations. In the subclusters the stellar density is very high, so many (about a third, including most of the high-mass objects) objects have close encounters of 100 AU or less, which would truncate their disks, and possibly harden binary systems and disrupt planetary systems. Again, this calculation has been criticized on the grounds that, had the simulation been carried to longer times, the star formation efficiency would be far larger than observed.

A calculation with similar physical assumptions but much higher numerical resolution [37] starts with a cloud of 500 $M_\odot$, temperature 10 K, and size 0.4 pc. The calculation provides very detailed results on the properties of the IMF and of the stellar multiple systems (Chap. 6). However, after only 1.5 initial free-fall times the stars and brown dwarfs contain 191 $M_\odot$. Had the calculation been continued it would again have produced a high star-formation efficiency.

A hydrodynamics simulation with additional effects [319, 385], was performed on a 128^3 grid, with a magnetic field (treated as ideal MHD) included, and with feedback effects resulting from the assumed bipolar outflows generated by the forming stars. The cloud is supercritical, with mass to flux ratio about 2 times the critical value. The initial turbulence decays, and is not driven externally, but

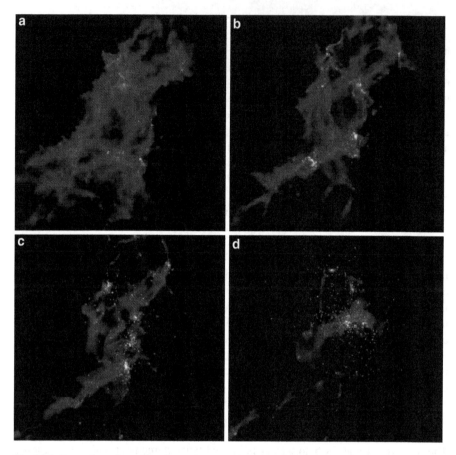

Fig. 2.14 Formation of a stellar cluster from a turbulent molecular cloud region. The grey scale indicates the log of the column density, ranging from $0.025\,\mathrm{g\,cm^{-2}}$ (*dark*) to $250\,\mathrm{g\,cm^{-2}}$ (*light*) [79]. The *light dots* indicate stars. The frames are 1 pc on a side. The evolution of the system of $1{,}000\,\mathrm{M_\odot}$ is plotted at times of 1.0 (**a**), 1.4 (**b**), 1.8 (**c**), and 2.4 (**d**), in units of the initial free-fall time of 1.9×10^5 yr. Reproduced by permission of John Wiley and Sons Ltd. from I. A. Bonnell, M. R. Bate, S. G. Vine: *MNRAS* **343**, 413 (2003). © 2003 Royal Astronomical Society

once stars form, it is assumed that they generate bipolar outflows, with momentum proportional to the current stellar mass and with a typical velocity of about $50\,\mathrm{km\,s^{-1}}$. The injection of momentum re-energizes the turbulence from within, and with the parameters chosen, the cluster-forming cloud remains in near virial equilibrium, and the formation of subsequent stars is delayed. In the standard simulation, the cloud had a radius of about 1.5 pc and a mass of $\sim 1{,}000\,\mathrm{M_\odot}$. After 2 initial free-fall times, or roughly 1 Myr, the star-formation efficiency was only 3%. The model contains a number of arbitrary but reasonable parameters, and as one would expect, as the strength of the outflow (as measured by the outflow momentum per unit stellar mass) decreases, the efficiency increases, in inverse

2.8 Turbulence and Star Formation

proportion. The effect of increasing the magnetic field is to reduce the efficiency. If the magnetic field is reduced to a negligible value, the efficiency increases by a factor of ≈ 2 relative to the standard case.

A further physical effect that must be considered is the feedback on the collapsing cloud resulting from the luminosity of the protostars that have formed within it. In numerical simulations, first, the luminosities, which are derived mainly from accretion, must be estimated, and second, the transfer of this radiation outward through the cloud must be calculated. In the previous calculations discussed above, radiative transfer was not considered, but simple approximations for the temperature of the gas were used (Chap. 3). The addition of this effect to three-dimensional numerical codes results in a significant enhancement of required computing time, so the simulations cannot be followed as far as those with simpler physics. An initial calculation of this type [38] shows that in a cloud of $50\,M_\odot$ after 1.4 initial free-fall times the number of stars formed is reduced by a factor 2 compared to the calculation without radiative transfer, and the number of brown dwarfs is reduced by a factor of about 5. However the total mass of stars and brown dwarfs formed is not significantly changed. A further simulation [424], with the same cloud mass and including magnetic fields, shows that the effect of radiation transfer is to reduce the efficiency slightly, and that the combined effects of radiation transfer and an initial magnetic field with a mass to flux ratio 3 times critical (similar to the observed value) gives an efficiency per free-fall time (see below) of about 10%. Without the magnetic field the efficiency is a factor 2–3 higher. However the calculation was carried only about 1.3 initial free-fall times. The physical effect of the magnetic field is to slow down the collapse on large scales and thus to reduce the rate of star formation. Radiative transfer effects do not become important until protostars have formed, and the effect is to heat the surroundings, increasing the Jeans mass and suppressing further star formation locally, on small scales.

How are these numerical results on star formation efficiency to be compared with observations? The answer to this question proves to be difficult when one considers the details. First we consider the various possibilities of defining the efficiency, going beyond the global and very general definition given in (1.4).

For a single molecular cloud core that forms a single stellar system the efficiency is

$$\epsilon_{\text{core}} = \frac{M_*}{M_{\text{core}}} \quad (2.53)$$

where the part of the core that doesn't accrete onto the star within it is assumed to have been ejected by the effects of stellar outflows. In an analytic model of collimated bipolar protostellar outflows [346], the results show an ϵ_{core} of about 25% for spherical cores and about 75% for cores with a high degree of flattening. These results are roughly consistent with observations [15] that show that the initial mass function of cores in a star forming region has the same shape as that for stars, but is displaced upwards by a factor of about 3 (Fig. 2.6). In principle, the radiation from the protostar could also affect the efficiency; this effect turns out to be important for high-mass stars (Chap. 5) but not for low-mass stars.

For a cluster-forming clump, the total efficiency, defined when star formation is complete, is

$$\epsilon_{\text{tot}} = \frac{M_*}{M_{\text{clump,init}}} \quad (2.54)$$

where M_* is the total mass in stars and $M_{\text{clump,init}}$ is the original total mass of the clump. This quantity is difficult to determine observationally, because by the time all the stars have formed, much of the remaining gas has been dissipated. Thus the observed efficiency is based on the current values of star mass and clump mass

$$\epsilon_{\text{clump}} = \frac{M_*}{M_* + M_{\text{clump}}} \quad (2.55)$$

where M_{clump} is now the current clump mass, not counting that which has already been dissipated. Thus this efficiency is a function of time, with values ranging from zero to one. It is difficult to determine for a given observed cluster what its evolutionary history was and to compare with theory, where the typical calculation is stopped at 1 or 2 initial free-fall times. Observed values of ϵ_{clump} in embedded clusters, still in the process of formation, range from 8 to 33% [289]. Even smaller values (3 to 6%) are obtained for five nearby star-formation regions based on infrared observations by Spitzer [164]. It is possible that these efficiencies correlate with the age of the system, as expected. The average value is higher than the overall efficiency for a molecular cloud as a whole, where it is estimated to be roughly 2%.

A more meaningful comparison with observations can be obtained through use of still another definition of the efficiency [283]. The dimensionless star formation rate per free-fall time, equivalent to the efficiency per free-fall time, is defined in terms of the density of the object considered:

$$\epsilon_{\text{ff}} = SFR_{\text{ff}} = \frac{\dot{M}_* t_{\text{ff}}(\rho)}{M(>\rho)} \quad (2.56)$$

where $M(>\rho)$ is the mass within a given volume with density greater than a threshhold value ρ, and $t_{\text{ff}}(\rho)$ is the free-fall time at that density. The quantities on the right-hand side can be estimated from observations; note that the uncertain molecular cloud lifetime does not enter. For densities varying from 10^2 to 10^5 cm^{-3}, observations show that the quantity ϵ_{ff} does not vary significantly, and on the average falls in the range 1–3%. By way of comparison, the overall efficiency per free-fall time, averaged over all the molecular clouds in the Galaxy, is about 1%. The numerical simulations involving hydrodynamics with decaying or driven turbulence, which start in the same density range, give much higher values, closer to 20–30% (per free-fall time). In the model with a magnetic field and regeneration of turbulence through outflows [385] the computed efficiency per free-fall time is about 3%. The implication is that clusters form relatively slowly, over many free-fall times. The model with a magnetic field and radiative transfer, although preliminary, gives a value of about 10%.

Although many aspects of the star formation problem still remain unresolved, the introduction of the theory and simulations involving supersonic turbulence have resulted in progress. The observational results over a wide range of scales show that the velocity dispersion scales with the square root of the size of the region; this fact is a strong argument for the existence of supersonic turbulence. Although it is no longer believed that most star formation occurs in magnetically subcritical cores and is controlled by the ambipolar diffusion time, still the magnetic fields must have some influence on the overall process. The magnetic energy in molecular clouds is known to be comparable to the turbulent kinetic energy. Thus the combination of turbulence and magnetic fields in numerical simulations is necessary, although difficult. Eventually, inclusion of even more physical effects in such simulations, along with simultaneous treatment of a wide range of scales, from the size of a whole molecular cloud (parsecs) down to a few solar radii, should result in further progress on this key problem.

2.9 Induced Star Formation

There is some evidence to suggest that star formation may have been induced, not only by turbulent effects, as mentioned in the previous section, but by other kinds of shocks. Induced (or "triggered") star formation simply means that interstellar clouds with masses originally less than their Jeans mass are compressed by some external agent, resulting in a reduction of their Jeans mass to the point where they are forced into collapse. On large scales, galaxy-galaxy mergers as well as galactic spiral arms can induce such compressions. In this section we consider the smaller-scale processes of star formation induced by supernova shock waves, by expanding HII regions, and by cloud–cloud collisions. Although there is some observational evidence that induced star formation has occurred, it is still thought that most star formation occurs by the "spontaneous" processes described in the previous sections.

The supernova shock trigger for star formation is supported by the argument that meteoritic material in the solar system is known [311] to have had live (radioactive) ^{26}Al, which has a half-life of only 0.7 Myr, in it at the time of solidification, as well as a number of other so-called extinct radioactivities [193], including ^{60}Fe, with a half-life of 1.5 Myr. The ^{26}Al decays to ^{26}Mg which is found in meteorites in excess of the normal isotope ratio ^{26}Mg/ ^{24}Mg. Since the ^{26}Al was presumably produced in supernovae, some material in the ejecta must have travelled to a molecular cloud, been injected into a cloud core, evolved to the onset of collapse, collapsed into a disk, and solidified, all in a time of order 1 Myr. Both the magnetic diffusion picture and the turbulent star formation picture would have difficulty explaining this short time scale.

However it is possible that the collapse of molecular cloud material could have been induced on a shorter time scale by the same event that produced the ^{26}Al, namely the supernova shock wave [104]. Numerous calculations, for example [86, 493, 501], have been made of the interaction between clouds and shocks of various kinds. For example, consider a supernova which is set off with an energy of 10^{51} ergs, 10 parsecs away from a pre-existing relatively dense cloud of radius 1 parsec. The shock hits the cloud at 1,000 km/s. A transmitted shock is sent into the cloud, which moves more slowly than the shock outside the cloud. The main shock wraps around the cloud and meets with itself at the back, sending another shock back into the cloud and also one headed away from the cloud. Effects caused by the shock include acceleration of the cloud, stripping of the outer edge of the cloud, and compression of the interior. The shock that wraps around the outside of the cloud induces Kelvin–Helmholtz and other instabilities which tend to shred the cloud. If the cooling time behind the shock is long compared to the time for the shock to cross the cloud, then the cloud will reexpand after passage of the shock and will not be forced into collapse. If the cooling time is short, there is a chance that star formation can be induced, but it still must happen before the cloud is shredded. In general it seems to be difficult for star formation to be induced by such an event. Nevertheless there is observational evidence supporting this picture: a detailed study of the Upper Scorpius OB association [419] strongly suggests that star formation there was initiated by a supernova shock.

The following argument [157] illustrates in a simple way how to get conditions favorable for shock-induced star formation. Note that the way the Jeans length is written in (2.62) implies that if the cloud $R > R_{\text{Jeans}}$ it will be unstable to collapse even before the shock hits it.

To estimate the likelihood, we can compare the gravitational collapse time of the cloud at the post-shock compressed density to the time it takes for the shock to destroy the cloud through Rayleigh–Taylor and related instabilities. The cloud destruction occurs by the interaction of its surface gas with the fast post-shock flow that propagates just outside the cloud. The velocity shear generates the instabilities, and numerical simulations [255, 392] show that the time for the cloud to be torn apart is about 3 times longer than the time for the shock to propagate through the inside of the cloud. The destruction time is

$$t_{\text{dest}} \approx \frac{3R}{v_s(\rho_0/\rho_c)^{1/2}} \qquad (2.57)$$

where the cloud has initial density ρ_c and radius R, and the interstellar gas outside the cloud has density ρ_0. The propagation speed of the shock outside the cloud is v_s and that inside the cloud [351] is the denominator on the right-hand side of (2.57), in the limit of a strong shock.

Let c_c and c_{cA} represent, respectively, the sound speed in the cloud before and after the shock hits it. Let ρ_{cA} be the compressed density in the cloud behind the shock. The free-fall time of the compressed region of the cloud is

$$t_{\text{ff}} \approx \frac{1}{(G\rho_{cA})^{1/2}}. \qquad (2.58)$$

2.9 Induced Star Formation

The shock momentum equation in the frame of the shock reads

$$P_1 + \rho_1 v_1^2 = P_2 + \rho_2 v_2^2 \tag{2.59}$$

where the subscript 1 refers to the gas into which the shock is moving, and 2 refers to the gas behind the shock (the shock equations are discussed in more detail in Sect. 3.4). In the limit of a strong shock, P_1 is small compared to $\rho_1 v_1^2$ and $\rho_2 v_2^2$ is small compared with P_2. Thus, for the shock moving through the cloud, in the above notation

$$\rho_{cA} c_{cA}^2 \approx \rho_0 v_s^2. \tag{2.60}$$

Combining these results we obtain

$$\frac{t_{dest}}{t_{ff}} \approx \frac{3R(G\rho_c)^{1/2}}{c_{cA}}. \tag{2.61}$$

Let λ be the ratio of the cloud radius to its Jeans length before the shock hits:

$$\lambda = \frac{R}{R_{Jeans}} \approx \frac{R(G\rho_c)^{1/2}}{c_c} \tag{2.62}$$

so that

$$\frac{t_{dest}}{t_{ff}} \approx 3\lambda \frac{c_c}{c_{cA}}. \tag{2.63}$$

When this ratio is larger than 1, the cloud can collapse before it is torn apart by the passing shock. Therefore two conditions favor induced star formation in this situation: (1) R is large compared to R_{Jeans}, but if this is true, the cloud will be unstable to collapse even without the shock. Thus for actual induced star formation the quantity 3λ will be of order unity. (2) The cloud cools appreciably after the shock goes through it. If the shock is approximately isothermal, as many types of interstellar shocks are, then $c_c \approx c_{cA}$ and collapse could be possible. Note that the shock velocity does not appear explicitly in (2.63), but actually it is important because a relatively slow shock allows more time for cooling.

A series of simulations [86, 88, 89, 529] suggests that the supernova trigger is a viable explanation of the existence of at least some of the short-lived radioactive isotopes. The initial condition is a centrally condensed isothermal (10 K) self-gravitating Bonnor–Ebert sphere of 1–2 $M_\odot$, representing a molecular cloud core with mass slightly below its Jeans mass. It is hit by a mild shock wave, at say 25 km/s, which does not shred and destroy the cloud. The idea is that the supernova was far enough away and that there was sufficient intervening interstellar material to slow the shock to this velocity. A supernova can produce both ^{26}Al and ^{60}Fe; the shock could impact a nearby molecular cloud, inducing collapse and injecting some of the ^{26}Al into the collapsing cloud material. The simulations show that a portion of the original cloud material is compressed sufficiently so it becomes marginally unstable to collapse, while the outer regions of the cloud appear to escape. Thus the general conclusion from such calculations is that in order to get the cloud to collapse, first, it must be fairly close to its Jeans mass anyway, second, the shock must not be too fast nor too slow (5–70 km/s), and third, cooling must be important, so that the layer behind the shock can cool to close to the original cloud temperature.

An important question connected with the shocked cloud is whether the radioactive material behind the shock front actually gets injected into the compressed material that is going to form the star and disk. High-resolution numerical simulations [88, 89, 528, 529] indicate that Rayleigh–Taylor instability occurring at the boundary between the cloud and the external material can mix some of the matter containing supernova-produced radioactivities into the collapsing region. It has not actually been proved that a sufficient amount of radioactive material can be introduced into the inner few AU of the disk, but the results shown in Fig. 2.15 indicate that injection into the collapsing cloud on larger scales does occur.

The main difficulty in the hypothesis of supernova-triggered solar-system formation is that a single supernova does not produce the correct abundance ratios of all of the dozen or so extinct radioactivities that have been detected in meteorites. Thus alternate explanations of the radioactive isotopes have been proposed. At least some of these isotopes could have been produced in the solar nebula itself [463] by energetic particles produced in solar flares impacting dust grains. However the problem is that ^{60}Fe cannot be produced this way; it apparently requires the supernova environment. Energetic particles in the pre-collapse molecular cloud could also produce some of these isotopes. Another variation on the model [401] involves a pre-existing solar nebula disk (rather than a molecular cloud core) which is impacted by a shock from a supernova less than 1 parsec away. If the radioactive isotopes are incorporated into grains of roughly micron size or larger, they could penetrate into the disk. There are other stellar sources besides supernovae that can produce the radioactive isotopes; two important ones are asymptotic giant stars, and massive $60 M_\odot$ Wolf–Rayet stars. Both of these types of objects can eject the isotopes in stellar winds which can then trigger star formation. Although the supernova trigger is the favored mechanism for the formation of the solar system, the other processes just mentioned could well have made contributions to the radioactive inventory in the meteorites as well.

Expanding HII regions around massive stars could also serve as triggers for star formation. The expansion of an HII region is supersonic with respect to the surrounding cool gas, and it drives a shock, sweeping up the circumstellar gas into a dense shell. The shock precedes the ionization front, and once the shell between them has swept up sufficient mass, it becomes dense and cool, leading to instability and gravitational collapse [162]. The general picture is illustrated in Fig. 2.16. Further work [548] strengthened the case that the fragments that are likely to form in the dense, cool shell will be massive. As such they could generate their own HII regions and initiate a further wave of star formation. Particular examples of star formation observed at the edge of HII regions are Sharpless 104 [133] and RCW79 [570]; an image of the latter region is shown in Fig. 2.17.

2.9 Induced Star Formation

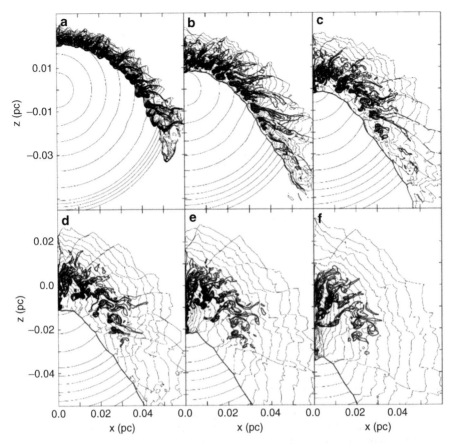

Fig. 2.15 Supernova shock wave hitting a cool cloud of $1\,M_\odot$. The shock is shown at time (**a**) 22,000 years, (**b**) 44,000 years, (**c**) 66,000 years, (**d**) 88,000 years, (**e**) 110,000 years, and (**f**) 132,000 years after it first encounters the cloud. *Thin lines:* contours of equal density, separated by a factor 1.5 in density, with a maximum value of $7.93 \times 10^{-16}\,\mathrm{g\,cm^{-3}}$; *thick lines:* contours of equal density of the material behind the shock wave that is being injected into the cloud. The shock-compressed layer is developing a Rayleigh–Taylor instability. The size of one zone in this numerical simulation is 19 AU. The coordinates give the distance scale in units of parsecs on the *x*- and *z*-axes. Reproduced by permission of the AAS from [529]. © 2002 The American Astronomical Society

The image was obtained by Spitzer at $8\,\mu$m and shows that there is a dust ring around the edge of the HII region. The region has been observed at numerous wavelengths. In the millimeter continuum at 1.2 mm a number of dust condensations are observed, practically coinciding with the edge of the HII region. Their masses are 50 to several hundred $M_\odot$. Near infrared photometry and Hα observations show that in the dust condensation near the south-east edge of the HII region there is a compact HII region and several Class I protostars, indicating that a cluster has formed with at least one massive star. There is good evidence that the large HII

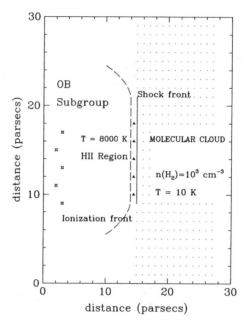

Fig. 2.16 Sketch of the HII region associated with a group of OB stars interacting with a molecular cloud. The stars are originally outside the cloud. After a time of 2 Myr, the ionization front and its preceding shock front have travelled about 10 pc, and the dense layer between them is forced into gravitational collapse, forming new, relatively massive stars. The layer between the shock front and the ionization front has a density of about $n(H_2) \approx 10^5$ cm^{-3} and a temperature of 100 K. The triangles, representing maser sources, infrared sources, or compact continuum sources, are the observed signposts of the massive stars, which themselves are obscured. Once these new stars evolve, their own HII regions can propagate even farther into the cloud, setting off a new wave of star formation, thus inducing a process known as *propagating* star formation. Adapted from [162]

region has expanded into the interstellar medium and swept up a compressed layer of gas and dust between the ionization front and the shock front, and when the layer became massive and dense enough it cooled and fragmented, producing new stars (the "collect and collapse" model). The age of the large HII region is about 1.7 Myr. Analytic models of an expanding HII region in a molecular cloud [548] can be used to deduce that the dense shell fragmented 10^5 years ago, which is consistent with the ages of the protostars and the compact HII region in the clump near the south-east edge.

The numerical simulation of such an expansion [130] with around 10^6 SPH particles showed that an initially uniform-density cloud at 10 K and molecular density 100 cm^{-3}, with an O star turned on at the center, was driven to the point of fragmentation after a time of about 3 Myr, a radius of the HII region of about 10 pc, and a typical fragment mass of $20 M_\odot$. These findings are fairly consistent with analytical results [548]. The implications of the "collect and collapse" model are that the stars in the daughter cluster, formed in the dense shell, should all have

2.9 Induced Star Formation

Fig. 2.17 Induced star formation. In the interior of the expanding bubble is an HII region, powered by a massive star. The Spitzer image of RCW79 shown here is taken at 8 μm and indicates emission from dust particles. Two groups of newly-formed stars are suspected to have formed in the dust shell, one near angle 200° relative to the center of the frame, and the other near angle 45°. The radius of the HII region is about 6.4 parsecs. The direction north is up, east to the left. Picture credit: NASA/JPL-Caltech/E.Churchwell [University of Wisconsin-Madison]

about the same age, and that this age should be distinct from that of the original set of stars that generated the HII region. Observational evidence in favor of this picture of propagating star formation has been obtained through estimates of the ages of spatially separated subgroups in an OB star association such as Sco Cen or Orion OB1 [55, 159].

Cloud–cloud collisions, which are generally supersonic, can compress interstellar material and, under the right conditions, induce star formation. It has been suggested [497] that such collisions on the scale of clouds with masses $\approx 5 \times 10^5\,M_\odot$ could explain the global star formation rate in galaxies, and observational evidence has been found in our Galaxy suggesting that events on these scales have occurred [445]. However, clouds exist on all scales, and numerical simulations of the process typically consider clouds in much lower mass ranges (10–1,000 $M_\odot$) where numerical issues are less severe. However compression induced by such a collision does not guarantee star formation. For example, if a shock is produced, converting kinetic energy of relative cloud motion into heat, one would expect the cloud to be disrupted if the kinetic energy were on the order of the gravitational binding energy of one of the clouds. If, however, the heat of compression is radiated away so that the shock is close to isothermal, one would expect that star formation could be possible. However even in this case, early numerical simulations [305] suggested that the clouds needed to be on the verge of Jeans instability for collapse to be induced.

A number of parameters are involved in such a simulation, including the mass ratio of the two clouds, their temperature, the impact parameter in an off-center collision, and the Mach number, which is the ratio of the relative velocity of collision to the post-shock sound speed. The full parameter space has not been explored; however an interesting example is shown in Fig. 2.18. Two clouds of $10\,M_\odot$ each, with radius 0.22 parsec and temperature 35 K, collide off-center with a relative velocity of $1\,\mathrm{km\,s^{-1}}$. They are Bonnor–Ebert spheres that are not unstable to collapse. At the initial temperature and density, the interstellar heating/cooling balance (Sect. 2.3) leads to a decrease of temperature on compression; thus after shock compression the material cools down to 10 K, a situation favorable to allowing at least part of the clouds to collapse. The end result, shown in the lower panel of the figure, is a single protostar of mass $0.7\,M_\odot$ and radius 90 AU. The angular momentum of the material provided by the off-center collision results in a disk around the protostar, which develops spiral arms. The fate of the disk is unclear, but it could either fragment, producing a low-mass companion, or simply transport angular momentum outward, allowing disk material to accrete onto the protostar. The mean density of the protostar is $\approx 10^{-13}\,\mathrm{g\,cm^{-3}}$ so it is entering the adiabatic collapse phase. A wide range of outcomes is possible with such simulations, from no star formation to formation of several fragments to a binary with a circumbinary disk.

Radiation-driven implosion is a mechanism in which an expanding HII region envelops a dense globule and forces it into collapse by increasing the pressure at the surface [257]. The globules can either be pre-existing or formed by dynamical instabilities generated by the expansion of the HII region into a molecular cloud [159]. Cooling of the compressed cloud material is again key to the forcing of collapse. The heating of the compressed cloud will oppose collapse unless the excess thermal energy is radiated away. Near- isothermal conditions are then favorable for triggering star formation. It would seem counterintuitive for ionizing radiation, which normally tends to disrupt molecular cloud material, to actually aid in generating collapse. But under the right conditions it can happen. For example, numerical simulations [161, 197] of ionizing radiation impinging on a pre-existing turbulent clumpy molecular cloud results in the ionization of the less-dense regions but the compression of the higher-density clumps, producing pillar-like structures similar to those observed, and induced gravitational collapse near the tips of the pillars. A further consequence is an increase in the turbulent kinetic energy in the cloud; thus the ionizing radiation can serve as a driving mechanism (Sect. 2.8). An example of the outcome of a numerical simulation of this process is shown in Fig. 2.19.

2.9 Induced Star Formation

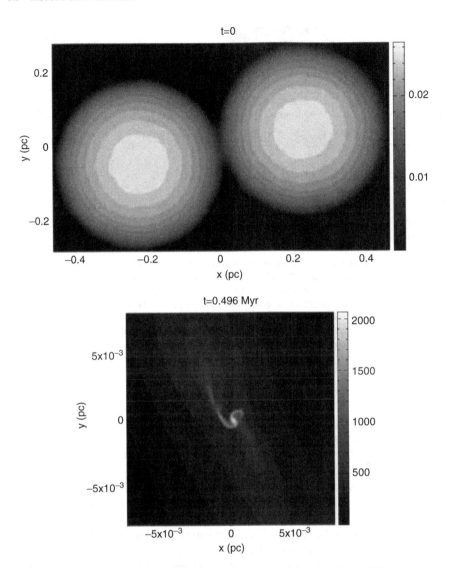

Fig. 2.18 Induced star formation by a cloud–cloud collision calculated with an SPH code in three space dimensions. Two clouds, each with $10\,M_\odot$, collide off-center at 1 km/s. The initial velocity of the clouds is parallel to the x-axis. The upper panel gives the initial condition, and the grey scale gives the column density, integrated along the z-direction, in $\mathrm{g\,cm^{-2}}$, ranging from 1.0×10^{-3} to 2.69×10^{-2} in sixteen equal logarithmic intervals. The lower panel (on a greatly reduced scale) shows the star-disk system that has formed after 0.496 Myr, where the logarithmic grey scale ranges from 0.24 to $2.04 \times 10^3\,\mathrm{g\,cm^{-2}}$. Reproduced by permission of John Wiley and Sons Ltd. from S. Kitsionas, A. P. Whitworth: *MNRAS* **378**, 507 (2007). © Royal Astronomical Society

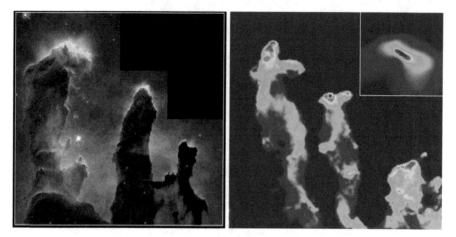

Fig. 2.19 Induced star formation by radiation-driven implosion. *Left:* Optical photograph of the Eagle Nebula (M16). Credit: NASA/ESA/Space Telescope Science Institute/ASU (J. Hester and P. Scowen). *Right:* detail from a three-dimensional numerical (SPH) simulation of ionizing radiation impinging on a clumpy molecular cloud. The surface density ranges from log $\Sigma = -2.7$ (*red*) to -5.5 (*dark blue*), where Σ is the surface density in $\mathrm{g\,cm^{-2}}$. *Right panel* reproduced by permission of the AAS from [197]. © The American Astronomical Society. The *inset* (not previously published) shows a protostellar disk that has formed in one of the pillars. Composite figure courtesy M. Gritschneder

2.10 Summary

Magnetically controlled star formation. If prestellar cores are magnetically subcritical, that is, they have less mass than the magnetic Jeans mass, they can contract in quasi-equilibrium, and bring the densest central regions to the point of collapse through the process of ambipolar diffusion, which increases the mass-to-flux ratio. The time scale is relatively long, about 10 free-fall times, but it could be shortened in a turbulent region. The main problem with this scenario is that numerous cores have been observed to be supercritical, in which case ambipolar diffusion will still occur but will not control the rate of star formation.

Turbulence-controlled star formation. Observed line widths in interstellar clouds are interpreted in terms of supersonic turbulence. The complicated shock patterns randomly generate highly-compressed regions, which, if they last long enough can reach the point of instability to collapse. On the other hand, on more global scales the turbulence is the primary mechanism that supports most molecular cloud regions against gravitational collapse. The advantage of the turbulence picture is that the time scale for star formation is on the order of 1 Myr, which can explain numerous observations. However, simulations often give efficiencies of star formation that are too high, unless very special properties of the turbulence are assumed. Another problem is that turbulence decays on the time scale of one crossing time, and it has been difficult to identify a mechanism that can continuously regenerate it. Also, the

effects of the magnetic field must be included in numerical turbulent simulations. Nevertheless turbulence is now thought to be the major effect in the determination of the star formation rate and efficiency, and the properties of the resulting stars.
Induced star formation. Supernovae shocks, cloud–cloud collisions, and the sweeping up of dense shells and clumps of gas by massive-star winds and HII regions can induce star formation, and there is considerable observational evidence that these processes occur. Substantial cooling behind shock fronts is a requirement for star formation. It is thought that only a relatively small fraction of the observed star-formation rate can be accounted for by induced processes. However it appears that such an event is necessary to explain the presence in the Solar System of the decay products of extinct radioactive isotopes such as ^{26}Al and ^{60}Fe.

2.11 Appendix to Chap. 2: Note on Numerical Methods

The numerical results discussed in this book, on problems of hydrodynamics and magnetohydrodynamics in 2 or 3 space dimensions, are mostly based on two basic techniques. The first involves expressing the differential equations as finite differences on an Eulerian grid (fixed in space), as discussed, for example, by [435,493]. The second, known as smoothed particle hydrodynamics (SPH), is based on representation of the fluid by a set of particles, each with a given mass, which move around in space under the effect of the various forces involved, such as those from gravity and the pressure gradient. There is no grid, and the method is Lagrangian in nature. The basic properties of the method are reviewed by [368,422]. A direct comparison of the two methods on the same hydrodynamic collapse problem is shown in our Figs. 6.11 and 6.12.

The advancement of physical quantities in time in both cases is based on an *explicit* numerical scheme, in which a quantity at the advanced time t^{n+1} is based entirely on known quantities at the previous time t^n and the assumed time interval Δt. Thus, for a simple example of an equation in one space dimension

$$\frac{\partial y(x,t)}{\partial t} = W(y,x,t) \qquad (2.64)$$

where y is a physical variable such as density, the explicit time-differencing scheme would be

$$y(t^{n+1}) = y(t^n) + \Delta t\, W(t^n). \qquad (2.65)$$

An alternative to this scheme, known as an *implicit* method, can be written

$$y(t^{n+1}) = y(t^n) + \Delta t\, W(t^{n+1}) \qquad (2.66)$$

where generally the solution has to be iterated to self-consistency. This method is used in one-dimensional calculations of stellar evolution.

In any explicit scheme, Δt is limited by two important effects. The first is the Courant–Friedrichs–Lewy (CFL) condition [121] which states that if a grid zone has a width Δx and the fluid is moving through it with a sound speed c_s and a velocity $|v|$, then Δt is limited to a fraction of the crossing time $C_0 \Delta x/(c_s + |v|)$. Here C_0, the Courant number, must be <1 and is usually taken to be about 0.5. The actual time step is taken to be the minimum CFL condition over the entire grid. A numerical instability arises if the condition is not met. In the (gridless) SPH method, the resolution element Δx is replaced by the *smoothing length h* (see below). The second limitation is needed if viscosity, either physical or artificial, is present [see (4.18)–(4.21)]. Then the limit [411] is $\Delta t < (\Delta x)^2/(4\nu)$ where ν is the kinematic viscosity.

In star-formation problems involving collapse from an initially extended configuration of relatively uniform density into a very centrally condensed configuration, a grid-based solution with uniform grid spacing Δx soon becomes inadequate to represent the flow. Much of the mass flows into the central zone and becomes unresolved. Similar problems arise if the computational problem involves fragmentation. The difficulty may be overcome by rezoning the grid in regions that have developed, for example, higher than average density, a procedure known as *adaptive mesh refinement* [50, 256]. The idea is for the code to automatically generate (or delete) finer grids to follow the details of small-scale structures that develop. A three-dimensional zone that needs refinement can be divided into 8 zones, so that the overall spatial resolution is increased by a factor 2. The time-step restrictions must be followed on the refined grid, so that several time steps on a refined grid may be necessary for each time step on the overlying coarser grid. A refined zone can be further refined if needed; in extreme collapse problems, 40 or more levels of refinement are needed if the spatial scale reduction per level, for reasons of accuracy, is limited to a factor 2. For collapse problems, such a procedure is certainly more efficient than providing a single very large uniform grid with the grid spacing needed for the highest-density regions. In SPH calculations, no such refinements are necessary; the spatial resolution is self-adapting as particles move into regions of higher density.

To represent the density or another physical variable in smoothed particle hydrodynamics, one assumes that a particle is not a point mass, but rather a smoothed-out distribution of density over a small volume. Thus, if a particle j momentarily has a position in space $\mathbf{r}_j$, then the contribution of that particle to the overall density at some radius $\mathbf{r}$ is given by

$$\rho_j(\mathbf{r}) = \frac{m_j}{h^3 \pi^{3/2}} \exp[-(|\mathbf{r} - \mathbf{r}_j|/h)^2] \qquad (2.67)$$

where m_j is the particle mass and h is the smoothing length. Then the total density at $\mathbf{r}$ is

$$\rho(\mathbf{r}) = \sum_{j=1}^{N} \rho_j(\mathbf{r}). \qquad (2.68)$$

Usually N does not have to include the full set of particles, because only close-by particles make a significant contribution to $\rho(\mathbf{r})$. The Gaussian form of the smoothing function is given just as an example; typically the function can have other forms and is cut off at about $2h$. To determine h, which is different for each particle, one requires that the smoothing volume has to include about 50 other particles. Thus regions of higher density automatically develop higher spatial resolution.

The number of zones in a grid-based calculation or the number of particles in an SPH calculation must obey an accuracy requirement known as the Jeans condition [42, 511]. At all points in the domain and at all times the local Jeans length (2.16) must be resolved. In a grid calculation, that means that, locally, there must be at least 4 zones per Jeans length. In SPH, the requirement is that the Jeans length must contain several smoothing lengths. If one ignores this requirement, one runs the risk of observing numerical fragmentation under conditions where physical fragmentation would not occur. If the initial number of particles in an SPH simulation is insufficient to meet this requirement at later times, particle splitting can be employed, with one particle replaced by 8 particles, to improve the resolution.

From the resolution requirement and the CFL condition, it is clear that, for example, in the high-density centers of rapidly collapsing regions, the required time step can be very short, in fact so short that it becomes impossible to follow the simulation for a reasonable amount of time. To deal with this problem, either on a grid [279] or in SPH [41], *sink particles* are introduced. A volume with a given radius is defined (say 1 AU); once the mean density within such a volume exceeds a pre-determined limit, then the volume is replaced by a single particle with the same mass and momentum. Either on a grid or in SPH, the particle is followed as a Lagrangian object, which moves through space under the influence of the gravity of all external matter. The particle can also accrete additional mass and momentum from the surrounding region. Material inside the sink is not resolved; however the sink can be used as a source of energy input from the accretion luminosity of a presumed protostar that has formed within it.

2.12 Problems

1. A spherical cloud of $2\,M_\odot$ has a uniform density of $5 \times 10^{-19}\,\mathrm{g\,cm^{-3}}$, uniform temperature of $10\,\mathrm{K}$, and uniform composition of molecular hydrogen. It is rotating with uniform angular velocity $\Omega = 2 \times 10^{-14}\,\mathrm{rad\,s^{-1}}$ and is threaded by a uniform magnetic field of 2×10^{-5} Gauss.

   (a) Calculate the gravitational energy, the thermal energy, the rotational energy, and the magnetic energy. Find out if the cloud is unstable to collapse.
   (b) The cloud collapses with conservation of mass, angular momentum, and magnetic flux. How do magnetic energy and rotational energy scale with radius (compare with the scaling of the gravitational energy)? The collapse is isothermal. What happens to the total internal energy?

(c) Will the collapse stop under these assumptions? If so, what stops it and what is the final radius? (Think of physical simplifications).
(d) Is the total energy of the cloud conserved during the collapse? Why?

2. The aim of this problem is to derive the Jeans length in the same way that Sir James Jeans did it, and to identify where a physical inconsistency occurs in this method.

The equations of continuity and momentum are

$$\frac{\partial \rho}{\partial t} + \mathbf{v} \cdot \nabla \rho = -\rho \nabla \cdot \mathbf{v} \tag{2.69}$$

$$\rho \left(\frac{\partial \mathbf{v}}{\partial t} + \mathbf{v} \cdot \nabla \mathbf{v} \right) = -\nabla P - \rho \nabla \phi \tag{2.70}$$

where ρ is the density, P is the pressure, $\mathbf{v}$ is the velocity, and ϕ is the gravitational potential.

Now assume that velocity, density, and ϕ can be represented by a mean state plus a small fluctuation:

$$\rho = \rho_0 + \delta\rho; \quad \mathbf{v} = \mathbf{v}_0 + \delta\mathbf{v}; \quad \phi = \phi_0 + \delta\phi. \tag{2.71}$$

Also assume that the mean state is in equilibrium and is isothermal, so $\mathbf{v}_0 = 0$, $\partial \rho_0 / \partial t = 0$, and $P = c_s^2 \rho$ where c_s is the constant sound speed.

(a) Expand and linearize the equations to get two expressions relating $\delta\rho$ and $\delta\mathbf{v}$.
(b) Then make the simplification that ρ_0 is constant and use the linearized Poisson equation $\nabla^2 (\delta\phi) = 4\pi G \delta\rho$ to get an expression for $\delta\rho$ as a function of time

$$\frac{\partial^2}{\partial t^2} (\delta\rho) = \rho_0 \left(4\pi G \delta\rho + \frac{c_s^2}{\rho_0} \nabla^2 \delta\rho \right). \tag{2.72}$$

(c) Now assume that the perturbed density $\delta\rho$ has a plane-wave solution

$$\delta\rho = K' \exp[i(\omega t + kx)] \tag{2.73}$$

where K' is a constant, and find the dispersion relation between ω and k. If ω^2 is positive, the solution is oscillatory. If ω^2 is negative, the disturbance grows exponentially. The critical case corresponds to the Jeans limit. Obtain the Jeans length and the Jeans mass and compare with values obtained by simpler means (using energies, for example).
(d) What is physically inconsistent about the assumptions made?

3. Express (2.10) in terms of velocities: sound velocity, Alfvén velocity, rotational velocity, mean turbulent velocity. Show that the equation for the magnetic Jeans mass is roughly equivalent to the condition that the free-fall time of the sphere be comparable to the crossing time of an Alfvén wave.

2.12 Problems

4. (a) Use the shock jump conditions for an isothermal shock [(3.37) and (3.38)] to show that the density ratio across the shock is given by the square of the Mach number v_1/c_s, where c_s is constant across the shock.
   (b) A shock is moving at Mach 6 with respect to a spherical clump of radius 2 pc, mass 225 $M_\odot$, and sound speed 0.6 km/s. The shock is isothermal. Would the clump be unstable to collapse before the shock hit it? The gas external to the clump is hotter than that in the clump but in pressure balance with it. What is the Mach number of the shock outside the clump? After the shock hits the clump, calculate the crossing time and the free-fall time of the compressed layer. Is induced star formation likely? What if the shock were Mach 12? Would that make a difference?

5. Consider a uniform-density sphere of radius R, mean column density (particles cm^{-2}) N, particle mass m, uniform magnetic field B, and full-width half-maximum line width ΔV. These are all observable quantities. Make the approximation that the line width arises purely from turbulence.

   (a) Calculate the magnetic energy density (energy per unit volume), the kinetic energy density (in turbulence), and the gravitational energy density in terms of the above quantities.
   (b) Assume a simple model of a cloud core in which all three energy densities are equal. Show that $B \propto (\Delta V)^2/R$; find the constant of proportionality. Assume the cloud is all molecular hydrogen.
   (c) A core in Orion A has a measured line width of 1.7 km s^{-1} in H_2CO, a radius of 0.18 pc, and a measured magnetic field strength $B = 250\,\mu G$. A core in Orion KL has a line width of 2.6 km s^{-1} in NH_3, a radius of 0.047 pc, and $B = 3{,}000\,\mu G$. Do these observations agree with the model or not? For the first case in this problem calculate the column density and the visual extinction (in magnitudes) in the core.
   (d) Show also that, in the model, $B \propto \Delta V n^{1/2}$, where n is the density in particles per cm^3. Does this relation agree with observations?

Chapter 3
Protostar Collapse

Once magnetic forces have become dynamically unimportant, turbulence has decayed, and rotational effects have become unimportant relative to gravity in the core of a molecular cloud, and the mass of the core exceeds the thermal Jeans mass, gravitational collapse proceeds. Some of the questions that can be asked include (1) what are the initial conditions for collapse? (2) What is the physics that induces and maintains collapse? (3) What is the role of magnetic fields during collapse? (4) Can the embedded infrared sources (Class I objects) be identified with the stage of evolution just after disk formation? (5) What is the observational evidence for infall? (6) What is the origin of bipolar outflows? (7) Once a disk forms, how is angular momentum transported? (8) Will the core fragment or form a single star? (9) What fraction of the core mass ends up in the star? (10) How is the high-mass star formation problem solved? In fact, for high mass star formation, turbulence is important in the initial conditions, and the problem is considered in more detail in Chap. 5. The present chapter applies in general to low-mass star formation; however the physical regimes of the collapse itself, as described next, apply to all masses.

Protostellar evolution can be divided into three phases. During the first, isothermal collapse, the gas is optically thin to the infrared radiation of the grains. Instability to collapse is a consequence of the fact that the released gravitational energy is essentially all radiated away, so that the thermal energy stays well below the gravitational energy. Heating of the gas by compression and cosmic rays is balanced by grain and molecular cooling, with a resulting temperature near 10 K. Although this equilibrium temperature is not exactly constant as a function of density, it is convenient to assume, without significant error, that the gas is isothermal at 10 K over the density range $10^{-19}\,\mathrm{g\,cm^{-3}} < \rho < 10^{-13}\,\mathrm{g\,cm^{-3}}$. Once the gas collapses to the point where the density is high enough for the gas to become optically thick, which occurs first in the high-density center, the second, adiabatic, phase begins. Much of the released gravitational energy goes into heat, the pressure increases rapidly, and the collapse slows down. However, once the central temperature has increased to about 2,000 K, the molecular hydrogen dissociates and instability to collapse resumes, because much of the released gravitational energy must go into dissociation energy rather than into an increase in thermal pressure.

The collapse continues until dissociation is complete at the center, at which time a small amount of material comes into hydrostatic equilibrium there, forming the "stellar" core. The third phase of evolution involves the accretion of the remainder of the collapsing cloud onto the core, through an accretion shock at its outer edge. These same general phases are encountered for spherical protostars as well as for rotating protostars or protostars with turbulent initial conditions. In the rotating case there may be accretion onto a disk as well. We first describe in a little more detail the three main phases in the spherical case, and in later chapters we show some numerical results for rotating and turbulent protostars.

3.1 Protostellar Initial Conditions

The standard initial conditions for a low-mass spherical protostar in a molecular cloud core can be characterized as follows:

- The core is composed of a solar mixture of elements, with about 70% hydrogen by mass; however the hydrogen is in molecular form, so the overall mean molecular weight per free particle μ, in atomic mass units, is about 2.37
- The core has negligible infall motion
- The core has a mass of about 1 to a few solar masses and a radius of 0.05–0.1 parsec
- The core has a distribution of density that could be (a) uniform, (b) a power law with $\rho \propto r^{-n}$, where observations typically give n in the range 1–2, or (c) a Bonnor-Ebert sphere (see below) with a relatively flat density distribution in the center, but approaching $\rho \propto r^{-2}$ in the outer regions, a distribution that matches some observations [20, 24, 539]
- The core is nearly isothermal at T $\approx$ 10 K
- The main parameters are α = thermal energy/|gravitational energy| and the form of the density distribution.

Observations show that cores are near virial equilibrium with $\alpha \approx 0.4$. Magnetic effects and turbulent effects may be of some importance in a fraction of low-mass cores, but for simplicity they are not considered here.

The theoretical Bonnor-Ebert sphere is simply an isothermal equilibrium structure, in which gravity, plus an assumed external pressure P_{surf}, are balanced by the internal pressure gradient. The equations are

$$\frac{Gm}{r^2} + \frac{1}{\rho}\frac{dP}{dr} = 0. \tag{3.1}$$

$$\frac{dm}{dr} = 4\pi r^2 \rho \tag{3.2}$$

$$P = c_s^2 \rho \tag{3.3}$$

3.1 Protostellar Initial Conditions

Here m is the total mass interior to a given spherical surface at radius r, ρ is the mass density at r, P is the pressure, $c_s = (R_g T/\mu)^{1/2}$ is the sound speed, and μ is defined above. The first equation expresses force balance on a mass element (hydrostatic equilibrium), the second gives the mass (dm) of a spherical shell at radius r, and the third is the ideal gas equation of state. The equations can be combined to produce

$$\frac{1}{r^2}\frac{d}{dr}\left(r^2 c_s^2 \frac{d\ln\rho}{dr}\right) = -4\pi G\rho. \tag{3.4}$$

Using dimensionless variables

$$-u = \ln(\rho/\rho_0), \tag{3.5}$$

where ρ_0 is the central density, and

$$\xi = \frac{r}{c_s}(4\pi G\rho_0)^{1/2} \tag{3.6}$$

one obtains the equation

$$\frac{1}{\xi^2}\frac{d}{d\xi}\left(\xi^2 \frac{du}{d\xi}\right) = e^{-u} \tag{3.7}$$

with boundary conditions at the center $u(0) = 0$ and $(du/d\xi)(0) = 0$. The parameter in the solution is the outer dimensionless radius ξ_1, where the integration outward from the center is arbitrarily stopped. Note that at ξ_1 the density is not, in general, zero, and the external pressure P_{surf} matches the internal pressure at that point.

Once the outward integration has reached ξ_1 the total mass enclosed is

$$M(\xi_1) = 4\pi \int_0^{\xi_1} \rho_0 e^{-u} \alpha_B^3 \xi^2 d\xi = \left(\frac{1}{4\pi\rho_0}\right)^{1/2}\left(\frac{R_g T}{\mu G}\right)^{3/2} \xi_1^2 \left(\frac{du}{d\xi}\right)_{\xi=\xi_1}, \tag{3.8}$$

where $r = \alpha_B \xi$ and

$$\alpha_B = \frac{c_s}{(4\pi G\rho_0)^{1/2}}. \tag{3.9}$$

Equation (3.7) is used to perform the integration.

The basic properties of the solutions to (3.7) are found as follows. Assume that the total mass M and temperature T are fixed. Then for a given ξ_1 the mass equation (3.8) determines ρ_0, and the relation $R = \alpha_B \xi_1$ determines the outer radius R:

$$R = \frac{GM(\xi_1)}{c_s^2 \xi_1 (du/d\xi)_{\xi=\xi_1}}. \tag{3.10}$$

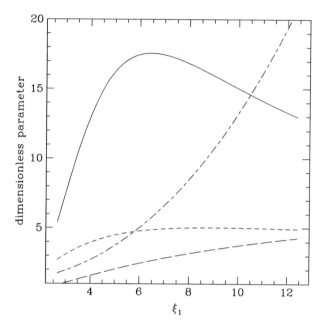

Fig. 3.1 Properties of solutions for the Bonnor-Ebert sphere, as a function of ξ_1, the dimensionless radius at which the integration is cut off. *Solid curve:* the dimensionless surface pressure, *long-dashed curve:* $u(\xi_1)$, *short-dashed curve:* the quantity $\xi_1 (du/d\xi)_{\xi=\xi_1}$, *long-dashed–short-dashed curve:* the degree of central concentration $\rho_0/\bar{\rho}$

The outer pressure is obtained from the ideal gas equation

$$P_{\text{surf}} = \frac{c_s^8 \xi_1^4 (du/d\xi)_{\xi=\xi_1}^2 \exp(-u_1)}{4\pi G^3 M(\xi_1)^2}, \qquad (3.11)$$

where $u_1 = u(\xi_1)$. For small ξ_1 the density distribution is relatively flat, the radius is large, and the equilibrium is stable. As ξ_1 is increased note that $\xi_1 (du/d\xi)_{\xi=\xi_1}$ increases (Fig. 3.1) so the radius decreases (out to about $\xi_1 = 9$) and gravity becomes more important. Also, the ratio of central to mean density $\rho_0/\bar{\rho}$ increases. As the plot shows, for a critical value $\xi_{1,crit} = 6.5$ the external pressure reaches its maximum; there the equilibrium is marginally stable. Noting that at $\xi_{1,crit}$ we have $(du/d\xi)_{\xi=\xi_1} = 0.375$, the corresponding radius and pressure are

$$R_{\text{crit}} = 0.41 \frac{GM\mu}{R_g T}, \quad P_{\text{crit}} = 1.40 \frac{c_s^8}{G^3 M^2}. \qquad (3.12)$$

It is clear that R_{crit} is almost exactly the Jeans length derived from the energy argument. At the critical ξ_1 the ratio $\rho_0/\bar{\rho}$ is 5.8, and the ratio of central density to density at the outer edge is 14. From (3.12) the critical Bonnor-Ebert mass can be

3.1 Protostellar Initial Conditions

written, for a fixed c_s and external pressure

$$M_{\text{crit}} = 1.18 \frac{c_s^4}{P_{\text{crit}}^{1/2} G^{3/2}} = 1.82 \left(\frac{n}{10^4 \text{cm}^{-3}}\right)^{-1/2} \left(\frac{T}{10 \text{ K}}\right)^{3/2} M_\odot, \quad (3.13)$$

where n is the mean number density of particles in a cloud. Given observational estimates for n and T in a given interstellar cloud, the critical mass can be determined and the actual cloud masses compared with the critical value to see if they are stable or unstable to collapse. Equation (3.13) is also useful if the external pressure in a region with starless cores is known. Given the sound speed within a core, and identifiying the external pressure with the surface pressure, the critical mass is

$$M_{\text{crit}} = 1.5 \left(\frac{c_s}{0.2 \text{ km s}^{-1}}\right)^4 \left(\frac{P_{\text{surf}}/k}{10^5 \text{ K cm}^{-3}}\right)^{-1/2} M_\odot. \quad (3.14)$$

This equation illustrates two important points. First, under typical conditions in molecular cloud cores, masses above about $1.5 \, M_\odot$ are unstable to collapse, while lower-mass clouds are not. Thus masses at the lower end of the observed stellar mass range are apparently difficult to form. Second, an increase in the external pressure results in a reduction of the critical mass. However in order to get a core of say $0.1 \, M_\odot$ to collapse a rather large increase is needed.

We now discuss briefly the reason for instability. For small ξ_1 the equilibrium is stable. Gravity is not important compared to the external pressure in determining the equilibrium, and an increase in the external pressure, for a given mass and c_s, results in a reduction in size, compression, and an increase in the internal pressure to match the external pressure. For $\xi_1 > \xi_{1,crit}$ the equilibrium is unstable. The external pressure is no longer important, compared with gravity, in determining the equilibrium. A small perturbation, for example a decrease in radius at fixed mass and c_s, results in the gravity force becoming larger than the restoring pressure force. A simple way of looking at the situation involves the total energies. The decrease in radius results in a more negative gravitational energy, but the total internal energy doesn't change with radius. The resulting net inward force will compress the cloud even further, so the situation is unstable. Another way of expressing the situation is to quote the well-known thermodynamic result [506] that if the equation of state of a gas is written $P = \text{constant} \times \rho^\gamma$, and the gas sphere is in equilibrium against the force of gravity, then the equilibrium is unstable if $\gamma < 4/3$. In the isothermal case $\gamma = 1$. (Note that the argument does not apply when the external pressure is significant in maintaining the equilibrium.) Again, the physical reason is when the protostar is compressed, the released gravitational energy does not go into heat and the associated increase in pressure, but instead is radiated away through the optically thin medium.

The limit of $\xi_1 \to \infty$ corresponds to the *singular isothermal sphere*, which has $\rho \propto r^{-2}$, an infinite outer radius, an infinite central density, and is an unstable equilibrium.

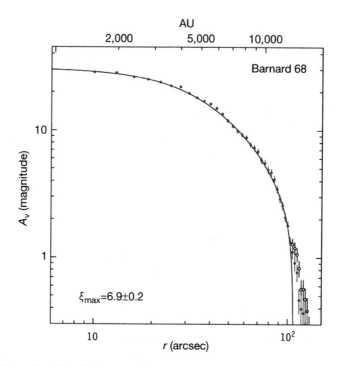

Fig. 3.2 Observed radial profile (*points*) of the molecular cloud core Barnard 68. The azimuthally averaged visual extinction, which is related to the gas column density, is plotted as a function of distance from the cloud center. *Solid line:* the corresponding profile of a theoretical Bonnor-Ebert sphere with $\xi_{max}(=\xi_1) = 6.9$. Reprinted by permission from Macmillan Publishers Ltd., J. F. Alves, C. J. Lada, E. A. Lada: *Nature* **409**, 159 (2001). © 2001 Nature Publishing Group

These theoretical curves can be compared with observational spatially resolved density distributions for some cores without embedded infrared sources (known as *starless* cores or *prestellar* cores), which are thought to represent conditions just prior to the onset of collapse. An example is shown in Fig. 3.2. Here the dust extinction of background giant stars in the near infrared (H and K bands) by the prestellar core Barnard 68, is used to determine the density profile in the core, which has about $2\,M_\odot$, a radius of 12,500 AU, and a mean density of $1.5 \times 10^{-19}\,\mathrm{g\,cm^{-3}}$. From the reddening of the background stars one can deduce the total dust extinction, and therefore N_{H2} at about 1,000 different points across the core [14]. Note that the data represent the projection on the plane of the sky of the actual three-dimensional density distribution $\rho(r)$. In Fig. 3.2, by astronomical convention, the data is plotted in terms of visual extinction, A_V, in magnitudes, and a theoretical Bonnor-Ebert profile is converted to those units. The dots give the data, while the solid line gives the best fit Bonnor-Ebert equilibrium sphere, which happens to have an outer dimensionless radius $\xi_1 = 6.9$, very close to the critical value (above which the sphere is unstable to collapse) of 6.5. Thus only a slight perturbation, such as cooling of the cloud, would initiate collapse. Note, however, the typical molecular cloud core

3.2 Isothermal Collapse

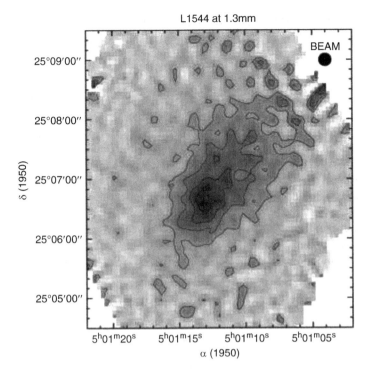

Fig. 3.3 A map of the 1.3 mm emission in the prestellar molecular cloud core L1544. The scale of the plot is approximately 0.1 pc. *Darker regions* correspond to higher flux density and therefore to higher column density of dust. The source is a relatively isolated one in the Taurus molecular cloud. Reproduced with permission from Wiley D. Ward-Thompson, F. Motte, P. André: *MNRAS* **305**, 143 (1999). © 2002 Royal Astronomical Society

is not spherical, as illustrated in Fig. 3.3 which shows a 1.3 mm map [538]. At this wavelength the dust is optically thin, and a measurement at a given point gives the dust column density, which can be converted to N_{H2}. The map shows an irregular density structure and non-spherical density maxima. Note also that other cores may have different density distributions, in some cases closer to the r^{-2} of the singular isothermal sphere, which is also used in theoretical calculations as a possible initial condition. Finally, note that the agreement of an observed cloud density distribution with that of a Bonnor-Ebert model does not necessarily imply that the actual cloud is in strict hydrostatic equilibrium. Simulations of turbulent interstellar clouds [335] show that Bonnor-Ebert-like profiles are often found, but they are transient and the corresponding clouds are not in equilibrium.

3.2 Isothermal Collapse

The free-fall time from the standard initial conditions falls in the range 1–2×10^5 yr. One can derive, from dimensional analysis, an approximate infall rate $\dot{M} \approx M_{\text{core}}/t_{\text{ff}}$. Using $t_{\text{ff}} \approx (G\rho)^{-1/2}$ and $\rho \propto M_{\text{core}}/R_{\text{core}}^3$ as well as the assumed initial

Jeans condition $R_{core} = 0.4G\mu M_{core}/(R_g T)$ one finds

$$\dot{M} \approx c_s^3/G. \tag{3.15}$$

For the singular isothermal sphere the proportionality constant is almost exactly unity and $\dot{M}$ is constant in time at a typical value of $2 \times 10^{-6} M_\odot$ yr^{-1} for $T_{init} = 10$ K. To see that $\dot{M}$ is constant for this case, consider the mass equation (3.2) with $\rho \propto r^{-2}$, leading to $dm/dr =$ constant, so $m \propto r$. The free-fall time $t_{ff} \propto (m/r^3)^{-1/2} \propto r$. Hence $\dot{M} \approx m/t_{ff} =$ constant in time.

As the protostar collapses the temperature remains approximately isothermal at 10 K over several orders of magnitude increase in density. Spherical collapse is governed by the equations of motion, mass conservation, and the isothermal equation of state. In the situation where an extreme degree of density contrast develops between center and outer edge, it is convenient to express the equations in terms of m as an independent variable and to use the Lagrangian time derivative, following the motion of mass elements:

$$\frac{Gm}{4\pi r^4} + \frac{dP}{dm} = -\frac{1}{4\pi r^2}\frac{d^2r}{dt^2} \tag{3.16}$$

$$\frac{dr}{dm} = (4\pi r^2 \rho)^{-1} \tag{3.17}$$

$$P = \frac{R_g}{\mu}\rho T. \tag{3.18}$$

The symbols are the same as those used in (3.1)–(3.3). Since the temperature T is fixed in time and space, these equations are sufficient to specify the solution for ρ, r, and P as functions of (m, t). The solution is obtained numerically, subject to the inner boundary condition $m = 0$ at $r = 0$ and the outer boundary condition that either the pressure is constant in time or that the radius is fixed with zero velocity at the outer surface.

Even if the cloud has initially uniform density (and therefore uniform free-fall time), during collapse the inner regions increase their pressure relative to the surface value, which is constrained by the outer boundary condition. A pressure gradient is thus set up; the boundary of the region that has the gradient propagates inward at the speed of sound relative to the velocity of the infalling material. In other words, material in the central regions continues to have constant density in space, but increasing in time, as long as the rarefaction wave generated at the outer edge has not had time to reach it. As long as the initial sound travel time $t_s = R/c_s$, where R is the total radius and c_s is the sound speed, is comparable to or shorter than the free-fall time, the entire configuration soon develops a gradient in pressure and density. The collapse is therefore somewhat retarded from a true free fall. Once the gradient is established, the evolution time becomes far shorter in the central regions than in the outer regions.

3.2 Isothermal Collapse

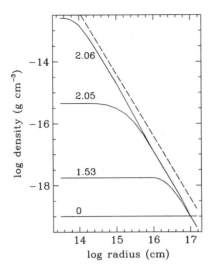

Fig. 3.4 Density vs. radius (*solid curves*) at four different times (in units of the initial free-fall time) during the collapse of an isothermal sphere starting from uniform density (*lowest curve*). The *dashed line* shows a profile with $\rho \propto r^{-2}$. After [299]

The solution [299], starting from near the Jeans limit for a protostar of 1 $M_\odot$ with initially uniform density, is illustrated in Fig. 3.4. After about 1.5 initial free-fall times, the central density has increased by several orders of magnitude, but the mass involved in the central peak is very small, only 10^{-3} of the total mass. The outer regions, which contain most of the mass, are collapsing slowly with respect to the central regions. Typical infall velocities increase to 2–3 times the sound speed, or roughly 0.5 km s^{-1} for $T = 10$ K. This general end result, a very centrally condensed collapsing configuration with $\rho \propto r^{-2}$ outside a central "plateau" occurs also when the initial condition is a Bonnor-Ebert density profile, but forced to be out of equilibrium, that is, the pressure force is assumed to be somewhat less than the gravitational force [68, 395]. In fact it can be shown, through an analytic similarity solution for the equations of isothermal collapse [299, 413] that the r^{-2} density profile is obtained at a given radius in the limit of long times after the start of collapse. In practice, for this condition to be satisfied it also must be true that the sound travel time through the initial sphere must not be much greater than the free-fall time, as indicated above. Once the pressure gradient is established, the collapse is no longer a true free fall; in fact the outward force resulting from the pressure gradient is about half that of gravity. Also as mentioned above, the r^{-2} profile indicates that the collapse evolves toward a steady state, with a constant mass accretion rate of material into the density peak. If one measures that accretion rate from the data in Fig. 3.4 one finds, at any radius, $m/t_{\rm ff} = 2 \times 10^{-6} M_\odot$ yr^{-1}. The infall velocities, given by $\dot{m} = 4\pi r^2 \rho v$ are then approximately constant, close to the sound speed.

Another possible starting point that has been considered is the singular isothermal sphere [456], with $\rho \propto r^{-2}$. The equilibrium is unstable and a small perturbation will result in collapse. The end result is the same as if the collapse started at constant density, but the process of collapse is different. The dynamical time is shortest at the center, so the collapse begins there, a situation known as *inside-out collapse*. The region that is collapsing spreads outward at the speed of sound, so the boundary between the collapsing region and the relatively static region is given by $R_b = c_s t$. Thus the outer boundary condition plays a less significant role than in the case of a collapse starting from constant density, and, in the singular case, the rarefaction wave spreads outwards rather than inwards. The accretion rate for the singular case is $\dot{M} \approx c_s^3/G = 1.9 \times 10^{-6} M_\odot \, \text{yr}^{-1}$ for $c_s = 0.2 \, \text{km s}^{-1}$, not very different from that obtained from Fig. 3.4. It is approximately constant with time as long as the r^{-2} profile is maintained; however on scales smaller than 10^{14} cm, at later phases of the collapse, this profile changes.

3.3 Adiabatic Collapse

At the beginning of the isothermal collapse, when the opacity (κ) from dust grains is below $1 \, \text{cm}^2 \, \text{g}^{-1}$, the density is about $10^{-19} \, \text{g cm}^{-3}$ and the size is about 10^{17} cm, the optical depth $\tau \approx \kappa \rho R \ll 1$. Once the optical depth in the central peak exceeds unity, which corresponds to a density of about $10^{-13} \, \text{g cm}^{-3}$, the energy released by the compression of the gas can no longer be radiated freely by the dust, and the isothermal approximation no longer holds. The heating of the gas results in a temperature gradient. The gas rapidly becomes optically thick because the opacity increases with temperature, and the standard radiative diffusion equation, used in stellar interiors calculations (e.g. [119]) becomes an adequate approximation for relating the radiative flux F_r and the temperature gradient (see Chap. 8):

$$F_r = -\frac{16\pi a c r^2}{3\kappa_R} T^3 \frac{dT}{dm}, \tag{3.19}$$

where κ_R is the Rosseland mean opacity (defined in Chap. 8) in $\text{cm}^2 \, \text{g}^{-1}$, c is the velocity of light, a is the radiation density constant, and F_r is the energy crossing a sphere at radius r per unit time per unit area, integrated over all frequencies. The opacity is dominated by dust grains at temperatures up to about 1,500–2,000 K and by molecules and atoms above that temperature range. Between temperatures of about 400 and 1,500 K the grains are composed of mainly silicates, iron, and their compounds. Between 400 and 175 K, volatile organics are also present. Below 175 K, ice grains are present along with the other species. Dust opacities and molecular and atomic opacities [175, 452] are plotted in Fig. 3.5. Below 175 K the opacity is dominated by ice grains, and the Rosseland mean increases as T^2. Between 175 K and about 1,700 K the opacity is dominated by silicates and iron grains. The sudden dip about 1,700 K is caused by the evaporation of the silicate

3.3 Adiabatic Collapse

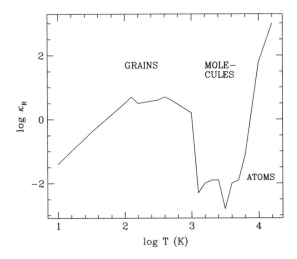

Fig. 3.5 An example of Rosseland mean opacity, in cm² g⁻¹, appropriate for a protostar with solar composition, plotted as a function of temperature. The density along the curve is given by $\rho = 10^{-19} T^3$. Dust grains dominate the opacity up to a temperature of about 1,500 K; atoms dominate beyond about 4,000 K. The dust grains consist of a standard mix of silicates, organics, amorphous ice, FeS, and iron. The ice grains evaporate at about 160 K. Data from [452]

and iron grains; above that temperature and up to about 3,500 K molecular opacity dominates with important contributors being H_2O, CH_4, NH_3, and CO.

Equations (3.16), (3.17), and (3.19) must now be solved along with the energy equation:

$$\frac{dL_r}{dm} = \left(-\frac{dE}{dt} - P\frac{dV'}{dt}\right), \quad (3.20)$$

where $L_r = 4\pi r^2 F_r$, $V' = 1/\rho$, and E is the internal energy per unit mass. The first term on the right-hand side is the rate of increase of internal energy, and the second is the rate at which the gas is compressed by the force of gravity. The equation is straightforwardly derived from the first law of thermodynamics. It turns out, however, that in the central regions of the protostar during the relatively short-lived phase of optically thick collapse, the radiative diffusion time, $t_{diff} \approx \kappa_R \rho (\Delta r)^2/c$, rapidly becomes much longer than the local free-fall time. The core actually follows a nearly adiabatic curve with slope 0.4 in the (log ρ, log T) plane, as would be expected for a gas of H_2 with $E = 2.5 R_g T/\mu$.

To obtain a solution to the four differential equations referred to in the previous paragraph, the equation of state (3.18) is needed to obtain P and E as functions of ρ, T, and composition. But now c_s is no longer constant. Also, as the temperature increases, the molecular hydrogen is subject to dissociation, and later on, to ionization. The equation of state presented here is somewhat simplified in that it assumes that helium is in atomic form and is not ionized in the temperature range considered and that the metals, which do not contribute significantly, have a mean atomic weight per particle of 16. Also, radiation pressure is not included. It is also a good approximation to assume that, as the temperature increases, dissociation of H_2 is completed before the ionization of hydrogen starts. Thus the degree of dissociation can be defined as

$$y = \frac{\rho(H)}{\rho(H) + \rho(H_2)} \quad (3.21)$$

and the degree of ionization is

$$x = \frac{\rho(H^+)}{\rho(H) + \rho(H^+)}. \quad (3.22)$$

Let the mass fractions of hydrogen, helium, and heavier elements be given by X, Y, and Z, respectively. Then the pressure [119] is given by

$$P = \frac{R_g}{\mu} \rho T, \quad (3.23)$$

where the mean atomic weight per free particle μ is given by

$$\mu^{-1} = [2X(1 + y + 2xy) + Y]/4 + Z/16. \quad (3.24)$$

Thus, if hydrogen is in molecular form, $\mu = 2.367$ for a standard solar composition of $X = 0.71$, $Y = 0.27$, $Z = 0.02$. A general expression for E, under the same assumptions, is

$$E = [X(1 - y)E_{H2}/R_g + 1.5X(1 + x)y + 0.375Y + 0.09375Z]R_g T$$
$$+ X(1.3 \times 10^{13} x + 2.14 \times 10^{12}) y. \quad (3.25)$$

In this equation, the terms on the right-hand side that are multiplied by $R_g T$ represent the thermal energy of a free particle per unit mass, $1.5 R_g T/\mu$, and the remaining terms, in x and y, refer, respectively, to the ionization energy and dissociation energy of hydrogen. The internal energy per unit mass of molecular hydrogen is

$$E_{H2}(T) = \frac{3}{2}\frac{R_g T}{2} \quad (3.26)$$

only for $T < 80$ K, where only translational degrees of freedom are excited. Above that temperature, rotational degrees of freedom begin to become excited, and in the range $300 < T < 1{,}000$ K

$$E_{H2}(T) = \frac{5}{2}\frac{R_g T}{2}. \quad (3.27)$$

Between 80 and 300 K, the degree of excitation must be calculated in detail, as given, for example, by [56]. Above 1,000 K the vibrational levels also become

3.3 Adiabatic Collapse

excited, contributing additional degrees of freedom; however the dissociation of the molecules occurs before this excitation is complete.

The dissociation energy per molecule for H_2 is 4.48 eV, while the thermal energy of a molecule is $\frac{3}{2}kT \approx 0.25$ eV at 2,000 K, the temperature where dissociation starts. Thus once dissociation starts, almost all the increased heat from compression goes into dissociation rather than into thermal pressure, and the pressure gradient can no longer balance gravity, leading to renewed collapse. The dissociation can be calculated [515] from

$$\frac{[P(H)]^2}{P(H_2)} = K(T) = 3.49 \times 10^8 \exp(-52490/T). \tag{3.28}$$

Using the ideal gas equation of state on the left-hand side we obtain

$$\frac{y^2}{1-y} = \frac{2.11}{\rho X} \exp\left(-\frac{52490}{T}\right) \tag{3.29}$$

which can be solved for y as a function of (ρ, T).

Under the assumption that dissociation of hydrogen is complete before ionization starts, and that hydrogen is the only species that is undergoing ionization, one can similarly solve for the degree of ionization x. From the Saha ionization equation [251] and (3.22) one obtains the quadratic equation

$$\frac{x^2}{1-x} = \frac{4.0105 \times 10^{-9}}{X\rho} T^{3/2} \exp\left(-\frac{157600}{T}\right). \tag{3.30}$$

The evolution of the center of the protostar during this phase is sketched in Fig. 3.6. The material becomes optically thick and begins to heat near point A in the figure. After a short period of readjustment, during which the rotational degrees of freedom are excited in the H_2 molecules, the collapse approaches an adiabat. The slope in the diagram follows d log T/d log $\rho = (\gamma - 1) \approx 0.42$; for pure molecular hydrogen with 5 degrees of freedom it would be 0.4 (γ is the ratio of specific heats C_p/C_v). As heating progresses, the force produced by the pressure gradient exceeds that of gravity, and the collapse slows down. Near point B in the diagram a small amount of mass in the center ($10^{-2} M_\odot$) approaches hydrostatic equilibrium; this region is known as the *first core*. It starts with $T \approx 170$ K and $\rho \approx 2 \times 10^{-10}$ g cm^{-3}. Outside the core, infall velocities become supersonic with respect to the slowly compressing core material, and a shock wave forms at its outer edge, at a radius of about 4 AU.

Before the core has had a chance to accumulate much mass, it compresses to a density of about 10^{-8} g cm^{-3} and $T = 1,600$ K, where the H_2 begins to dissociate (point C). The dissociation energy per molecule is so large compared with its thermal energy that most of the gravitational energy released on compression goes into dissociation energy rather than into heating of the gas. The increase in temperature slows noticeably, and γ falls to about 1.1. The collapse, temporarily

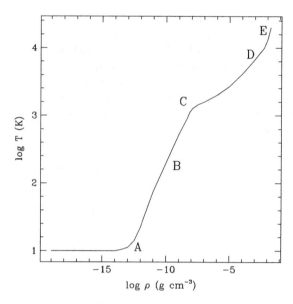

Fig. 3.6 Schematic diagram of the evolution of the center of a protostar during the isothermal phase (horizontal portion of *curve*) and the adiabatic phase. Salient points, as described in the text, are marked by *letters*. This *curve* represents the transition of gas from an interstellar cloud to a star

halted in the first core, begins again in the center of that core, and velocities eventually approach free fall. Collapse continues over several more orders of magnitude in density, until the H_2 is mostly dissociated in the very center (about 8,000 K; point D). The curve steepens, approaching d log T/d log ρ = 0.67, as would be expected for a gas of neutral H and He with $\gamma = 5/3$.

As a result of the rapidly increasing pressure at the center, the material in this second collapse regains equilibrium after the dissociation of molecular hydrogen has been completed and the central temperature has increased to 20,000 K (point E). The density by this time has increased to about 10^{-2} g cm^{-3}, so the physical conditions are approaching those in stars; however the initial mass of the *stellar* core is only $10^{-3} M_\odot$. Most of the mass of the protostar at this time is still in the outer isothermal region, at densities of 10^{-18}–10^{-19} g cm^{-3}, 17 orders of magnitude lower than that in the center.

The second hydrostatic core also develops a shock front at its edge, which has a radius of only a few $R_\odot$, and it gradually builds up by accretion through the shock front and becomes a star. The first shock front disappears relatively soon in the process of erosion of the first core because of the collapse occurring at its center. As the core temperature increases, the hydrogen begins to ionize, at temperatures considerably higher than those in the ionization zones of stellar atmospheres, because of the much higher density here. One might expect that a third collapse would occur because of the reduction in γ resulting from hydrogen ionization,

However the increase in degree of ionization with temperature is much slower than in low-density atmospheres, and furthermore the equation of state under these conditions is beginning to be non-ideal, that is, interactions between the particles must be considered. It turns out that γ never falls below 4/3, so the second core remains in hydrostatic equilibrium in spite of ionization. The complicated nature of the equation of state requires detailed calculations, whose results are usually presented in tabular form [446].

3.4 Accretion Phase

The third phase of evolution involves the collapse of the remaining protostellar material onto the second (stellar) core. The main source of energy of the protostar is the infall kinetic energy of material falling into the accretion shock at the outer edge of the stellar core. Just behind the shock, that energy is almost entirely converted into radiation, which flows back out through the shock and into the infalling material. An additional source of energy is the gravitational contraction of the core itself. Thus the total luminosity can be written

$$L = L_{\text{acc}} + L_{\text{int}}. \tag{3.31}$$

For low mass stars during the main accretion phase, L is almost completely supplied by L_{acc}. The infall kinetic energy per unit mass into the core is $\frac{1}{2}u^2 = GM_{\text{core}}/R_{\text{core}}$. Multiply by $\dot{M}$ to get the energy inflow per second. Assuming this is all converted to radiation in the shock,

$$L_{\text{acc}} \approx \frac{GM_{\text{core}}\dot{M}}{R_{\text{core}}}. \tag{3.32}$$

In the cool, dusty, optically-thick collapsing layers outside the core, this total radiation is absorbed and then re-radiated in the infrared region of the spectrum. The protostar becomes an observable infrared source. At the time of initial formation of the core, much of the material is still in the isothermal phase with density not much higher than it was originally. Thus, the main accretion phase lasts at least another two initial free-fall times, for a total low-mass protostellar lifetime of a few times 10^5 yr.

There are several important time scales associated with protostellar collapse.

- The free fall time

$$t_{\text{ff}} = \left(\frac{3\pi}{32G\rho}\right)^{1/2} \tag{3.33}$$

which ranges from less than 0.01 yr just outside the core to 2×10^5 yr near the outer edge of the protostar

- The Kelvin-Helmholtz contraction time of the core [251]

$$t_{KH} \approx \frac{GM^2}{RL_{int}} \qquad (3.34)$$

which gives the time for an object of mass M in hydrostatic equilibrium to contract from infinity to radius R, at an average luminosity L_{int}. It is also approximately the time for an object to achieve "thermal balance" between energy generated by contraction and energy radiated according to the Virial Theorem. Here L_{int} is the energy liberated from the interior of the equilibrium core, not counting the accretion luminosity generated at the very surface. For a core of $0.5\,M_\odot$ at $3\,R_\odot$ and $L_{int} = 1\,L_\odot$, typical of the main accretion phase, this time is about 3 Myr.
- The accretion time scale $t_{acc} = M_{core}/\dot{M}$ which, for a core of $0.5\,M_\odot$ is about 2×10^5 yr. If $t_{acc} \ll t_{KH}$, which is the normal situation here, the core is compressed adiabatically, little radiation escapes from the interior, and $L \approx L_{acc}$. If the inequality is reversed, the star evolves through the Hertzsprung–Russell diagram while accreting, and $L \approx L_{int}$.
- The radiative diffusion time scale [457]

$$t_{diff} \approx \frac{3\kappa_R \rho (\Delta r)^2}{c} \qquad (3.35)$$

which represents the time for photons to diffuse by a random walk over a distance Δr in a medium with Rosseland mean opacity κ_R (c is the velocity of light). During the adiabatic collapse phase, $t_{diff} > t_{ff}$, so the radiation released from the warm material is not able to escape but is carried along with the infalling material. In the main accretion phase, the situation is different. The radiated energy is released at the core boundary and propagates outward. Here the relevant time scales are t_{acc}, roughly the production time scale of radiation, and t_{diff} in the dusty infalling envelope. Since $t_{diff} \ll t_{acc}$, practically all the radiated energy produced at the core passes through the envelope. Thus the observed luminosity is very close to L, although its spectral energy distribution is quite different from that at the core.

The r^{-2} density profile set up in the isothermal collapse region suggests that the accretion rate should be roughly constant in time: $\dot{M} = \text{const} \times c_s^3/G$. In fact that statement appears to be in conflict with observations. A Class 0 protostar is defined observationally (Chap. 1) as one with less than half of its total mass in the central stellar core; a Class I protostar has more than half of its mass in the core. Yet the observed ratio by number of Class 0 vs. Class 1 objects is about 1 to 5–10 (subject to observational and theoretical uncertainties in both the core mass and the envelope mass), which can be explained only if the accretion rate drops off significantly with time. What is responsible for the drop is so far not explained. It is possible that the r^{-2} profile is not realized in the central regions because the

3.4 Accretion Phase

sound travel time is somewhat longer than the free-fall time. That situation could occur for reasonable initial conditions. Also, at later stages the accretion rate is determined by disk physics, so the argument regarding the r^{-2} profile may not apply. Furthermore, this spherically symmetric solution does not take into account winds and outflows that could limit the amount of mass from the molecular cloud core that can be accreted onto the star.

It is reasonable to assume, however, that over short intervals of time a steady state collapse is set up, with $\dot{M}$ constant in space [482, 555]. In steady state, physical properties at a given radius vary only very slowly with time, so in the Eulerian hydrodynamic equations the partial time derivative $\partial/\partial t = 0$ at all points in space. Once the core has formed and its gravity has become significant, the density distribution in the infalling envelope, especially in the region near the core, is modified.

$$\rho(r) = \frac{\dot{M}}{4\pi r^2 v} = \frac{\dot{M} r^{-3/2}}{4\pi (2GM_{\text{core}})^{1/2}}, \qquad (3.36)$$

where v is the infall velocity. Note that if $\dot{M}$ is constant with r, the velocities become highly supersonic and follow a profile $v \propto r^{-1/2}$ as expected for free fall, while the density follows $\rho \propto r^{-3/2}$.

A somewhat simplified solution for a model of a protostar involves dividing it into three parts: the hydrostatic structure of the core, the shock wave, and the steady-state infalling envelope. The standard stellar structure equations apply to the core, the time-independent hydrodynamic equations apply to the infalling envelope, and the two regions are joined by the shock jump conditions [482]. The shock relations must include the terms representing radiative energy flow through the shock. Let physical quantities ahead of the shock in the infalling region be labelled with subscript "1" and those behind the shock in the core be labelled "2". The temperature is represented by T, the radiative flux by F, and the velocity by v. Between regions 1 and 2 exists a very narrow "relaxation layer". In that layer the infall is drastically decelerated and most of the kinetic energy is converted into heat; the resulting temperature is approximately $kT_{\text{relax}} \approx \frac{1}{2}\mu m_u v_1^2 \approx GM_{\text{core}}\mu m_u / R_{\text{core}}$ where μm_u is the mean mass of a particle. For a core of $0.5\,M_\odot$ and radius $3\,R_\odot$, $T_{\text{relax}} \approx 5 \times 10^6$ K. But the region ahead of the shock is optically thin, so most of this energy is radiated away ahead of the shock. In fact, so much energy is radiated that the shock is practically isothermal: $T_2 \approx T_1$.

The shock jump conditions for a radiating shock are, in the frame of reference of the shock [482]

$$\rho_2 v_2 = \rho_1 v_1 \qquad (3.37)$$

$$P_2 + \rho_2 v_2^2 = P_1 + \rho_1 v_1^2 \qquad (3.38)$$

$$L_2 - \dot{M}(w_2 + v_2^2/2) = L_1 - \dot{M}(w_1 + v_1^2/2), \qquad (3.39)$$

where $L = 4\pi r^2 F$ and $w = E + P/\rho$, the specific enthalpy. The first equation is simply a consequence of the fact that the mass per unit area per unit time (ρv)

entering the shock must be the same as that leaving it. The second equation is obtained from the momentum equation

$$\rho\left(\frac{\partial \mathbf{v}}{\partial t} + \mathbf{v} \cdot \nabla \mathbf{v}\right) = -\nabla P. \tag{3.40}$$

The velocity of the shock itself, in the frame of reference chosen, is zero, and the flow through the shock can be considered to be in steady state. Thus the time derivative in the momentum equation vanishes, and (3.38) is derived from (3.40) simply by integrating across the shock. (In fact in the protostar accretion problem the shock is practically stationary in space.) The third equation states that the increase in fluid energy per unit time, as a result of passing through the shock, equals the rate at which pressure forces do work on the gas, minus the net amount of energy per unit time lost by radiation (L_1–L_2). More details on the properties of this equation may be found in [298, 476]. Note that in the form given it applies to a spherical accretion flow.

The time scale for accretion ($M_{core}/\dot{M}$) becomes much longer than the time for radiation to diffuse from the shock front to the outer edge of the protostar. Thus the protostar begins to radiate from the surface, known as the dust photosphere, where the dust optical depth is unity, which corresponds typically to a temperature of 100–300 K. At the shock front, practically all the inflowing kinetic energy is converted to heat, and practically all this energy is radiated out into the optically thin region just ahead of the shock. Farther out, the radiation is absorbed, re-radiated, and thermalized in the optically thick dusty infalling region, so that the observable radiation lies in the mid-infrared. Very little net energy is absorbed in these layers, so the observable luminosity is given to a good approximation by the total (mostly accretion) luminosity of the core. The energy radiated out from the shock, L_1, can be approximated as arising from a photospheric layer just behind the shock front, with

$$L_1 \approx 4\pi R_{core}^2 \sigma_B T_{eff}^4 \tag{3.41}$$

so the approximate temperature of the radiation from the core is

$$T_{eff} \approx \left(\frac{GM_{core}\dot{M}}{4\pi\sigma_B R_{core}^3}\right)^{1/4}, \tag{3.42}$$

which, for a core at 3 $R_\odot$, $\dot{M} = 3 \times 10^{-6} M_\odot$ yr^{-1}, and $M_{core} = 0.6 M_\odot$ gives $T_{eff} \approx 7,000$ K, which produces radiation in the optical.

The typical structure of a protostar during this phase, from the core out to the dust photosphere, is shown schematically in Fig. 3.7. Material infalls through the outer optically thin envelope region, then through the optically thick dust envelope. Once the temperature has increased above 1,500 K so that the dust has evaporated, molecular opacity dominates and there is a drop of two orders of magnitude in opacity by the time the temperature reaches 2,000 K. The opacity then rapidly increases again

3.5 Comparison with Observations

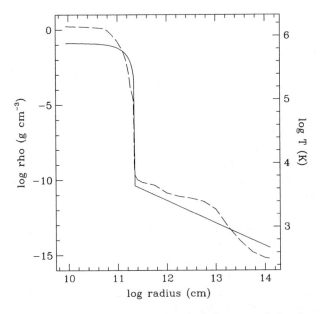

Fig. 3.7 A schematic diagram of the structure of a spherical protostar during the main accretion phase, for a core mass $0.48\,M_\odot$ and a steady-state envelope, infalling at the rate $10^{-5}\,M_\odot\,\text{yr}^{-1}$. Temperature (*dashed curve; right scale*) and density (*solid curve; left scale*) are plotted as a function of radius. The accretion shock at the outer edge of the stellar core is located at 2.2×10^{11} cm. Deuterium burning is occurring in the core. The temperature just behind the shock is 8,000 K. The dust destruction front is located at 9.5×10^{12} cm. Interior to that point the hydrogen has been dissociated. The dust photosphere is located at 1.25×10^{14} cm. Data from [483, 484]

to higher T, but still this region is only marginally optically thick. The material is decelerated in the shock, its kinetic energy is converted to radiation, and the radiation flows back outward through the same regions. Calculations show that the radius of the shock front remains nearly constant at 2–3 $R_\odot$, and the dust destruction front also remains at a nearly constant distance from the core, about 1 AU but dependent on the accretion rate. The dust photosphere is the "visible" surface for a distant observer; for a given frequency ν it occurs at an inward integrated optical depth $\tau_\nu = \int \kappa_\nu \rho dz = 2/3$, where z is the distance inward from the edge of the protostar. For typical parameters the "mean" photosphere, averaged over frequency, is at about 10 AU. However its radial position is strongly dependent on frequency.

3.5 Comparison with Observations

The main points of comparison between observed protostars and models are (1) the position in the infrared Hertzsprung–Russell diagram, (2) the spectral energy distribution, and (3) profiles of spectral lines indicative of infall motions. In most cases departures from assumed spherical symmetry are required to explain the observations: these could involve a rotating infall region, and/or the presence of a disk and/or an outflow cavity in the vicinity of the poles. Thus the comparisons discussed here do allow for the possibility of non-spherical effects. More detailed discussion of theoretical models of rotating protostars and disks follows in the next chapter.

Normal stars in hydrostatic equilibrium are observable in the Hertzsprung–Russell diagram with surface temperatures down to about 3,000 K, not cooler [209, 251]. The observable surface of a protostar, however, corresponds to much cooler temperatures. The reason for these low temperatures is that the outer layers of protostars are collapsing, and most importantly, are extended enough and cool enough so that dust particles exist which result in a high optical depth for the infalling region. The observable surface of a 1 $M_\odot$ protostar starts its evolution at cool temperatures of order 10 K, progresses leftward (to higher temperatures) in the H–R diagram, and ends, once the infalling envelope has been accreted, in the region of pre-main-sequence stars which are in hydrostatic equilibrium with surface temperature in the range 3,000–4,000 K.

A number of evolutionary tracks have been published based on various levels of approximation. Calculations have been carried out [482] based on the steady-state approximation discussed above, in which the protostar is divided into core, shock, and infalling envelope. Other work [299, 563] represents full spherically symmetric solutions of the equations of hydrodynamics and radiative transfer. Another approach [378] is to compare observations with approximate collapse models of a star-disk-envelope system in a (T_{bol}, L_{bol}) diagram, where T_{bol} is the temperature of the black body with the same mean observed frequency as the protostar (Fig. 3.8). This diagram is designed as a close analog to the H–R diagram. The quantity T_{bol} is used [382] because the effective temperature for a protostar is not a well-defined quantity. The temperature one observes at the "surface" depends significantly on frequency. But one can average over frequency by defining a flux-weighted mean:

$$T_{bol} = 1.25 \times 10^{-11} <\nu> \quad \text{where} \quad <\nu> = \frac{\int F_\nu \nu d\nu}{\int F_\nu d\nu} \quad \text{and} \quad (3.43)$$

$$L_{bol} = 4\pi D^2 \int F_\nu d\nu \quad (3.44)$$

where F_ν is the flux density, i.e. the energy per unit area per unit time per unit frequency interval radiated by the protostar and received at the earth, and D is the distance. The constant is defined by the requirement that if the spectrum were a black body with the same mean frequency, the black body temperature would equal T_{bol}.

A wide variety of models is possible, because approximations to the full three-dimensional evolution of protostellar collapse have to be made. Examples of two different types of models are shown in Fig. 3.8. The models from [378] are based

3.5 Comparison with Observations

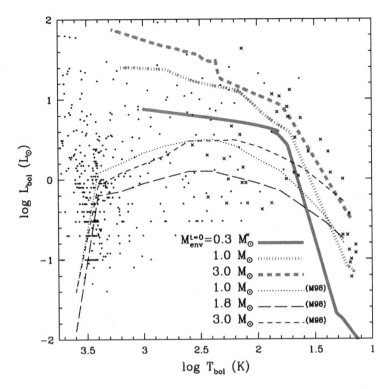

Fig. 3.8 Theoretical evolutionary tracks for protostars of different masses through an analog of the Hertzsprung–Russell diagram. The total luminosity, integrated over frequency, in solar units, is plotted as a function of bolometric temperature. The *dots* and *crosses* correspond to a selection of observed objects in nearby star-forming regions. *Heavy lines:* tracks from [569] with constant masses of 3.0, 1.0, and 0.3 $M_\odot$ (*top* to *bottom*). *Light lines:* tracks from [378] with initial masses as indicated. In these cases the masses are assumed to decrease with time. For initial masses 1.0 and 3.0 $M_\odot$, the final mass is 0.5 $M_\odot$ and the initial envelope temperature is 10 K. For the initial mass 1.8 $M_\odot$, the final mass is 0.3 $M_\odot$ and the initial envelope temperature is 8 K. Reproduced, by permission of the AAS, from [569]. © 2005 The American Astronomical Society

upon several simplifiying assumptions. They take into account the presence of a stellar core, a circumstellar disk, and an infalling envelope. The accretion rate is assumed to start off at the standard value for the singular isothermal sphere, then to decline exponentially with a time constant t_*. This assumption is reasonable, since Class II objects (T Tauri stars) are known to have accretion rates far below c_s^3/G, where c_s refers to the sound speed in the molecular cloud core. The envelope mass is also assumed to decrease exponentially in time from an initial value $M_{e,0}$, with a time constant t_e. This assumption takes into account the dissipation of the envelope by the jets and winds generated near the central star. The curves with initial masses of 1.0 and 3.0 $M_\odot$ have a final mass of only 0.5 $M_\odot$. The curve with initial mass of 1.8 $M_\odot$ has a final mass of 0.3 $M_\odot$. The luminosity along the tracks first increases, then decreases mildly, and the tracks pass roughly through the median of the observed points. The tracks suggest that at least some of the observations can be explained by models with final masses in the range 0.3–0.7 solar masses, with a time scale of 0.5 Myr for the protostar phase (Class 0 and Class I). However, there are numerous observed objects with luminosities higher or lower than those of the calculated tracks. There are a number of parameters in this model: the initial mass, the initial temperature, the time scale for the reduction in $\dot{M}$, and the time scale for envelope dissipation. The nearly vertically downward portions of the tracks at the end of the evolution are the Hayashi tracks. They are defined, for a given mass star with a given composition, as the locus of fully convective hydrostatic models, down which low-mass stars evolve during the pre-main-sequence contraction (see Chap. 8 for more details). The tops of these nearly vertical tracks correspond to the assumed end of accretion, and the stars then contract at constant mass to the zero-age main sequence (ZAMS) where nuclear burning takes over as the energy source. As one can see from the figure, the objects reach the top of their Hayashi tracks at about 1 $L_\odot$.

The second set of tracks shown in Fig. 3.8 assume a constant accretion rate, as appropriate for an initial condition of a singular isothermal sphere [569]. The mass of the protostar remains constant in time. Again, the model consists of a stellar core, a surrounding disk, and an infalling envelope. The results show that the luminosity increases continuously with time for a given mass, and the tracks skirt the upper envelope of the observed points, giving a typical luminosity much higher than the observed average.

That seemingly arbitrary assumptions have to be introduced to reduce the luminosities of the theoretical protostars to values near the median of the observations of protostars (Fig. 3.8) is an illustration of the so-called *luminosity problem* [249]. Suppose a protostar accretes at a constant rate of $\dot{M} = c_s^3/G = 1.6 \times 10^{-6} M_\odot \text{ yr}^{-1}$, where c_s corresponds to a temperature of 10 K in the molecular cloud core. Then the starlike core reaches 0.5 $M_\odot$ in 3×10^5 yr, in reasonable agreement with the lifetime of the protostellar phase. However there are two points of disagreement with the observations. First, the accretion luminosity (3.32) should increase with time as the core mass increases, because R_{core} is practically constant in time at about 3 $R_\odot$. But this behavior is not in agreement with the trend of the observations. Second, once the protostar reaches 0.5 $M_\odot$, its accretion luminosity is calculated to be over 8 $L_\odot$, well above typical observed values. In fact, the observations of Class I objects show an even larger spread in luminosities than can be accounted for by models in the expected mass range (Fig. 3.8), especially on the low-luminosity side [163]. Various solutions have been proposed to the luminosity problem; a promising one [146] is that the accretion rate from the infalling envelope onto the disk is relatively constant, but the accretion rate from disk to star is neither constant nor smoothly decreasing with time, but involves short episodes of very rapid accretion separated by relatively long periods of slow or negligible accretion. This assumption is compatible with the expected behavior of a gravitationally unstable disk. Calculations with such an assumption give a better agreement between the full range of observed luminosities and the evolutionary tracks, but as yet there is no fully satisfactory explanation to the luminosity problem.

3.5 Comparison with Observations

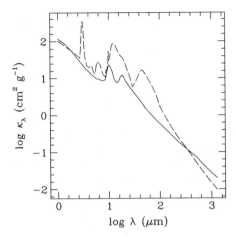

Fig. 3.9 Dust opacity as a function of wavelength. *Solid line:* dust grains composed of silicates and graphite without ice mantles. *Dashed line:* grains composed of silicates and graphite with thick ice mantles. Data from Ossenkopf and Henning [399]. The opacity is expressed per gram of gas–dust mixture, assuming a gas-to-dust ratio of 100, by number. These grain opacities are slightly different from those derived from standard interstellar grains [140]. In the case plotted, the grains have been allowed to coagulate for 10^5 years at a molecular-cloud-core density of 10^6 cm^{-3}

A full hydrodynamic solution for the protostellar collapse, without arbitrary assumptions but in spherical symmetry, not accounting for the presence of a disk, has been presented for the case of $1\,M_\odot$ [563]. The initial condition is based on a fully three-dimensional simulation of the turbulent interstellar medium, in which protostellar cores are produced by compression induced by transient shock waves (Chap. 2). A cluster containing a substantial number of such cores was produced, and the core containing closest to a solar mass was then followed in detail with a one-dimensional spherically symmetric calculation. The mass accretion rate onto this core was obtained from the 3-D solution. The Hayashi track is reached in a time of order 10^5 yr and the luminosity across most of the infrared H–R diagram and at the top of the Hayashi track is over $10\,L_\odot$, noticeably higher than that which the corresponding mass would have in Fig. 3.8 and again illustrating the luminosity problem.

A second method of comparison with observations can be made through the spectral energy distribution. The key to the continuous spectrum is the behavior of the dust opacity as a function of wavelength, which is shown in Fig. 3.9 for two different assumptions regarding the grain properties [399]. The grain opacity can vary significantly depending on grain properties, such as fluffiness, composition, and size distribution. Note that the opacity is orders of magnitude larger at 1 μm than at 100 μm.

Following Hartmann [202] we can estimate the radiation properties of a protostar by considering a simple spherical model. First, the relevant part of the mass may be considered to be free-falling at a rate $\dot{M}$ onto a core of mass M_{core}. The density distribution in the envelope is given by (3.36). We can then integrate inward to obtain the optical depth as a function of radius and wavelength, assuming the dust opacity is independent of radius down to the dust evaporation temperature:

$$\tau_\lambda = \frac{\kappa_\lambda \dot{M} r^{-1/2}}{2\pi (2GM_{\text{core}})^{1/2}}, \qquad (3.45)$$

where κ_λ is the opacity at λ in cm^2 g^{-1}. Assuming that the stellar core at the center of the protostar has characteristics something like those of T Tauri stars, a characteristic wavelength of emission from the core is 1 μm. At that wavelength, using the opacity from Fig. 3.9 one can solve for the optical depth, integrated inward to an inner radius of r_{in}:

$$\tau(1\ \mu m) \approx \frac{4.36 \times 10^4 \dot{M}}{(M_{\text{core}} r_{\text{in}})^{1/2}} \approx 76 \qquad (3.46)$$

for a core of 0.6 $M_\odot$, dust evaporation point of 10^{13} cm, and $\dot{M} = 3 \times 10^{-6} M_\odot$ yr^{-1}. Thus the 1 μm radiation is strongly absorbed in the envelope and the core is not observable at that wavelength. Now we ask where the radius of the photosphere is (the characteristic radius where most of the radiation at wavelength λ is radiated to space) and find it is very dependent on λ. Defining the photospheric radius r_λ where $\tau_\lambda = 2/3$, we find

$$r_\lambda = \frac{9\kappa_\lambda^2 \dot{M}^2}{32\pi^2 GM_{\text{core}}}. \qquad (3.47)$$

The plot in Fig. 3.10 shows r_λ as a function of λ for $\dot{M} = 3 \times 10^{-6} M_\odot$ yr^{-1} and $M_{\text{core}} = 0.6 M_\odot$. Note that it is at about 20 AU in the range 10–30 μm but on the order of stellar radii at 1,000 μm. Typical protostars are optically thin at millimeter wavelengths.

We now can estimate the temperature at the radiating photosphere by defining a mean radius r_m and a mean temperature T_m from the black-body relation:

$$L = 4\pi r_m^2 \sigma_B T_m^4, \qquad (3.48)$$

where L is the actual luminosity of the source and r_m is the radius with optical depth $\tau_m = 2/3$. To obtain r_m one needs an average opacity to put into (3.47). The temperature range for protostellar envelopes is 30–1,500 K, with a corresponding wavelength range of 2–100 μm. If one simply averages log κ over that range for the solid curve in Fig. 3.9 one obtains $\kappa_m \approx 5$ cm^2 g^{-1}, which is in good agreement with the Rosseland (8.40) and Planck (5.13) mean opacities in this temperature range [452]. For the standard case ($M_{\text{core}} = 0.6 M_\odot$ and $\dot{M} = 3 \times 10^{-6} M_\odot$ yr^{-1}) one obtains $R_m \approx 20$ AU and $T_m = 177$ K, indicating that indeed protostars inhabit the infrared portion of the H–R diagram.

Figure 3.11 shows the spectral energy distribution of a particular Class I protostar. One of the main features of this distribution is that it is broader than that of a black body at a single temperature, a result of the fact that the "photosphere"

3.5 Comparison with Observations   117

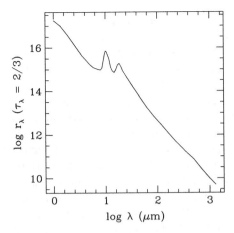

Fig. 3.10 The radius, in cm, of the dust photosphere of a protostar as a function of wavelength. The opacities used are from the *solid curve* in Fig. 3.9

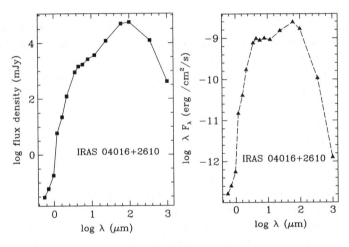

Fig. 3.11 Observational data for the spectral energy distribution of a Class I protostar. Data from the compilations of [248] and [438]. (*Left*): the flux density F_λ in millijanskys is plotted against wavelength. (1 Jansky = 10^{-26} watts m^{-2} Hz^{-1}). (*Right*): the same data are plotted with λF_λ as a function of λ, where F_λ is expressed per unit wavelength interval

exhibits a range of temperatures, depending on wavelength. A number of simplified models have been produced that reproduce such spectra, consisting, in general, of a central stellar source with a given core mass, a surrounding disk, and an infalling optically thick dusty envelope. A number of parameters are involved, for example, the core mass, $\dot{M}$, the rotational angular velocity of the initial molecular cloud core that collapsed to form the protostar, and the viewing angle with respect to the rotation axis. There is not a unique fit, and the results depend on the assumptions made in the model, but this particular source can be modelled with approximate parameters $\dot{M} = 10^{-6} \, M_\odot \, \mathrm{yr}^{-1}$, mass $0.5 \, M_\odot$, $\Omega_{\mathrm{cloud}} = 10^{-14} \, \mathrm{s}^{-1}$ and inclination of $60°$. The estimated total luminosity L_{bol} of the object is $3.7 \, L_\odot$.

The general properties of Class 0 protostars [19] are as follows: the ratio of the submillimeter luminosity L_{submm}, $\lambda > 350 \, \mu\mathrm{m}$, to the total bolometric luminosity is greater than 5×10^{-3}; the spectral energy distribution (Fig. 1.18) is relatively narrow, with characteristic black-body temperatures of 15–30 K; there is generally undetectable near-infrared emission shortward of $5 \, \mu\mathrm{m}$; the mass of the infalling envelope is greater than that of the stellar core; and $T_{\mathrm{bol}} < 80$ K. An object assigned to Class 0 may not necessarily conform to all these criteria. The quantity $L_{\mathrm{submm}}/L_{\mathrm{bol}}$ is considered a more reliable indicator for Class 0 than is T_{bol}. However the general interpretation of the class is that it represents the very earliest phase of protostellar accretion. Estimates of the time scale for residence in the class range from a few times 10^4 yr [179] to 10^5 yr [164]. That time scale is consistent with the relatively small number of Class 0 objects, only $\sim$10–25% relative to Class I objects. All Class 0 objects have strong bipolar outflows which create low-density cavities in the polar directions in the infalling envelope. Thus it is not possible to model these envelopes as spherically symmetric. Typical disk-envelope models [504] include a bipolar cavity whose geometry is constrained to fit the observations in particular cases. Some near and mid-infrared flux is observed, which represents light from the central object or disk beamed toward the poles, and scattered in the cavity toward the observer. Because of the difficulty of modelling these sources, protostellar parameters derived from the models, such as accretion rates of disk onto star, $\dot{M}$ of the infalling envelope, M_{core}, and M_{final} (the expected mass after completion of accretion) often vary by up to an order of magnitude, resulting from different model assumptions.

The quantities that can be determined directly from observations are

- T_{bol}, given a well-sampled spectral energy distribution
- L_{bol}, by integration over the SED and by adoption of a distance
- $L_{\mathrm{submm}}/L_{\mathrm{bol}}$, from the radiation observed at $\lambda > 350 \, \mu\mathrm{m}$
- M_{env}, the infalling envelope mass, directly from the flux density at 1.3 mm and the assumed distance. At this wavelength the envelope is optically thin. The basic method for determining the mass is given in Sect. 4.2.

Errors in luminosity could be up to 50%, in T_{bol} by ± 10 K, and in mass by a factor 3.

The quantity L_{submm} is a measure of the envelope mass, which is optically thin at these wavelengths. L_{bol} is a measure of the stellar core mass through (3.32). The quantity R_{core} is usually set to about $3 \, R_\odot$, and $\dot{M}$ is estimated from the sound speed in the region around the protostar, typically a few times $10^{-6} \, M_\odot \, \mathrm{yr}^{-1}$. Empirically it is found [18] that $M_{\mathrm{env}}/M_{\mathrm{core}}$ exceeds unity when $L_{\mathrm{submm}}/L_{\mathrm{bol}} > 5 \times 10^{-3}$, approximately.

A sample of data for Class 0 objects [178] is presented in Fig. 3.12. Envelope masses have been determined for these objects; they fall in the range 0.16–$6 \, M_\odot$. Evolutionary tracks have been calculated [19, 179, 378, 469, 569]. These approximate models vary considerably in their assumptions, in particular the assumed accretion

3.5 Comparison with Observations

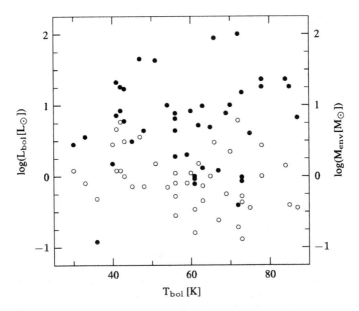

Fig. 3.12 Luminosities log L_{bol} (*filled circles*) and envelope masses log M_{env} (*open circles*) as a function of T_{bol} of a selection of protostars of Class 0 and Class 0/1 within a distance of 500 parsec. The quantities are plotted in solar units. Class 0 objects chosen satisfy all three of the following criteria: there is no detection of radiation with $\lambda < 5\,\mu m$, the ratio $L_{submm}/L_{bol} > 0.005$, and $T_{bol} < 80\,K$. Objects which do not satisfy only one out of the three criteria are classified as Class 0/1. Reproduced by permission of the AAS from [178]. © The American Astronomical Society

rate as a function of time. They generally go through the observed points, but the derived properties vary considerably from model to model. In general, the derived ages are in the range 10^4–10^5 yr and final masses are most probably in the range 0.5 to a few $M_\odot$. For example the specific object RNO 15 has measured $T_{bol} < 73\,K$, $L_{bol} = 15\,L_\odot$, $M_{env} = 0.43\,M_\odot$, and $L_{submm}/L_{bol} = 0.013$. Note that the mean luminosity of the Class 0 objects is higher than that of the Class I's, indicating a higher accretion rate.

The third method of comparison with observations is the analysis of Doppler shifts in spectral lines to detect evidence for infall. The collapse velocities are high only very close to the core. On larger scales, which are much more likely to be observable, the velocities are generally low, about 1 km s^{-1} for the free-fall velocity at 1,000 AU from a core of 0.5 $M_\odot$. An additional complication is the fact that a protostar exhibits not only infall motions but also rotation, outflow, and turbulence. Therefore it has proved to be very difficult to actually detect evidence of infall in protostars of about a solar mass, but in a few cases that evidence is quite convincing. It is important because it allows a direct estimate of the infall accretion rate.

The basic idea is to detect "infall asymmetry" in a spectral line, usually in the radio or millimeter region where the dust optical depth is small and one can see far into the protostar. The required conditions are: the line has to be somewhat optically

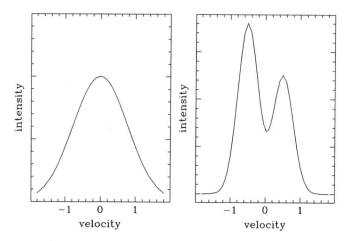

Fig. 3.13 Qualitative sketches of emission line profiles from a collapsing cloud in the optically thin case (*left*) and the moderately optically thick case (*right*). Intensities and velocities are on arbitrary linear scales. Velocity is measured relative to that of the center of mass of the cloud. A negative velocity corresponds to a blue shift in the spectrum. The separation of the two peaks gives a typical infall velocity. The far wings of the line, near velocities of -1 and $+1$, give the highest observable infall velocity

thick, and the degree of excitation of the line must increase with increasing depth into the cloud. If temperature and density increase toward the center of the cloud, this condition is likely to be met.

Consider, first, a simplified example of a spherical collapsing cloud, with highest density near the center ($\rho \propto r^{-1.5}$), and infall velocity increasing inwards. For a line that is optically thin, the line profile will be composed of contributions from the far hemisphere (infalling toward the observer, so blueshifted) and the near hemisphere (infalling away from the observer, so redshifted). The line will be broadened because of the different Doppler shifts arising from emission in the various layers, and we neglect the additional broadening resulting from thermal or turbulent motions. Most of the mass is in the outer regions, with small or zero velocity in the line of sight to the observer. Most of the emission will be near the rest wavelength of the line, and the line profile will be symmetric about that wavelength. The inner layers with high infall velocity have relatively little mass and will contribute to the line profile in the wings, but at reduced intensity. It is shown by Hartmann ([202], pp. 72–73), using more detailed geometrical and radiative transfer arguments, that an optically thin line is symmetric about zero velocity and has the general shape indicated in the left-hand panel of Fig. 3.13. However an expanding envelope with a similar density and velocity distribution would produce the same line profile, and therefore an optically thin line cannot be used to distinguish between collapse and expansion.

For the case of an optically thick line, the situation is shown in the right-hand panel of Fig. 3.13. Strong absorption, from the overlying static material (Fig. 3.14), takes place around zero velocity, and the blue-shifted intensity peak is stronger than the red-shifted one. In the case of expansion, the situation would be reversed, with

3.5 Comparison with Observations

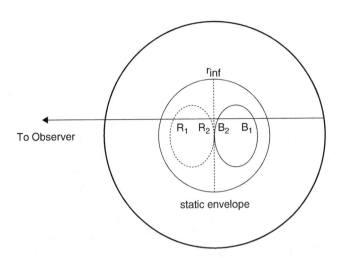

Fig. 3.14 Aid in the explanation of infall asymmetry. The spherical protostar is divided into a static region (outside r_{inf}) and an infalling region. The *dashed line* corresponds to the locus of points with a constant projected velocity in the observer's line-of-sight (v_{los}) in the red-shifted portion of the cloud. The *solid line* corresponds to the same absolute value of v_{los} in the blue-shifted portion. The points R_1, R_2, B_1, B_2 are the points of intersection of an observer's line of sight, offset from the center of the cloud, with these curves. Velocity is measured relative to that of the center of mass of the cloud. Reproduced, by permission of the AAS, from [572]. © The American Astronomical Society

the red-shifted peak stronger. This is a complicated problem in line transfer in a medium with velocity gradients, and the details of the line profile will depend on the density, velocity, and excitation distributions. However this profile is a definite indicator of collapse. To see physically what is happening, one can make the approximation that the velocity varies rapidly with distance (the *Sobolev* or Large Velocity Gradient approximation). Thus in a line with unshifted central wavelength λ_0, the velocity in the line of sight that contributes emission at a slightly displaced wavelength λ is

$$v_{\text{los}} = c(\lambda - \lambda_0)/\lambda_0 . \tag{3.49}$$

This emission arises from the very small amount of material that has velocity very close to v_{los}. The inner parts of these curves (R2, B2 in Fig. 3.14) have higher excitation than the outer parts (R1, B1) so the line emission will be more intense. However the light from R2 will be self- absorbed at R1 along the line of sight to the observer. So the observed radiation will come primarily from R1. On the other hand, the observer will primarily see light from B2 on the far side, which is more intense than that from R1 because of higher excitation. And there will be no self absorption. So the line profile will be asymmetric, with stronger radiation on the blue side than on the red side. Again, the dip between the two peaks is caused by self-absorption by non-moving or very slowly-moving material near the outer edge of the cloud. Typical molecular lines used for these studies include those of CS, H_2CO, and $C^{18}O$.

It is now claimed that infall has been detected in a number of objects [381] of which the best candidates are B335 [572], IRAS 16293-2422 [535,571], and L1527 [379, 573]. The first two are both very cold objects, as indicated by their spectral energy distributions . The near fit to black-body curves at low temperature (~ 30 K) is a characteristic of a Class 0 source. IRAS 16293 is a binary, and L1527 is Class 0/1. B335 has a low rotational velocity ($\Omega \sim 10^{-14}$), and low turbulence, thus it is considered an ideal object for detection of infall. The other two sources have significant rotation. However B335 also has a bipolar outflow almost in the plane of the sky, which complicates the situation. B335 has been modelled [572] by an "inside-out" collapse model, which starts from the singular isothermal sphere, and at a given time has an infalling central region and a static outer region (Fig. 3.14). The line profiles of several different transitions in CS were compared with theoretical line profiles. The strengths and profiles of optically thin and optically thick lines are qualitatively matched by the profiles in the left and right panels, respectively, in Fig. 3.13. A fit to the data shows that the outer edge of the infall region is at a radius of 0.04 pc, the core mass is about $0.4 M_\odot$, and $\dot{M} \approx 3 \times 10^{-6} M_\odot$ yr^{-1}. However, not all observed features [554] agree with this model, and other effects, including the outflow, must be taken into consideration. Thus it is quite clear that infall has been detected, but it is not clear that the "inside-out" model is the correct one. A similar comparison was made for IRAS 16293-2422 [571], except in this case rotation had to be included in the radiative transfer model. Again, the fits provide good evidence for infall. with a deduced $\dot{M} = 3 \times 10^{-5} M_\odot$ yr^{-1} onto a central mass (binary + circumbinary disk) of $2.9 M_\odot$. The data are also consistent with rotation, with a "centrifugal radius" (where the rotational velocity becomes comparable to the sound speed) of 610 AU, close to the orbital separation of the binary.

Observations of several other objects [231] provide estimates of infall $\dot{M} \approx 4\pi\rho r^2 v_{\text{infall}}$ for both Class 0 and Class I objects. In the Taurus star-forming region, $\dot{M}$ of a few times $10^{-6} M_\odot$ yr^{-1} was found in both classes. For the Taurus objects, these dynamical values of $\dot{M}$ were consistent with another rough estimate obtained from L_{bol} and an assumed accretion time. However such estimates are oversimplified, as in most cases the effects of rotation and a disk have to be taken into account. In cluster sources, of which IRAS 16293 is an example, $\dot{M}$ values an order of magnitude or more higher are found [537]. It turns out to be difficult to fit most observations with a pure spherical "inside-out" model. Collapse models starting from a Bonnor-Ebert sphere are better, but even these have to be modified to include rotation, outflow, and possibly magnetic fields.

3.6 Summary

A protostar collapses because, first, its mass is greater than the local Jeans mass, and, second, it is optically thin to the infrared radiation generated by heating from gravitational compression. As collapse proceeds, gravitational energy becomes more negative while thermal energy stays about the same, and the object becomes more and more gravitationally bound. If collapse starts with mass near the (thermal) Jeans mass, the infall results in a high degree of central concentration, with a small amount of material near the center having density much higher than the average density. This central core becomes optically thick as a result of dust opacity when its average density is around 10^{-13} g cm^{-3}. It then heats up, the collapse is slowed down by the increasing pressure gradient, and a small core forms in hydrostatic equilibrium. Upon further compression, the center of the core heats to ~2,000 K, the molecular hydrogen dissociates, and a second collapse is induced. It ends with the formation of a "stellar" core, in equilibrium, with a density of about 10^{-2} g cm^{-3} and a temperature of 20,000 K.

At this time most of the protostellar mass is still at relatively low density. The main part of the protostellar evolution consists of the accretion of the remaining material onto this core. The luminosity is provided by the conversion of infall kinetic energy into heat and then into radiation at the accretion shock. This radiation is absorbed in the dusty infalling envelope, and reradiated in the infrared, where it can be observed. A typical observed low-mass protostar has a "photospheric" temperature, best defined by $T_{\rm bol}$, of a few hundred K for a Class I source and less than 80 K for a Class 0 source, and a luminosity of order 1 L$_\odot$ for a Class I source and somewhat higher for Class 0. A protostar is usually identified by its presence in a star-forming region, a value of $T_{\rm bol} <$ 500–600 K, a spectral energy distribution that fits that of an optically thick dusty envelope, and, in a few cases, direct detection of infall velocity by analysis of spectral line profiles. The accretion time for a protostar is typically a few times 10^5 years.

3.7 Problems

The first two problems are based on the Bonnor-Ebert sphere, a model for molecular cloud cores that fits some observations and is a standard initial condition for protostellar collapse.

1. Consider a spherical isothermal cloud at temperature T, mean molecular weight μ, mass M, radius R, surface pressure $P_{\rm surf}$, and isothermal sound speed $c_s^2 = R_g T/\mu$. Assume the sphere has uniform density. Using the equilibrium Virial Theorem, including thermal, gravitational, and surface effects (2.6), calculate $P_{\rm surf}$ as a function of R. Show there is a maximum value of $P_{\rm surf}$:

$$P_{s,\rm max} = \frac{3.15 c_s^8}{G^3 M^2} . \tag{3.50}$$

Calculate the corresponding radius and compare it with the Jeans length derived from the simple energy condition for a sphere. The equilibria are unstable for R less than this critical value.

2. The Bonnor-Ebert sphere actually does not have uniform density but requires a density gradient to stay in equilibrium. In this problem you will calculate the equilibrium structure of the sphere.

   (a) Integrate numerically the equations for mass distribution and hydrostatic equilibrium, starting at the center with an assumed density ρ_0. Pick a sound speed and a mass; note that typical core temperatures are 10 K and masses are one to a few $M_\odot$. Stop the integration when you reach your total mass. Calculate P_{surf}.
   (b) Show that the Virial Theorem is satisfied. Note that the density is not uniform.
   (c) Varying the assumed value of ρ_0, calculate several models, keeping M and c_s constant, and find the surface pressure maximum $P_{s,\text{max}}$. Compare its value with that in Problem 1. List $P_{\text{surf}}(R)$ for a few models around the peak.
   (d) Plot the density distribution $\rho(r)$ for the model with $P_{s,\text{max}}$. What is the ratio of central-to-mean density? In dimensionless units, what is ξ_1?
   (e) Calculate the radius of the model with $P_{s,\text{max}}$ and compare it with the Jeans length, as in Problem 1.

3. Assume a power-law solution in (3.4) i.e. $\rho = Cr^{-n}$. Find physically reasonable values of n and C. Find m (which is the mass interior to r) as a function of r. If the sphere is cut off at a finite value of total mass M and radius R, what is the gravitational potential energy E_{grav}?

4. Along the curve in Fig. 3.6 find the minimum Jeans mass before point B. If the cloud is able to fragment into smaller pieces during the collapse (it is not clear that it actually will), this mass represents the smallest possible fragment. Fragmentation is very unlikely during the adiabatic collapse and dissociation phases [36].

5. In this problem you will calculate the approximate evolution of a protostar through the infrared H–R diagram.

   (a) Assume that the protostellar core accretes mass at the rate

   $$\dot{M} = \dot{M}_0(1 - t/\tau), \qquad (3.51)$$

   where $\dot{M}_0$ is the initial accretion rate (say $2 \times 10^{-6}\, M_\odot\, \text{yr}^{-1}$), t is the time, and τ is the time when accretion stops (say 0.5 Myr, but you can pick your own parameters). Calculate the mass as a function of time and the final mass.
   (b) Obtain a formula for the accretion luminosity of the core as a function of time. Assume that the core radius is constant at $3\, R_\odot$.
   (c) Produce a table of L_{acc}, M, and $\dot{M}$ at 15–20 different times starting at 0.05τ and ending at 0.95τ.
   (d) At each time, calculate the approximate mean photospheric temperature and plot the evolution in the H–R diagram.
   (e) At each time find the approximate wavelength at maximum intensity in the observed continuous spectrum.

3.7 Problems

6. A protostar of $1\,M_\odot$ starts collapse at the Jeans limit from a cloud core at 10 K. The cloud is composed of 72% H and 28% He by mass, and at the beginning all the hydrogen is in the form of H_2. Collapse ends on the Hayashi track as a fully convective star in hydrostatic equilibrium with all of the hydrogen and all of the helium ionized and with the equation of state of an ideal gas. The convective star can be represented by a polytrope of index $n = 1.5$ whose total gravitational energy is given by

$$E_{grav} = -\frac{3}{5-n}\frac{GM^2}{R}. \qquad (3.52)$$

   (a) Assume no energy is radiated during the collapse. What is the radius of the final star? What is the mean internal temperature of the final star? The dissociation potential of H_2 is 4.5 eV per molecule, the ionization potential of hydrogen is 13.6 eV, and the sum of the two ionization potentials of He is 79 eV per atom.

   (b) Assume that the star accretes at a constant rate $\dot M$ and that all of the accretion energy is radiated at the shock, which has a average radius 1.2× the final stellar radius. What is the final stellar radius and mean internal temperature? Note: this is a simple energy budget problem.

Chapter 4
Rotating Protostars and Accretion Disks

In the previous chapter, the idealized case of the spherically symmetric collapsing protostar was discussed. However it was clear that observations of most protostars are not consistent with such a simple model, and additional effects must be included. This chapter considers the effects of rotation. The angular momentum problem was introduced in Chap. 1: even the relatively small-scale molecular cloud cores have far too much angular momentum to be able to collapse to stellar dimensions. Various physical effects can contribute to the solution of this problem. Two of them which are relevant for the protostellar phase are

1. Fragmentation of the core into a binary or multiple system, with much of the angular momentum going into orbital motion
2. Collapse of the cloud into a central stellar object surrounded by an orbiting disk, which contains most of the angular momentum.

Investigation of the first solution involves three-dimensional hydrodynamic simulations, which are further discussed in Chap. 6. The disk solution, which is the subject of this chapter, can be treated for the most part with assumed axial symmetry in the collapsing cloud and in the disk structure; thus two-dimensional hydrodynamic simulations are sufficient. But if the disk becomes gravitationally unstable, its structure is no longer axisymmetric, and three-dimensional simulations are required for a detailed description of this phase of its evolution.

A number of issues may be identified with regard to the formation and evolution of disks. Among them are

- What are the initial conditions for the formation of a star-disk system?
- What is the predicted appearance of the emergent spectrum during the various phases of disk evolution?
- What are the important mechanisms for angular momentum transport in a disk? At what stage is each dominant? What are the time scales?
- What are the implications of the physical properties of newly formed disks, and the angular momentum transport processes, with regard to planet formation?

Disk evolution may be divided into three stages. The first is the formation stage, lasting 2–5×10^5 yr, in which the disk structure is built up from the infall of material from the protostellar cloud. This stage may be identified with the observed objects known as Class 0 and Class I sources (Chap. 3). The initial mass of the disk may be relatively high, so that it becomes gravitationally unstable right after formation and undergoes a rapid process of mass accretion onto the central star. It is possible that FU Orionis objects, which have been observed to undergo rapid increases in luminosity, are in a transition stage when the disk mass is still relatively large. The second stage is known as the viscous stage, during which internally generated torques, for example, those arising from turbulent viscosity in the presence of a magnetic field, result in redistribution of the angular momentum in the disk. The disk then evolves, with both accretion of matter onto the star and the spreading of the outer regions of the disk. This stage is usually identified with Class II objects, which exhibit a photospheric spectrum along with infrared excess because of dust emission in the disk, and whose lifetime is $\approx 10^6$–10^7 yr. However, disk evolution also takes place while gas is still accreting onto it from the molecular cloud core (Class I objects). The viscous stage is also associated with the formation of giant planets. The final stage is known as the clearing stage, during which the disk either (1) is blown away by the action of irradiation from the central star or external sources, (2) is blown away by a stellar wind, (3) accretes onto the star, (4) is accumulated into protoplanets, or (5) is disrupted by external encounters. The first stage is closely connected with protostellar collapse, as discussed in Chap. 3. After showing an example of that process, we concentrate on the observational evidence for disks and the basic physics of disk evolution.

4.1 Disk Formation

The requirement of significant transfer of angular momentum out from the material that eventually ends up in the star can be solved by various physical processes operating at different phases of the evolution. Examples of these processes include magnetic braking in the molecular cloud phase (Chap. 2), formation of disks and binaries during the protostar collapse, and stellar winds, gravitational instability, or magnetorotational instability after the disk has formed. In this section we look at the collapse of rotating protostars in the axisymmetric (2-dimensional) approximation, without magnetic fields, which can produce disks (but not binaries).

The formation of a disk from a molecular cloud core brings in another aspect of the angular momentum problem. Even if the total angular momenta of the core and of the star-disk system are the same, the *distributions* of angular momentum are very different. Angular momentum transport is required during the transition from protostar to star-disk system, as is illustrated in Fig. 4.1. The upper curves show the distribution of specific angular momentum, as a function of interior (cylindrical) mass fraction, in a uniform-density cloud core and a centrally condensed cloud core with the density distribution of the singular isothermal sphere, both with solid

4.1 Disk Formation

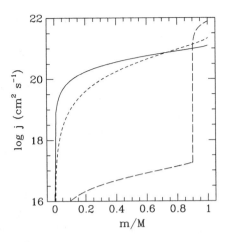

Fig. 4.1 Specific angular momentum as a function of the mass fraction inside a given cylinder about the rotation axis. All three curves have total mass $1\,M_\odot$ and total angular momentum $10^{54}\,\mathrm{g\,cm^2\,s^{-1}}$. *Solid curve*: molecular cloud core with uniform density and uniform angular velocity; *short-dashed curve*: cloud core with an r^{-2} density distribution and uniform angular velocity; *long-dashed curve*: a T Tauri star of $0.9\,M_\odot$ with surface rotational velocity of $20\,\mathrm{km\,s^{-1}}$, uniform angular velocity, and radius $3\,R_\odot$, surrounded by a disk of $0.1\,M_\odot$ and radius 30,000 AU. After [64]

body rotation. The total angular momentum is normalized to be consistent with observations of molecular cloud cores [191]. The long-dashed curve shows the distribution of angular momentum in a model of a star-disk system with the *same* total mass and total angular momentum as in the clouds. The star has 0.9 solar masses and is rotating uniformly with a typically observed surface rotational speed of 20 km/s. The disk has 0.1 solar masses and is Keplerian. Clearly some process had to transfer a lot of angular momentum from the inner regions to the outer regions. And there is an additional problem. The typical observed disk has a radius of 100–1,000 AU, not 30,000 AU as the disk shown in the figure must have to contain the correct total angular momentum. Thus even though the *specific* angular momentum ($R^2\Omega$) of the outer edge of a cloud core ($10^{21}\,\mathrm{cm^2\,s^{-1}}$) is about the same as that of the outer edge of a disk of 500 AU, the total angular momentum in the disk shown in the figure is far too large to be consistent with observations. Thus, substantial angular momentum has to be removed during the transition from cloud core to star plus disk.

The initial conditions for rotating cloud collapse are the same as those given at the beginning of Sect. 3.1, with the addition of a rotational parameter. Neglecting for the moment turbulent effects, the cloud is assumed to have a uniform angular velocity Ω. The rotational parameter is generally called β, the ratio of rotational energy to the absolute value of gravitational energy:

$$\beta = \frac{E_{\rm rot}}{|E_{\rm grav}|} = \frac{1}{3}\frac{R^3\Omega^2}{GM}, \tag{4.1}$$

where the last equality holds for a uniformly rotating sphere of uniform density. Observationally, a rotating cloud displays a linear change in the Doppler shift of an optically thin spectral line with position as one scans across the cloud (i. e. one edge of the cloud has a velocity toward the observer, the other edge away from the observer). In a number of cloud cores [191] the characteristic velocity pattern has been observed, and the typical value of β is 0.02–0.03. Thus rotation is not very important in the initial structure of the cloud, but if angular momentum is conserved during the collapse, it becomes much more important later on. In fact, once collapse starts, loss of angular momentum of the cloud as a result of magnetic braking is likely to be negligible, because the infall time becomes short compared with the time for an Alfvén wave to cross the cloud.

Rotating collapse can be treated in two different ways, first, by an approximate analytic solution [502] known as the Terebey–Shu–Cassen (TSC) solution, and, second, by a full numerical solution of the equations of hydrodynamics and radiation transport. The TSC solution, while convenient, makes some simplifying assumptions. As originally formulated by Ulrich [524], the outer radius of the spherical infalling cloud is R_c, and the cloud is initially rotating with uniform angular velocity Ω. Material falls toward a central object of mass M, in free fall with conservation of angular momentum, with a constant accretion rate $\dot{M}$. The gravitational effect of other material in the infalling envelope is neglected, as are pressure effects. It is convenient to define a "centrifugal radius" R_{ct} where material from the equator of the initial cloud at its outer edge (which has specific angular momentum $j = \Omega R_c^2$) reaches a balance between gravitational and centripetal effects, that is, a Keplerian orbit:

$$R_{ct} = \frac{\Omega^2 R_c^4}{GM}. \tag{4.2}$$

Material at other positions around the cloud surface, where the radius vector makes an initial angle θ_0 with the rotation axis, has smaller angular momentum and falls to the equatorial plane at smaller distances from the central object. The orbit of each individual mass element is calculated separately, assuming that it moves in the gravitational potential of the central mass M. If θ is the instantaneous angle between the radius vector of a mass element and the rotation axis, in the infall region, then at a given (r, θ) in polar coordinates, the density distribution can be derived

$$\rho(r, \theta) = \frac{\dot{M}}{4\pi (GMr^3)^{1/2}} \left(1 + \frac{\cos\theta}{\cos\theta_0}\right)^{-1/2} \left(\frac{\cos\theta}{\cos\theta_0} + \frac{2R_{ct}\cos^2\theta_0}{r}\right)^{-1} \tag{4.3}$$

where $\dot{M}$ and M are constants. The details of the derivation are given in [202,524]. This model is often used to represent infalling envelopes. To obtain θ_0 one uses the equation for the trajectory of a mass element starting at θ_0, which can be expressed in the form [524]

4.1 Disk Formation

$$r = \frac{R_{\text{ct}} \cos\theta_0 \sin^2\theta_0}{\cos\theta_0 - \cos\theta}. \tag{4.4}$$

Given (r, θ) this equation must be solved for θ_0. Thus the individual mass elements execute two-body orbits until an element hits the equatorial plane, where it meets its mirror particle arriving from the other side of the plane. The resulting shock strongly reduces the velocity component perpendicular to the plane, and as long as the angular momentum is large enough so that the element misses the central object, it joins the disk.

The TSC solution itself has two basic assumptions: (1) the overall cloud is slowly rotating, thus, for example, $R_{\text{ct}} \ll R_c$. Thus, on the larger scales the solution can be approximated. (2) The initial condition is close to that of the singular (equilibrium) isothermal sphere, with $\rho \propto r^{-2}$. The first assumption is supported by the fact that the rotational β is observed to be small in molecular cloud cores. In the inner regions, where $r \leq R_{\text{ct}}$ and where rotational effects are important, the model uses the Ulrich prescription for the gas flow as described above. This solution does not assume slow rotation; also, in the inner regions the assumptions of that analysis apply reasonably well. In the outer regions, a perturbation expansion is applied to obtain the rotating initial configuration and to describe the ensuing collapse. As in the spherical case, collapse starts at the center, the so-called "inside-out" collapse. The outer boundary of the region that is collapsing moves outward at the sound speed in the (assumed isothermal) structure. The outer region is static until the expansion wave reaches it.

The singular isothermal sphere has a constant mass accretion rate $\dot{M} = 0.975 c_s^3/G$, where c_s is the sound speed in the molecular cloud core. This feature can be used to estimate how fast the disk grows. Clearly, the central mass as a function of time is $M = 0.975 c_s^3 t/G$. In a time t that amount of mass arrived from within a radius R_x in the density distribution of the isothermal sphere:

$$M = \int_0^{R_x} \frac{4\pi r^2 c_s^2}{2\pi G r^2} dr \tag{4.5}$$

We can easily solve for $R_x = 0.4875 c_s t$. The specific angular momentum of material starting from R_x at an angle θ_0 with respect to the rotation axis is $j = R_x^2 \Omega \sin\theta_0$, and the "centrifugal" radius where that material ends up is $j^2/(GM)$ so

$$R_{\text{ct}} = .0579 c_s \Omega_0^2 t^3 \sin^2\theta_0. \tag{4.6}$$

Setting $\theta_0 = \pi/2$ to get the maximum radius, we find that the disk first forms when this radius exceeds the stellar radius, and that its outer edge expands as t^3.

The general procedure for using these approximate models is to set up parameterized models, make radiation transfer calculations based on those models, and compare with the spectral energy distributions of observed sources. An example of such a comparison is given in Fig. 1.6.

To study disk formation in the more general case, with less-restrictive approximations, where rotation is not necessarily small, where the self-gravity of the envelope is not neglected, and for arbitrary initial conditions, one must turn to detailed numerical solutions. The approximations are made that the collapse is axisymmetric, that is, there are no gradients in the azimuthal (ϕ) direction, and that there is symmetry with respect to the equatorial plane. For the moment we assume that magnetic fields are not coupled to the gas.

The general equations that are solved are shown below, where the two dimensions are the cylindrical radius R and the height above the midplane Z. The equations are set up on an Eulerian fixed grid. These equations are sometimes referred to as "$2\frac{1}{2}$-D" equations, because, although all variables are assumed to have zero derivatives in the ϕ-direction, the component of the equation of motion in that direction is included. The first equation is the equation of continuity, and the second and third are, respectively, the equations of motion in the R- and Z directions. In the R-component the rotational contribution has to be added to the right-hand side: $A^2/\rho R^3$, where $A = \rho R v_\phi$ is the angular momentum per unit volume and v_ϕ is the linear velocity component in the ϕ-direction. The quantities u and w are the R- and Z-components of the velocity $\mathbf{v}$, Φ is the gravitational potential, and P is the pressure. The fourth equation represents conservation of angular momentum. The fifth equation is the energy equation, where E is the internal energy per unit mass. In this approximation, the radiation energy density is not considered separately from the gas energy density. The last term in this equation is the divergence of the radiative flux, where the flux is

$$\mathbf{F} = \left(\frac{-c}{3\kappa_R \rho} \nabla a T^4 \right) \tag{4.7}$$

in an optically thick medium, where κ_R is the Rosseland mean opacity and a is the radiation density constant. This expression is the same as that used in stellar interiors, however in a protostar some regions of the problem may be optically thin. In the limit of optical thickness going to zero, it can be shown that the flux is given by

$$|\mathbf{F}| = c u_{\text{rad}} \tag{4.8}$$

where c is the velocity of light and u_{rad} is the radiation energy per unit volume, aT^4 for a black body. A combined approximate expression for the radiative flux, which is correct in the limits of very optically thin and very optically thick media, is called *flux-limited diffusion*. Thus the transition from the isothermal optically thin region of protostellar collapse to the optically thick adiabatic region can be approximately taken into account in a relatively simple way.

The modified equation reads

$$\mathbf{F} = \left(\frac{-c\lambda}{\kappa_R \rho} \nabla a T^4 \right) \tag{4.9}$$

4.1 Disk Formation

where the flux limiter λ must reduce to 1/3 in the optically thick limit. The usual way to calculate λ is [316] to define a quantity R_{lp}:

$$R_{\text{lp}} = \frac{|\nabla a T^4|}{\kappa_R \rho a T^4} \tag{4.10}$$

which is the ratio of the mean free path of a photon $(\kappa_R \rho)^{-1}$ to the scale height of the radiation energy density. Then

$$\lambda = \frac{2 + R_{\text{lp}}}{6 + 3R_{\text{lp}} + R_{\text{lp}}^2}. \tag{4.11}$$

Thus when the mean free path is short, $R_{\text{lp}} \to 0$ and $\lambda \to 1/3$. When the mean free path is long, $R_{\text{lp}} \to \infty$, $\lambda \to 1/R_{\text{lp}}$ and the flux goes to the limit acT^4, as required physically. Note that in this limit, (4.7) gives an unphysically large radiative flux. The temperature is low enough in the region being considered so that grains provide the main opacity source (Fig. 3.5). The last equation is the Poisson equation for the gravitational potential.

In the early part of the collapse, conservation of angular momentum is assumed. Once a disk forms and evolves, some kind of process of angular momentum redistribution must be included (see below).

$$\frac{\partial \rho}{\partial t} + \nabla \cdot (\rho \mathbf{v}) = 0 \tag{4.12}$$

$$\frac{\partial (\rho u)}{\partial t} + \nabla \cdot (\rho u \mathbf{v}) = -\rho \frac{\partial \Phi}{\partial R} - \frac{\partial P}{\partial R} + \frac{A^2}{\rho R^3} \tag{4.13}$$

$$\frac{\partial (\rho w)}{\partial t} + \nabla \cdot (\rho w \mathbf{v}) = -\rho \frac{\partial \Phi}{\partial Z} - \frac{\partial P}{\partial Z} \tag{4.14}$$

$$\frac{\partial A}{\partial t} + \nabla \cdot (A \mathbf{v}) = 0 \tag{4.15}$$

$$\frac{\partial (\rho E)}{\partial t} + \nabla \cdot (\rho E \mathbf{v}) = -P \nabla \cdot \mathbf{v} - \nabla \cdot \mathbf{F} \tag{4.16}$$

$$\nabla^2 \Phi = 4\pi G \rho. \tag{4.17}$$

These equations are solved in conjunction with the equation of state, as discussed in Chap. 3. For simplification, it is assumed that the grain temperature T_g and the gas temperature T are the same (Sect. 2.3). In addition it is necessary to consider the treatment of shocks, which are very likely to occur during disk formation. The standard numerical procedure is to use the Richtmyer–von Neumann [530] artificial viscosity to broaden shocks over a few zones while still obtaining the correct shock jump conditions (Sect. 3.4). In the $R-$ component of the momentum equation the derivative $\partial P/\partial R$ is replaced by $\partial/\partial R(P + Q_{RR})$ and in the $Z-$ component $\partial P/\partial Z$

is replaced by $\partial/\partial Z(P + Q_{ZZ})$ where

$$Q_{RR} = q^2 \rho (\Delta u)^2 \text{ if } \frac{\partial u}{\partial R} < 0 \tag{4.18}$$

$$Q_{RR} = 0 \text{ if } \frac{\partial u}{\partial R} > 0 \tag{4.19}$$

$$Q_{ZZ} = q^2 \rho (\Delta w)^2 \text{ if } \frac{\partial w}{\partial Z} < 0 \tag{4.20}$$

$$Q_{ZZ} = 0 \text{ if } \frac{\partial w}{\partial Z} > 0. \tag{4.21}$$

The quantities $\Delta u = u_{j+1,k} - u_{j,k}$ and $\Delta w = w_{j,k+1} - w_{j,k}$ are the velocity differences across the zone in the $R-$ and $Z-$ directions, respectively, and (j,k) are the zone indices in the two directions. The dimensionless constant q is of order unity; thus the Q's have dimensions of pressure. The artificial viscocity term is added only if the zone is compressing in a given direction. In terms of the actual kinematic viscosity coefficient, say in the $R-$ direction, $\nu = q^2 \Delta R |\Delta u|$ in units of cm^2 s^{-1}. The effect of the artificial viscosity is to dissipate kinetic energy of motion and convert it into heat, as is the situation in an actual shock. Thus an additional positive term must be added to the right-hand side of the energy equation to take into account the added heat:

$$-Q_{RR}\frac{\partial u}{\partial R} - Q_{ZZ}\frac{\partial w}{\partial Z}.$$

Care must be taken near the origin, because in a spherically converging flow (not usually the case in this collapse problem), artificial viscosity can be generated even if there is no shock [514].

The details of the numerical solution to these equations are discussed in [66]. The main problem in the calculation of the numerical collapse of a rotating protostar is the extreme range of length scales involved, from 10^{17} cm, the size of the cloud core, to 10^{11} cm, the size of the resulting star. Even if the star is resolved with only ten zones, the sound speed at a temperature of 10^6 K is 10^7 cm s^{-1}. The maximum time step in the numerical solution is determined by the Courant–Friedrichs–Lewy (CFL) condition [121]

$$\Delta t = \min\left(\frac{\Delta R}{c_s + |v|}, \frac{\Delta Z}{c_s + |v|}\right) \tag{4.22}$$

where c_s is the sound speed, v is the fluid velocity through a zone, and ΔR, ΔZ are the zone sizes. The requirement in the stellar zones is then $\Delta t < 10^3$ sec. The time for the entire cloud to fall onto the star and disk is at least one free-fall time from the initial density of the core, or 10^5 yr or 3.15×10^{12} sec. Thus 3 billion time steps would be required to do the simulation (after formation of a stellar core of

4.1 Disk Formation

reasonable mass). In fact there is an additional restriction on the time step [411] if viscosity is present:

$$\Delta t < \frac{(\Delta x)^2}{4\nu} \quad (4.23)$$

where Δx is the zone size in either direction.

To avoid the problem of very short time steps in the central regions, they are not resolved but treated as a sink zone, into which mass, energy, and angular momentum flow. In this way the main disk formation region, say outside 10 AU, can be calculated in detail. The mass and angular momentum that fall into the central zone are used to calculate a rough equilibrium model for the core and to determine the accretion luminosity. This luminosity is fed back into the grid as a central boundary condition for the radiative transfer:

$$L_{core} = L_{int} + \frac{3}{4} \frac{GM_{core}\dot{M}}{R_{core}} \quad (4.24)$$

where L_{int} is the internal contraction luminosity of the central star and R_{core} is reasonably taken to be a few solar radii. The second term, the accretion luminosity, is multiplied by 3/4 to take into account approximately the fact that the infall onto the core is not spherically symmetric but is mediated by an accretion disk [10]. This central source generally dominates the energy input to the protostar and determines the temperature in the infalling material and the disk. The extreme degree of central concentration of a protostar can also be dealt with by a series of nested fixed grids, with spatial resolution increasing toward the center by a factor of 2 for each grid. Thereby the inner regions are treated with high resolution, where it is needed, and the outer, low-density regions are represented by a relatively coarse grid.

An example of the solution of the full set of equations [565] is shown in Fig. 4.2. The initial rotating protostar has a temperature of 20 K and a density distribution $\rho \propto r^{-2}$ where r is the distance to the center. However it is not an equilibrium isothermal sphere; the force of gravity exceeds that of the pressure gradient by a considerable margin. The specific angular momentum $R^2\Omega$ at the outer edge is 2×10^{21} cm^2 s^{-1}, and $\beta \approx 0.01$. The numerical solution employed four nested grids, each with 124 by 124 zones in the (R, Z) plane.

Conservation of angular momentum is assumed. At a time of 7,000 yr, a disk of radius 260 AU is found. The interior of the disk is practically in hydrostatic equilibrium, and a shock wave is evident, as the dark band enclosing the disk, displaced somewhat from the surface of the actual equilibrium disk. This accretion shock is well resolved at about 10 density scale heights above the plane of the disk, and its location is determined by the point where the infall ram pressure, ρv^2, is matched by the gas pressure of the inner regions of the protostar. In fact there is a second shock. The component of velocity normal to the outer shock is dissipated, but the parallel component still results in infall. This inflow is deflected from radial motion by the mass of the inner disk and by effects of angular momentum, and it

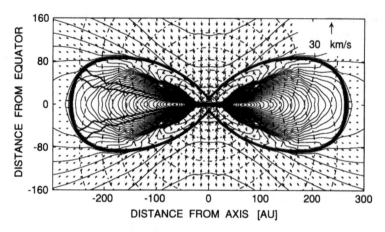

Fig. 4.2 Results from a two-dimensional hydrodynamic collapse calculation showing the structure of a newly-formed disk. *Dashed lines*: contours of equal temperature separated by $\Delta \log T = 0.1$ and with a minimum $T = 160$ K. *Solid lines*: contours of equal density with separation $\Delta \log \rho = 0.2$ and minimum value $\log \rho = -19.4$, in g cm^{-3}. *Arrows*, velocity vectors with length proportional to speed. Reproduced, by permission of the AAS, from [565]. © 1995 The American Astronomical Society

shocks again at the outer edge of the actual disk, then becoming part of the region of the disk that is near hydrostatic equilibrium.

Clearly this structure is more complicated than the theoretician's ideal "flat" disk. The vertical thickness is, however, somewhat exaggerated because of the logarithmic contour lines; in fact most of the mass of the disk lies interior to the two pairs of two diagonal (solid) lines in the left-hand side of the diagram, which correspond to 1 and 2 density scale heights, respectively. The heating of this disk, and therefore its vertical thickness, is controlled primarily by the luminosity of the central source. The central unresolved star has a mass of $2.7 \, M_\odot$ and a luminosity of $30 \, L_\odot$. The residual infalling envelope is optically thin along the rotation axis, so that the central object could more easily be seen in this direction than along the equatorial plane.

It is clear from the theoretical collapse calculations, also in the case of lower masses, that once the disk forms its mass becomes comparable to that of the central object, and it quickly becomes gravitationally unstable. Thus the evolution can no longer be calculated under the assumption of conservation of angular momentum of each mass element. The Toomre Q parameter is an indicator of gravitational instability, and it is defined by [508]

$$Q = \frac{\kappa c_s}{\pi G \Sigma} \tag{4.25}$$

4.1 Disk Formation

where Σ is the surface density (mass per unit area) and κ is the so-called epicyclic frequency

$$\kappa^2 = \frac{2\Omega}{R}\frac{d}{dR}(R^2\Omega) \qquad (4.26)$$

which reduces to $\kappa = \Omega$ for Keplerian rotation. Strictly speaking, if Q is greater than ≈ 1, the material is locally stable to an axisymmetric gravitational perturbation (presumably it is unstable if $Q < 1$). The axisymmetric instability would lead to a dense ring-like configuration. Q is also a useful indicator of non-axisymmetric gravitational instabilities, which tend to occur in the range $Q = 1.3 - 1.5$, that is, earlier in the evolution of the disk than the axisymmetric ones. The derivation and explanation of the condition for axisymmetric instability in a disk, based upon a linear perturbation analysis, is given in Shu [458], his Chap. 11.

If a disk's mass is a substantial fraction (say 0.3 or more) of that of the central star, it is likely to be gravitationally unstable, but even if the disk mass is small compared with that of the star, the disk can be gravitationally unstable if it is cold enough. The approximate argument runs as follows: set $Q = 1$, corresponding to the marginal condition for gravitational instability. Then use the (approximate) Keplerian $\Omega = (GM_*/R^3)^{1/2}$ and the (approximate) $\Sigma = M_{disk}/(\pi R^2)$ and solve for M_{disk}.

$$\frac{M_{disk}}{M_*} \approx \frac{c_s}{\Omega R}. \qquad (4.27)$$

Typically $c_s/(\Omega R) \sim 0.1$, suggesting that even a low-mass disk could be gravitationally unstable, but the disks typically observed around young stars have masses $\approx 0.01\,M_*$ and are probably stable. It has been shown [323] that for $M_{disk} \ll M_*$ the evolutionary time scale is given by $t_{grav} \approx \Omega^{-1}(M_*/M_{disk})^2$, so that a low mass disk evolves very slowly, over say 100 rotation periods if, in fact, it is gravitationally unstable. But, as mentioned earlier, M_{disk} is expected to be comparable to M_* during the early infall phases just after disk formation, and the time scale for the gravitational instability is a few orbital periods.

What is likely to happen as a result of the non-axisymmetric instability is the generation of spiral waves, which result in a gravitational torque on the material and the transport of angular momentum outward from the center. Full 3-D numerical simulations [306] show that this is in fact the outcome for $Q \approx 1.3$. For lower values of Q, near 1, the instability is suspected to result in fragmentation, that is, the formation of gravitationally bound subcondensations in the disk. As a disk builds up in mass through infall, the main effect on Q is the increase in Σ, so Q evolves from high values to low values, reaching 1.3 before reaching 1. As long as the disk accretion time is long compared with the dynamical time, spiral waves and angular momentum transport will start before the fragmentation regime is reached. The heating resulting from spiral wave shocks would tend to stabilize Q, so fragmentation may not occur. The disk would transfer material to the central star, reduce its mass, and eventually stabilize.

In fact there are two conditions that are required for a disk to fragment. The first is a Q-value close to unity in some region of the disk. The second [182] is that the cooling time of the disk must be less than about an orbital period; otherwise in fact the disk will heat up as a result of the instability and remain stable to fragmentation. The relatively massive disks that form initially as a result of protostar collapse are quite optically thick in the vertical direction, and their time scale for radiation transport and cooling is too long to allow fragmentation. Thus they will in fact develop spiral arms and transfer mass inward to join the central star. An exception may occur in the outer regions, which remain optically thin with short cooling times.

In any case numerical treatment of the non-axisymmetric gravitational instability requires a 3-D solution. The key modification to the above equations is in the equation of conservation of angular momentum (4.15), which is replaced by

$$\frac{\partial A}{\partial t} + \nabla \cdot (A\mathbf{v}) = -\left(\rho \frac{\partial \Phi}{\partial \phi} + \frac{\partial P}{\partial \phi}\right). \qquad (4.28)$$

The variation in gravitational potential and pressure in the $\phi-$ direction provides a net torque that transfers angular momentum outward in the disk. Numerous 3-D simulations have been carried out, with varying results but for example [74] show that even in a relatively low-mass disk it is difficult to meet the criteria for fragmentation. Thus the typical disk that results from the protostar collapse during the phase when it is gravitationally unstable [360] may qualitatively have the spiral-arm structure as shown in Fig. 4.3.

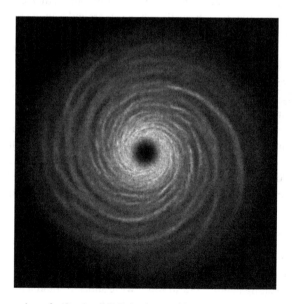

Fig. 4.3 A face-on view of a simulated disk that is unstable to the generation of spiral waves. The radius of the disk is about 25 AU. The grey scale represents the surface density in $g\,cm^{-2}$, ranging from $\log \Sigma = -2$ (*dark*) to $+4$ (*white*). Reproduced by permission of John Wiley and Sons from F. Meru, M. R. Bate: *MNRAS* **406**, 2279 (2010). © 2010 Royal Astronomical Society

4.1 Disk Formation

The difficulty with the otherwise high-quality 3-dimensional simulations that have been carried out is that hydrodynamics plus radiation transfer places high demand on computer resources. Thus typical simulations can follow the gravitationally unstable disk for a few thousand years, depending on assumed disk size and spatial resolution, short compared with the total duration of the gravitationally unstable phase. The question has been investigated whether the gravitational instability can be approximated by a pseudo-viscosity, making it possible to treat the problem in 1 or 2 space dimensions, which is numerically more tractable. The general opinion is that it is not a valid approach [307] because the gravitational instability is of global, rather than local, nature. It may, however, be feasible to use such an approach [27] in the particular situation when a disk is marginally gravitationally unstable, that is the minimum Q value hovers around the critical value above which the instability is not of importance. In the disk formation picture, the disk may in fact be marginally unstable, with the increase in Σ caused by mass accretion balanced by heating caused by the instability itself. Another possibly appropriate situation [74] is a disk where heating from the instability is balanced by radiative cooling and the cooling time is related to the orbital frequency by $t_{\text{cool}}\Omega = \text{constant}$.

To use a viscosity (or pseudo-viscosity) in 2D or 3D solutions of the evolution of a disk, one must solve the full Navier–Stokes equations for a compressible fluid [298]. In cylindrical coordinates with no gradients in the ϕ-direction, the 3 equations of motion and energy equation then become

$$\frac{\partial(\rho u)}{\partial t} + \nabla \cdot (\rho u \mathbf{v}) = -\rho \frac{\partial \Phi}{\partial R} - \frac{\partial P}{\partial R} + \frac{A^2}{\rho R^3}$$
$$+ \frac{1}{R}\frac{\partial}{\partial R}(R\tau_{RR}) + \frac{\partial \tau_{RZ}}{\partial Z} - \frac{\tau_{\phi\phi}}{R} \quad (4.29)$$

$$\frac{\partial(\rho w)}{\partial t} + \nabla \cdot (\rho w \mathbf{v}) = -\rho \frac{\partial \Phi}{\partial Z} - \frac{\partial P}{\partial Z}$$
$$+ \frac{1}{R}\frac{\partial}{\partial R}(R\tau_{RZ}) + \frac{\partial \tau_{ZZ}}{\partial Z} \quad (4.30)$$

$$\frac{\partial A}{\partial t} + \nabla \cdot (A\mathbf{v}) = \frac{1}{R}\frac{\partial}{\partial R}(R^2 \tau_{\phi R}) + R\frac{\partial \tau_{\phi Z}}{\partial Z} \quad (4.31)$$

$$\frac{\partial(\rho E)}{\partial t} + \nabla \cdot (\rho E \mathbf{v}) = -P\nabla \cdot \mathbf{v} - \nabla \cdot \mathbf{F} + \rho E_{\text{diss}}. \quad (4.32)$$

If $\eta = \nu\rho$ is the coefficient of dynamic shear viscosity, then the components of the viscous stress tensor τ_{ij} are

$$\tau_{\phi R} = \tau_{R\phi} = \eta R \frac{\partial \Omega}{\partial R}$$

$$\tau_{\phi Z} = \tau_{Z\phi} = \eta \frac{\partial v_\phi}{\partial Z}$$

$$\tau_{RR} = 2\eta \frac{\partial u}{\partial R} - \frac{2}{3}\eta \nabla \cdot \mathbf{v}$$

$$\tau_{ZR} = \tau_{RZ} = \eta \left(\frac{\partial w}{\partial R} + \frac{\partial u}{\partial Z} \right)$$

$$\tau_{\phi\phi} = 2\eta \frac{u}{R} - \frac{2}{3}\eta \nabla \cdot \mathbf{v}$$

$$\tau_{ZZ} = 2\eta \frac{\partial w}{\partial Z} - \frac{2}{3}\eta \nabla \cdot \mathbf{v}, \tag{4.33}$$

and the viscous energy dissipation rate per unit mass is

$$E_{\mathrm{diss}} = 2\nu \left(\left|\frac{\partial u}{\partial R}\right|^2 + \left|\frac{\partial w}{\partial Z}\right|^2 + \left|\frac{u}{R}\right|^2 \right) - \frac{2}{3}\nu |\nabla \cdot \mathbf{v}|^2 + \nu \left|\frac{\partial u}{\partial Z} + \frac{\partial w}{\partial R}\right|^2 + \nu R^2 |\nabla \Omega|^2. \tag{4.34}$$

As (4.31) shows, angular momentum transport is generated by the shear in the ϕ-velocity in both the $R-$ and $Z-$ directions.

A calculation of disk formation [564] included angular momentum transport by spiral waves approximately in 2D through use of a pseudo-viscosity, although a correct treatment would involve 3D. The original cloud had 1 $M_\odot$, $\alpha = 0.39$, $\beta = 0.01$, and a radius of 6,667 AU. The initial density $\rho \propto r^{-2}$ where r is the spherical radius. Once the disk forms, a viscosity is added which mimics angular momentum transport by spiral waves, even though the calculation remains axisymmetric. The standard disk viscosity prescription is used

$$\nu = \alpha_{\mathrm{visc}} c_s H = \alpha_{\mathrm{visc}} \frac{c_s^2}{\Omega} \tag{4.35}$$

where H is the disk scale height and α_{visc} is a numerical constant of order 10^{-2} (see next section). The Navier–Stokes equations are solved in the disk. At each radial location in the disk, the Toomre Q is calculated, and the minimum value found. An assumed value of α_{visc} is taken to start. If the minimum Q is above the marginal stability value for non-axisymmetric gravitational instability (about 1.3) then α_{visc} is reduced throughout the whole disk, and angular momentum transport becomes less effective. If the minimum Q is below 1.3, then α_{visc} is increased. After a few time steps, the viscosity becomes self-regulated such that the disk is always near marginal gravitational instability. In the case of a molecular cloud core of 1 $M_\odot$, the regulated α_{visc} falls in the range 0.02–0.07, in rough agreement with the efficiency of angular momentum transport in a full 3D calculation [306] for a gravitationally unstable disk in a protostar of the same mass. The reduction of the problem to 2D, however, allows the calculation of the evolution of the protostar for a few times 10^5 years, a substantial fraction of its total lifetime. The accretion rate of disk material onto the star falls in the range $(2 - 0.5) \times 10^{-6}\, M_\odot\, \mathrm{yr}^{-1}$, decreasing with time.

4.1 Disk Formation

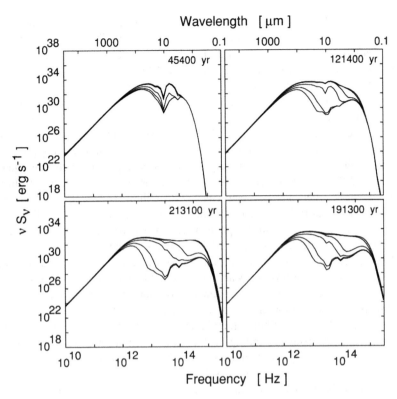

Fig. 4.4 Theoretical spectral energy distributions obtained by radiative transfer calculations based on the protostar models with initial molecular cloud core mass of 1 $M_\odot$. The quantity S_ν is the energy radiated, in erg s^{-1} Hz^{-1}, at the frequency ν. In each panel, the viewing angles, running from lower to upper curves, are 90°, 75°, 60°, 30°, 20°, and 0°, measured from the rotation axis. The evolution time is given in the upper right of each panel. The stellar core masses at the given times are, respectively, 0.29, 0.47, 0.55, and 0.57 $M_\odot$. The outer edges of the disks, which have similar structure to that shown in Fig. 4.2, are located, respectively, at 80, 240, 800, and 1,000 AU. Reproduced, by permission of the AAS, from [564]. © 1999 The American Astronomical Society

The calculation ends at 213,100 yr with the central star containing 0.567 $M_\odot$ and with most of the rest of the mass in the disk. The expected outer radius of the disk, based on conservation of angular momentum from the initial state, is about 500 AU; however because of angular momentum transport, the disk expands out to about 1,000 AU, although the outer parts are optically thin. Beyond this time, however, the time scale for the gravitational instability slows down considerably, to the point where the star is accreting mass very slowly. Other mechanisms for angular transport must be considered beyond this time, since the disk mass is still too large to be consistent with those observed around T Tauri stars (see next section).

Figure 4.4 shows the spectral energy distributions at four different times during the evolution of the 1 $M_\odot$ protostar, at various viewing angles. Most of the radiation originates at the central protostellar core and the hot regions of the disk close to it. The optically thick outer envelope absorbs and scatters much of the light, and most of the observable radiation is in the infrared. The equations indicate that the radiation transfer employed during the calculation of the collapse used a frequency-averaged opacity. However, at selected times during the evolution, the core/disk/envelope structure can be used to do a frequency-dependent calculation, by a solution of the equation of radiative transfer along various lines of sight passing through the object. The integration of the emerging radiation over the visible surface then gives the flux at each frequency. The results shown in the figure were based on 64 separate frequencies. During much of the lifetime of an accreting protostar, the infrared flux at say $10\,\mu$m can vary by orders of magnitude between equator and pole, and the total bolometric luminosity can vary by up to a factor 30. In the equator-on views, there are two peaks in the flux, the first arising from the cool, optically thick equatorial region, the second (around $1\,\mu$m) from indirect light from the central regions, emitted in the polar direction, where there is relatively little obscuring material, and scattered toward the observer. At the earliest time, the dip in the SED near $10\,\mu$m is absorption by a silicate dust feature. The figure indicates that near-infrared surveys for protostars are much more likely to pick up objects viewed pole-on than at other angles. For an inclination of more than 30° from the pole, the $2\,\mu$m flux is reduced by more than a factor 10 compared with the pole-on flux at the later times. This beaming effect, which indicates large differences in the observed luminosity of a protostar depending on the viewing angle, may contribute to the solution of the luminosity problem discussed in Chap. 3, where it was considered in the context of a spherically symmetric solution.

The various disk formation calculations can be summarized as follows:

- A centrally condensed cloud core collapses to a stellar core surrounded by a gravitationally unstable disk
- Spiral waves are an effective means of angular momentum transport during early phases of disk evolution
- It is possible that disks resulting from protostar collapse can be formed that are gravitationally unstable to the point of fragmentation, but it is not at all clear what factors control whether the outcome is in fact fragmentation or simply angular momentum transport by spiral waves
- After a few times 10^5 yr angular momentum transport by gravitational torques becomes ineffective, leaving a disk with about 35% of the system mass. Other mechanisms for angular momentum transport are required for the further evolution of the disk
- The increase in evolutionary time scale as the disk mass decreases in 2D simulations is consistent with nonaxisymmetric calculations of disk evolution with self-gravity
- The spectral appearance of a protostar is strongly dependent on viewing angle. The mid-infrared flux can vary by several orders of magnitude between pole-on and equator-on views

- Once the disk has formed, most of its mass lies between 100 and 1,000 AU
- The surface density distribution of the newly-formed disk is not a power law. It is nearly flat in the central regions, then drops rapidly farther out
- The temperature of the dust in the protostar drops roughly as the square root of the distance to the central object.

4.2 Observations of Disks

The previous section has described highly embedded disks in Class I protostars. Although our observational knowledge of the presence and properties of such disks in both Class 0 and Class I objects is rapidly developing [244], most of the observed data refer to Class II objects, that is, pre-main-sequence stars with surrounding disks of relatively low mass. The transition from the presumed massive disks that are generated during the formation stage to the low-mass Class II disks is not understood. But it is clear that many young stars are surrounded by disk-like structures [44, 51, 102]. We list and then discuss several key properties of disks and the methods that are used to deduce their presence. This section and the next describe the so-called "viscous" stage of disk evolution.

Disk observations span a wide range of wavelengths. The radial extent of these disks is in the range 10–1,000 AU, and the corresponding masses are roughly estimated to be 0.001–0.1 $M_\odot$. Central stars over a considerable mass range, from the brown dwarf region up to roughly 7 $M_\odot$, give indications that disks are present. The mass accretion rate from disk onto star is $\sim 10^{-8}$ $M_\odot$ yr^{-1} for stars with an age of 10^6 yr, and roughly decreases with age [102]. The lifetime of that phase of evolution when the disk radiates significantly in the infrared ranges from < 3 Myr to 10 Myr, for stars in the solar-mass range and below, with a median of about 3 Myr. The frequency of occurrence of disks around young stars with masses < 3 $M_\odot$ and with ages < 3 Myr is estimated to be 30–50%. Other significant properties are the distributions with radius of the gas surface density Σ_{gas} (R, t), the solid surface density Σ_{solid} (R, t) and the midplane temperature.

Disk sizes can be obtained best from direct imaging, of which there are many examples. The first announcement of a resolved disk around a young star [444] was based on radio maps of the ^{13}CO line emission of the star HL Tauri, which showed an elongated structure with a radius of about 2,000 AU and a mass estimated at 0.1 $M_\odot$. It was later determined that the velocities in that structure represent a rotating infall, so that the "disk" is not in equilibrium but is collapsing onto a small inner disk [211]. In a few other cases for disks resolved in the CO lines it was possible to observe the variation in orbital velocity as a function of distance to the star and to determine that the disk is rotating with a nearly keplerian velocity pattern (AM Aur [260], GG Tau [149]); here sizes of several hundred AU are indicated for disks near equilibrium. Disks can also be resolved in the radio continuum; a good example is the binary in L1551 IRS5 [440], imaged at the VLA at 7 mm (Fig. 1.10). In this case the disk size is only 10 AU, limited by the scale of the binary orbit.

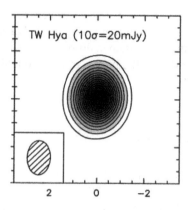

Fig. 4.5 A map of the dust continuum emission at 1.3 mm of the disk around the young star TW Hydrae. The distance scale is in arc seconds. The beam size is indicated in the lower-left corner, giving a resolution of about 1 arc sec. The observation was made by the Submillimeter Array on Mauna Kea, Hawaii. The disk outer edge is defined roughly by the radius that contains 95% of the observed emission. From these observations and theoretical models, the outer radius of the disk is determined to be 73 AU, and its mass 0.03 $M_\odot$. This disk is observed nearly face-on: the inclination of its plane to the plane of the sky is only 7°. The lowest contour corresponds to a flux level of 20 mJy and a confidence level of 10σ. Contours are separated by the same amount. Reproduced, by permission of the AAS, from [239]. © 2009 The American Astronomical Society

An HST image of the young star AB Aurigae (Fig. 1.15), using a coronagraph to mask the central star, shows a disk with a radius of about 1,300 AU. The coronagraph in the STIS instrument on the Hubble Space Telescope was used to image the disk around TW Hydrae, a nearby T Tauri star, in scattered starlight [437]. The disk has been resolved in the radio at the VLA at 7 mm [553] which represents emission from dust and also at 1.3 mm (Fig. 4.5). The scale of the emission region is ≈100 AU. Observations at higher spatial resolution (≈0.16 arc sec) at 7 mm wavelength [236] show evidence for a central "hole" in the dust emission out to about 4 AU. The presence of the hole was earlier suspected because of a deficit of flux in the spectral energy distribution at 2–20 μm, indicative of the lack of small dust [100] in the inner, warm regions of the disk. Nevertheless, there seems to be at least some gas in the region around 1 AU, as deduced from observations of CO [432]. This disk is of particular interest because it is nearby (51 pc) and it is thought to have properties similar to those of the primordial solar disk. The Hubble Space Telescope has provided other spectacular optical images of disks, particularly in the Orion region [350, 394]. Note in particular the flared structure of the disk (Fig. 1.7) associated with HH30 [97], which again has a scale of 200 AU.

"Debris disks" have also been resolved by HST; two good examples are the disk around the main-sequence A star Beta Pictoris (Fig. 1.16), of size 1,000 AU with an inner hole of 15 AU, and that around the main-sequence M star AU Mic [268], of size 200 AU with an inner hole of 12 AU. The first image of a debris disk, that of Beta Pictoris [468], was actually obtained with a telescope at Las Campanas Observatory in Chile through use of a coronagraph which blocked out

4.2 Observations of Disks

much of the direct starlight. The star had previously been observed in the early 1980s, by the Infrared Astronomical Satellite (IRAS), to have an infrared excess. This type of disk is quite distinct from the disks observed around young stars: they are found around main-sequence stars, the disks have very low mass, and very little gas is present. They are observed in scattered starlight from the small dust grains. Presumably these disks are the evolved remnants of the more massive, dusty, gaseous disks around young stars. Roughly 10–20% of main-sequence stars of solar type are observed to have such disks [509]. They are observed around stars with age range 10–1,000 Myr, and their dust masses, which roughly decline with age, are $0.1 - 10^{-4}\ M_\oplus$. However the underlying mass must be greater; the origin of the small dust grains is thought to be from collisions of larger objects in the meter-size range. The small dust particles, if not replenished by such collisions, would spiral into the star as a result of the Poynting–Robertson effect, be blown out by radiation pressure, or be destroyed by collisions.

These disks have been extensively studied by the Spitzer Space Telescope and have considerable importance in providing clues to the formation and evolution of planetary systems [363], but they will not be discussed in detail here.

Most of the known disks are not directly imaged but are detected indirectly. First, many young objects exhibit excess radiation at infrared, submillimeter, and millimeter wavelengths, above what is expected from a normal stellar photosphere. Spectral energy distributions derived from disk models agree with those observed, as discussed below. The disk radiation arises either from reprocessing of stellar radiation in the surface layers of the disk, or internal energy generation within the disk, for example by viscous dissipation. The amount of absorption of the radiation from the star implied by the infrared excess implies that if the absorbing material were spherically distributed, the star would not be optically visible. For a typical young star the visual extinction A_V would be about 500 magnitudes. The fact that the typical young star has $A_V < 3$ leads to the conclusion that the geometry of the circumstellar material must be highly non-spherical. The radiation arises from emission from dust particles. However, second, spectroscopy of relatively few young stars shows the presence of circumstellar gas, although this is much harder to detect than the dust. Third, a large fraction of the young stars with infrared excess also show ultraviolet excess radiation over that from a normal photosphere. This radiation is interpreted as arising from material falling from the disk onto the star, probably channelled along magnetic field lines, and producing shock waves on the stellar surface. We now discuss the estimation of disk masses, disk ages, and accretion rates onto the star, based on the above observations.

Disk masses [43] are obtained from long-wavelength continuum emission, in the millimeter range, to which the disks are optically thin. Figure 4.6 shows a histogram of the total disk mass, gas plus dust, in two star forming regions, Taurus and Ophiuchus [17, 45]. In the Orion region, the mass distribution is similar [338, 339]. The key equation for determining the disk mass is

$$F_\nu = \frac{L_\nu}{4\pi D^2} = \frac{1}{D^2} \int_{R_{\min}}^{R_{\max}} B_\nu[T(R)]\tau_\nu(R) 2\pi R dR \qquad (4.36)$$

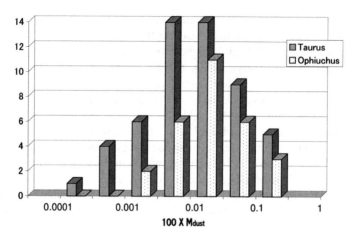

Fig. 4.6 Histogram of the estimated masses of disks, in solar masses, for Class II objects in the Taurus and Ophiuchus star-forming regions. Reproduced from [67]. Original data from [17, 45]. © Annual Reviews

where F_ν is the flux density received at earth (erg s^{-1} cm^{-2} Hz^{-1}), D is the distance to the source, R is the radial distance away from the central star, and τ_ν is the optical depth along the line of sight at frequency ν, for example, at 231 GHz ($\lambda = 1.3$ mm). This expression assumes an optically thin slab with a black-body source function B_ν which is a function only of distance from the star. The emergent intensity is $I_\nu = B_\nu \tau_\nu$, derived directly from the equation of transfer (it is multiplied by 4π to obtain the emission over all directions). One is observing the dust emission, integrated over the entire volume of the disk, which is not spatially resolved in the observation. Assume for simplicity that the disk is observed face-on. If the opacity, in cm^2 g^{-1} of dust, is $\kappa_{\nu,dust}$, then the optical depth is

$$\tau_\nu = \kappa_{\nu,dust} \Sigma_{solid}(R) \quad (4.37)$$

where $\Sigma_{solid}(R)$ is the dust surface density in g cm^{-2}. If one further assumes that the Planck function can be calculated in the Rayleigh-Jeans limit, so $B_\nu = 2k_B T \nu^2/c^2$, where k_B is the Boltzmann constant and c is the velocity of light, then the flux density becomes

$$F_\nu = \frac{\kappa_{\nu,dust} \cdot 2k_B \nu^2}{c^2 D^2} \int_{R_{min}}^{R_{max}} T(R) \Sigma_{solid}(R) 2\pi R \, dR \quad (4.38)$$

or

$$F_\nu \approx \frac{\kappa_{\nu,dust} \cdot 2k_B \nu^2 \langle T \rangle}{c^2 D^2} M_{dust} \quad (4.39)$$

4.2 Observations of Disks

where M_{dust} is the total dust mass and $\langle T \rangle$ is a suitable temperature average over the disk. This formula gives the basic expression for the disk mass. More generally [225]

$$M_{disk} = \frac{F_\nu D^2}{\kappa_\nu B_\nu(T_{dust})} \quad (4.40)$$

where T_{dust} is again a suitable average over the disk. This form of the equation can be used for other optically thin structures, such as protostellar envelopes or molecular cloud clumps. Here κ_ν is expressed in cm^2 per gram of material (gas plus dust), and therefore M is the total disk mass, gas plus dust. The assumption is usually made that the ratio of gas to dust by mass is about a factor 100.

Now one assumes that the temperature distribution and the surface density distribution can be written as power laws:

$$T(R) = T_0 \left(\frac{R}{R_{min}}\right)^{-q} \quad \text{and} \quad \Sigma(R) = \Sigma_0 \left(\frac{R}{R_{min}}\right)^{-p}, \quad (4.41)$$

where T_0 and Σ_0 are the values at the reference point R_{min}. Then one can calculate the average T (see [43] for details)

$$\langle T \rangle \approx T_0 \left(\frac{R_{max}}{R_{min}}\right)^{-q} \frac{2-p}{2-q-p} \quad (4.42)$$

where it has been assumed that $R_{min} \ll R_{max}$, which is reasonable since $R_{min} \approx 1$ AU for the dust while $R_{max} \approx 100$ AU. Standard temperature and density distributions give $q = 0.5$ and $p = 1$ so

$$\langle T \rangle \approx 2T(R_{max}) \quad (4.43)$$

and

$$F_\nu = \frac{\kappa_{\nu,dust} \cdot 4k_B \nu^2 T(R_{max})}{c^2 D^2} M_{dust}. \quad (4.44)$$

Typically the temperature at the outer edge of a disk is about 20 K. One still needs to provide an opacity law: a power law is usually assumed, but the dust opacity is still quite uncertain, by up to a factor 5:

$$\kappa_\nu = \kappa_0 (\nu/\nu_0)^\beta = \kappa_0 (\lambda/\lambda_0)^{-\beta}. \quad (4.45)$$

or, for example,

$$\kappa_\nu = 0.02(1.3\text{mm}/\lambda) \text{ cm}^2\text{g}^{-1}. \quad (4.46)$$

Here one has corrected for the fact that the total mass of the disk (gas plus dust) is about 100 times the dust mass, from which one obtains the final result

$$M_{disk} \approx 0.03 M_\odot \left(\frac{F_\nu}{1\text{Jy}}\right) \left(\frac{D}{100\text{pc}}\right)^2 \left(\frac{\lambda}{1.3\text{mm}}\right)^3 \left(\frac{50K}{\langle T \rangle}\right) \left(\frac{0.02}{\kappa_{1.3\text{mm}}}\right) \quad (4.47)$$

where the dust opacity is usually quoted to be uncertain by a factor of a few.

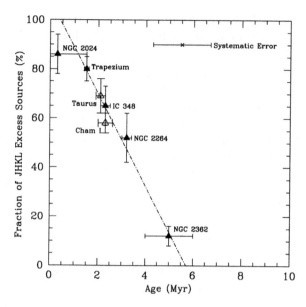

Fig. 4.7 The fraction of stars in a young cluster that show evidence of the presence of a circumstellar disk, as indicated by near infrared measurements in the J, H, K, and L bands. Ages of individual clusters are determined by fitting each source in the cluster to a given set of pre-main-sequence stellar evolutionary tracks, then taking the mean. The "systematic error" indicates the additional error that would be introduced by using different available sets of the evolutionary tracks. Reproduced by permission of the AAS from [199]. © 2001 The American Astronomical Society

Does the disk mass correlate with the mass of the central star? Some evidence indicates that it does [388]. Although there is a large scatter in disk masses for a given stellar mass (more than an order of magnitude, and there is an age spread as well) there is a trend suggesting $M_{disk} \propto M_*^{0.6}$. The typical disk mass is only a few percent of the stellar mass, well below the mass where gravitational instability would be important, as Fig. 4.6 indicates.

The lifetime of disks is an important constraint on the planet formation process. Figure 4.7 shows [199], for various clusters with a range of ages, the fraction of the stars in the cluster that have disks, as deduced from near-IR excess (2–4 µm). At an age of 3 Myr, only half of the stars have disks. Other data indicate, however, that disks can be present in stars of ages up to about 10 Myr (example: TW Hydrae; see above). However, disks have a temperature distribution, ranging from 1,500 K in the inner region to only 20 K in the very outer regions. The near-infrared emission arises primarily in the inner 0.1 AU of a typical disk; one would also like information at longer wavelengths.

Spitzer data, extending farther into the infrared, and therefore sampling regions farther out in the disks, suggest a similar trend [227, 403]. That is, the characteristic time scale for disks to dissipate is ≈5 Myr, and very few disks remain at 10 Myr.

4.2 Observations of Disks

A very large fraction of stars whose disks are not detected in the near infrared are also not detected at longer wavelengths, 3.6–70 μm, which strongly suggests that significant disks are not present out to 10–20 AU. Similar studies [21] extend the work out to the submillimeter, corresponding to distances of 50–100 AU. Again in the Taurus region, almost all disks that are not detected in the near IR are also not detected at these long wavelengths. The combined data indicate that there is a considerable spread in disk lifetimes, from 1 to 10 Myr.

However the IR and submillimeter observations detect emission from dust, not gas. What is more important is the mass of the gas as a function of time. It is very difficult to measure gas masses, although the presence of gas has been detected. There are only very preliminary indications that the decline of the gas mass roughly follows the decline in the infrared excess. Further information is expected from the Herschel satellite, which has the capability of detecting spectral lines in the far infrared.

It has long been known [219] that young stars have excess continuum radiation in the blue spectral region, where it fills in the photospheric absorption lines, and in the ultraviolet ($\lambda < 380$ nm). The excess radiation was associated with the boundary layer [332] through which disk material accretes onto the star, and later a detailed model was constructed [462] involving flow along stellar magnetic field lines, connecting star and disk, generally out of the plane of the disk. The hot radiation itself is interpreted to arise from shocks associated with disk material landing on the star, The ultraviolet observations can be used to determine approximate accretion rates ($\dot{M}_{disk}$) of disk gas onto the star (derived from $L_{acc} \approx GM\dot{M}_{disk}/R$). The accretion luminosity is derived from the excess radiation in the U photometric filter, above the expected photospheric value, and M and R for the star are derived from its position in the H–R diagram and comparison with theoretical evolutionary tracks. What is being measured is the accretion of gas. Data from a number of star-forming regions [99] are shown in Fig. 4.8. The data clearly show a decline of accretion rate with age, and $\dot{M} \approx 10^{-8} M_\odot$ yr^{-1} at 1 Myr. The solid line in this figure is not a least squares fit but rather the expected decline of $\dot{M}$ with age from a theoretical viscous "alpha" model (see below). The figure includes a range of masses; other observations [376, 389] indicate that there is a fairly strong mass dependence, $\dot{M} \propto M^2$, but with a large spread at any given mass. As a specific example, the object TW Hydrae is accreting at $5 \times 10^{-10} M_\odot$ yr^{-1} and has an age of about 10^7 yr. The determination of $\dot{M}$ is important because it helps to constrain the physical mechanism responsible for disk evolution. Clearly, these low rates imply that the star accretes only a small fraction of its total mass during its lifetime in the Class II phase.

The observed spectral energy distribution in objects with disks can provide information on the distribution of surface density and temperature. It is usual procedure to parametrize the disk surface density and temperature by power laws: $T \propto R^{-q}$ and $\Sigma \propto R^{-p}$, where R is the distance to the star. The power q is easy to determine from the observed SED, but the power p is more difficult since the spectrum is insensitive to it [260, 553]. Often it is simply assumed to be 1 or 1.5, as obtained from theoretical models, although such models in general do not

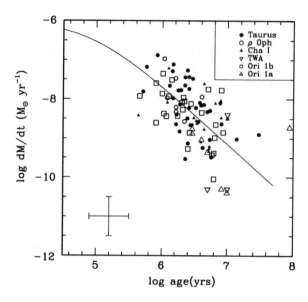

Fig. 4.8 Mass accretion rate of disk onto star as a function of age for a number of young objects in various star-forming regions, as indicated in the legend. The *solid line* is a theoretical disk evolution model [204]. Reproduced by permission of the AAS from [99]. © 2005 The American Astronomical Society

find $\Sigma(R)$ consistent with a single power law. If the disk is resolved, say in the millimeter spectral range, then the observed emission as a function of radius can be used to constrain p. In the case of the nearby TW Hydrae disk it is found that $p \approx 1$ and $q \approx 0.5$ [553]. Comparison of model SED's with observations is discussed in the next section.

4.3 Basic Theory of Disk Evolution

We assume that the evolution of disks is driven by gravitational instability during its earliest stages, when the disk is relatively massive. At later times, when much of the disk mass has been transferred to the star or lost in an outflow, viscous and/or magnetic torques control the evolution. This assumption applies for most Class II objects, whose observed disks have low mass. A useful approach, at this later phase, is to take cylindrical coordinates (R, Z, ϕ) and to assume that all variables are constant as a function of ϕ. Then the problem can be broken up into two one-dimensional subproblems, one for the evolution of the surface density, expressed just as a function of radius R (distance from star), and the second for the vertical structure of the disk (quantities as a function of Z) at a given R.

The basic equation for the radial evolution of a disk was derived [332] under the assumptions that the disk is vertically thin, is in Keplerian rotation at all radii,

4.3 Basic Theory of Disk Evolution

and has a negligible mass compared with that of the central object. The purpose of the following derivation (after [332]) is to show how viscosity and shear result in angular momentum transport and mass transport in the disk.

Define the total (gas plus dust) surface density $\Sigma(R)$ by $d\Sigma = \rho(Z)dZ$, where ρ is the mass per unit volume. The assumption of Keplerian rotation gives the angular velocity

$$\Omega = \left(\frac{GM}{R^3}\right)^{1/2}, \quad \text{and} \quad R\frac{\partial \Omega}{\partial R} = -\frac{3}{2}\Omega \qquad (4.48)$$

where M is the mass of the central star. The differential rotation produces a shear, and if viscosity is present, a frictional force is exerted. The frictional force per unit length around the circumference of an annulus is defined to be proportional to the viscosity times the rate of shear:

$$f = \nu \Sigma R \frac{\partial \Omega}{\partial R}. \qquad (4.49)$$

The above expression defines the viscosity coefficient ν which is in units of cm^2 s^{-1}. The torque exerted on an annulus by material just interior to it is given by

$$g = R \nu \Sigma \, 2\pi R (- R \frac{\partial \Omega}{\partial R}). \qquad (4.50)$$

The angular momentum of an annulus is increased by the torque exerted by material interior to it, and reduced by the torque it exerts on the material external to it. This is the basic mechanism for angular momentum transport in a disk. If the net torque on an element of mass is negative, it loses angular momentum and migrates inward in the disk. At the same time, angular momentum is transferred outwards. The actual viscosity has so far not been specified.

Setting the rate of change of angular momentum of an annulus equal to the net torque:

$$\dot{j} = \frac{D}{Dt}(jdm) = -dR\frac{\partial g}{\partial R} = dm\frac{Dj}{Dt} \qquad (4.51)$$

where the mass element, fixed in time, is $dm = 2\pi R \Sigma(R) dR$, and j is the specific angular momentum. (D/Dt is the Lagrangian derivative, following a mass element.) But

$$dm\frac{Dj}{Dt} = dm(\frac{\partial j}{\partial t} + \mathbf{v}\cdot \text{grad} j) = dm(v_R \frac{\partial j}{\partial R}) \qquad (4.52)$$

where v_R is the radial drift velocity. The specific angular momentum at a given radius is $\sqrt{GMR}$, independent of time, if the disk has negligible mass. Thus the equation for $\dot{j}$, divided by dR, becomes

$$2\pi R \Sigma v_R \frac{\partial j}{\partial R} = -\frac{\partial g}{\partial R}. \qquad (4.53)$$

Using $\frac{\partial j}{\partial R} = \frac{1}{2}\Omega R$ and the Keplerian Ω one can solve this equation for v_R:

$$v_R = -\frac{3}{\Sigma R^{1/2}}\frac{\partial}{\partial R}(\nu \Sigma R^{1/2}). \tag{4.54}$$

Using the continuity equation for the surface density

$$\frac{\partial \Sigma}{\partial t} + \frac{1}{R}\frac{\partial}{\partial R}(\Sigma R v_R) = 0, \tag{4.55}$$

one obtains the fundamental equation

$$\frac{\partial \Sigma}{\partial t} = \frac{3}{R}\frac{\partial}{\partial R}\left[R^{1/2}\frac{\partial}{\partial R}(\nu \Sigma R^{1/2})\right]. \tag{4.56}$$

This equation can be solved for $\Sigma(R,t)$ once an initial distribution is specified. The continuity equation can also be used to obtain the mass accretion rate as a function of R:

$$\dot{M} = -2\pi R \Sigma v_R. \tag{4.57}$$

However additional information must be provided to determine the viscosity ν. To obtain it we consider the vertical structure of the disk. At each radius, the disk is assumed to be in hydrostatic equilibrium, to be optically thick, and to be in vertical thermal balance, with the heat generated by viscous friction at a given radius balanced by radiative loss at the surface of the disk at the same radius. Thus, vertical, but not horizontal, transfer of heat is considered. Hydrostatic equilibrium requires that

$$\frac{1}{\rho}\frac{\partial P}{\partial Z} = -\frac{GMZ}{(R^2+Z^2)^{3/2}}. \tag{4.58}$$

The conservation of mass requires, as above, that $\partial \Sigma / \partial Z = \rho$. To write the energy equation, we define F as the vertical radiative flux, in erg cm^{-2} s^{-1}, and assume that all of the energy released by viscous dissipation is radiated in the vertical direction, so that

$$\frac{\partial F}{\partial Z} = \left(R\frac{\partial \Omega}{\partial R}\right)^2 \nu \rho \tag{4.59}$$

which essentially says that the divergence of the flux equals the energy generation rate per unit volume. The dissipation rate on the right-hand side has been derived by [298]. As a final relation we assume that the radiative transfer in the Z-direction is governed by the standard diffusion equation derived for stellar interiors, valid in material that is quite optically thick:

$$F = -\frac{4ac}{3\kappa_R \rho}T^3 \frac{\partial T}{\partial Z} \tag{4.60}$$

4.3 Basic Theory of Disk Evolution

where c is the velocity of light, a is the radiation density constant, and κ_R is the Rosseland mean opacity in cm^2 g^{-1}. Four equations have thus been set up for the gradients in T, F, Σ, and P.

The vertical structure equations can be simplified considerably into essentially a one-zone model, so that the evolution (4.56) can be solved as a function of only one spatial variable (R). If we assume that the ideal-gas equation holds, we have the pressure $P = R_g \rho T/\mu$ and the sound speed $c_s^2 = R_g T/\mu$. For a thin, vertically isothermal, disk, (4.58) simplifies to

$$\frac{c_s^2}{P} \frac{\partial P}{\partial Z} = -Z\Omega^2 \tag{4.61}$$

which has the solution

$$\frac{P_{surf}}{P_{cent}} \approx \exp\left(-\frac{Z^2 \Omega^2}{2c_s^2}\right) \tag{4.62}$$

where P on the left is evaluated at the surface of the disk and at the midplane. The vertical height in the disk at which the pressure has decreased by a factor e is known as the scale height H; we identify H with Z_{surf} and therefore

$$H \approx \sqrt{2} c_s / \Omega, \tag{4.63}$$

where c_s is evaluated at the midplane. This expression is commonly used in approximate disk analyses, usually without the $\sqrt{2}$.

The remaining equations are easily simplified. The mass equation $\partial \Sigma / \partial Z = \rho$ integrates to

$$\Sigma = \int \rho dZ \approx \rho_c \cdot 2H, \tag{4.64}$$

where ρ_c is the midplane density. The energy equation (4.59) becomes (using $dF \approx F$, the mass equation, and the Stefan–Boltzmann law)

$$2\sigma_B T_{\text{eff}}^4 = \frac{9}{4} \nu \Sigma \Omega^2, \tag{4.65}$$

where σ_B is the Stefan–Boltzmann constant and T_{eff} the surface temperature of the disk. If we define the optical depth from surface to midplane as $\tau_c = 0.5 \kappa_c \Sigma$, then it can be shown that in the limit of large τ_c the radiation equation (4.60) reduces to

$$T_c^4 \approx \frac{3}{4} \tau_c T_{\text{eff}}^4 \tag{4.66}$$

where T_c is the central (midplane) temperature.

The viscosity is usually calculated [453] according to the assumption

$$\nu \approx v_{turb} l \approx \alpha_{\text{visc}} c_s H \approx \alpha_{\text{visc}} \frac{c_s^2}{\Omega} \tag{4.67}$$

where v_{turb} is the average turbulent velocity, l is the typical turbulent length scale and α_{visc} is an arbitrary coefficient < 1, representing the ratio of turbulent speed to sound speed. The value of α_{visc} can be estimated through more detailed calculations of MHD turbulence (e. g. [26]) or estimated for gravitational instability [323]. Given the basic variables Σ and R, the reduced set of vertical equations can be then solved for ρ_c, T_c, ν, H, and T_{eff}. Thus, if the Σ distribution is given at some initial time, the evolution of Σ and other quantities in the disk can be solved for, in one space dimension, as a function of (R, t) with the use of (4.56).

The viscous time scale is given by

$$t_{visc} \approx \frac{R^2}{\nu} \approx \frac{1}{\alpha_{visc}\Omega} \left(\frac{R}{H}\right)^2 \tag{4.68}$$

which, for typical $R/H \approx 20$ and $\alpha_{visc} = .01$ gives several thousand orbital periods. This long time scale of course presents problems when one is calculating with a hydro code with time steps limited by the CFL condition (4.22).

The heat dissipated by viscosity in the disk must eventually be derived from the gravitational energy of the matter in the disk; thus some of the mass must flow inwards. However the action of viscosity also results in some transfer of angular momentum outwards; thus some mass must flow outwards to take up this angular momentum. Thus the radial drift velocity v_R is in general negative in the inner part of the disk, and positive beyond some radius in the outer part of the disk. Solutions of the evolution equation show that after a relatively short transient phase, the inner regions of the disk approach steady state, and it can be assumed that mass is flowing inward at a constant (as a function of radius) rate $\dot{M}$. If M is the mass of the central star, then the energy dissipation rate at a given radius is $\dot{E} = -GM\dot{M}/R$ and, differentiating with respect to R,

$$\Delta \dot{E} \approx \dot{M} \frac{GM}{R^2} \Delta R \tag{4.69}$$

which represents the energy dissipated per unit time as a mass element moves over a distance ΔR. Assuming that the energy dissipated there is radiated vertically and released at the surface of the disk, then the energy radiated is

$$\Delta \dot{E}_{rad} = 4\pi R \Delta R \sigma_B T_{eff}^4 \tag{4.70}$$

since the disk has two sides. Equating these, one obtains

$$T_{eff}^4 \approx GM\dot{M}/(4\pi R^3 \sigma_B).$$

Then $T_{eff} \propto R^{-3/4}$, a result which is independent of the prescription for the viscosity. A more accurate calculation, taking into account an inner boundary condition where the shear vanishes [425] gives

4.3 Basic Theory of Disk Evolution

$$T_{\text{eff}}^4 = \frac{3GM\dot{M}}{8\pi\sigma_B R^3}[1 - (R_{\text{in}}/R)^{0.5}] \quad (4.71)$$

where R_{in} is the radius of the inner edge of the disk. This boundary condition assumes that the disk extends almost all the way in to the star, and that just outside the star its angular velocity, increasing inwards to that point, reaches a maximum and rapidly declines to match Ω at the slowly rotating stellar surface. At the maximum there is no shear and no torque, which defines the location of R_{in}.

Specification of an inner boundary condition for the disk is problematical, because the star-disk interaction at the boundary layer is not fully understood. A reasonable scenario is that the disk is truncated near the star by the stellar magnetic field lines [462], roughly where the ram pressure associated with the accretion flow is balanced by the magnetic pressure. The truncation radius is given by

$$R_t = R_*\alpha_t \left(\frac{B_*^4 R_*^5}{GM_*\dot{M}^2}\right)^{1/7} \quad (4.72)$$

where the subscript $*$ refers to stellar properties, α_t is a dimensionless constant of order unity, and $\dot{M}$ is the accretion rate of the disk. For the typical magnetic field of a young star ($B_* \sim 1$ kilogauss) and the typical accretion rates deduced for these stars, the truncation point is a few stellar radii out, or approximately 0.05–0.1 AU. Inside that volume, known as the magnetosphere, completely different physics applies, as disk material is lifted out of the disk plane by magnetocentrifugal effects, and is channelled to the stellar surface, generally arriving at latitudes well above or below the equator.

One can integrate (4.71), assuming that the disk radiates as a black body at each radius, to get the total luminosity of the disk,

$$L = \int_{R_{\text{in}}}^{R_{\text{out}}} \sigma_B T_{\text{eff}}^4 4\pi R dR = \frac{GM\dot{M}}{2R_{\text{in}}} \quad (4.73)$$

which applies to the region of the disk that is in steady state. Only half of the total gravitational potential energy of the disk material is radiated. The rest is stored as rotational energy and is released when the orbiting disk material settles down onto the slowly rotating stellar surface. As an example, the disk provides about $0.1 L_\odot$ for a central mass of $1.0 M_\odot$, a stellar radius of $2 R_\odot$, and $\dot{M} = 1 \times 10^{-8} M_\odot$ yr^{-1}.

If $\dot{M}$ is too low, then the disk luminosity generated by internal viscous heating will be less important than that provided by re-radiation of stellar light from the disk. It turns out, by coincidence, that a flat disk without internal heating and with surface temperature determined by direct irradiation from the central star, also has $T_{\text{eff}} \propto R^{-3/4}$. The assumptions made are that the disk is infinitely thin and that it absorbs all the stellar radiation falling upon it (a "flat, black" disk). Then, to maintain thermal balance, it reradiates the same energy in the infrared. As one moves in the disk plane away from the star, the angle subtended by the star becomes smaller and smaller, and the vertical component of the stellar flux, which hits the disk, drops off as R^{-3} (see, for example [481]). If R_* and T_* are the stellar radius and surface temperature, respectively, the flux intercepted (and absorbed) by the disk is, assuming $R \gg R_*$,

$$F_d = \frac{\sigma T_*^4}{\pi} \frac{2}{3} \left(\frac{R_*}{R}\right)^3$$

Equating this flux to that radiated by the disk, σT_{eff}^4 gives

$$T_{\text{eff}} = \left(\frac{2}{3\pi}\right)^{1/4} T_* \left(\frac{R_*}{R}\right)^{3/4}. \tag{4.74}$$

A more accurate calculation shows that the disk intercepts and reradiates about 1/4 of the stellar luminosity [7], so in many cases this source will dominate the accretion luminosity. Setting the re-radiated disk luminosity equal to the internal luminosity and putting in numbers for $L_* = L_\odot$ and $R_* = 2 R_\odot$, for a solar-mass central star the critical $\dot{M} = 3.3 \times 10^{-8} M_\odot$ per year. Below that value, the re-radiation dominates and the disk is known as a "passive" disk.

However if $T_{\text{eff}} \propto R^{-3/4}$, then the sound speed along the surface of the disk $c_s \propto R^{-3/8}$. If the disk is vertically isothermal, then the simple vertical structure relations give

$$\frac{H}{R} \approx \frac{c_s}{\Omega R} \propto R^{-3/8} R^{1/2} \propto R^{1/8} \tag{4.75}$$

so the ratio of scale height to radius increases with radius, which is what is known as a "flared" disk. Because of flaring the disk will intercept more radiation than under the assumption of a flat disk, so the surface temperature will vary less rapidly than $R^{-3/4}$. Thus the radiation from the outer regions of typical disks tend to be dominated by stellar re-radiation rather than by internal heating.

The disk continuous spectrum can be obtained under the assumption that each annulus radiates as a black body at the local T_{eff}, so the luminosity as a function of wavelength L_λ is

$$L_\lambda = \int_{R_{\text{in}}}^{R_{\text{out}}} \pi B_\lambda[T(R)] 2\pi R dR. \tag{4.76}$$

The resulting spectrum, assuming $T_{\text{eff}} \propto R^{-3/4}$, is shown in Fig. 4.9, along with an assumed stellar black body with $T_* = 4{,}000$ K. The disk inner edge is truncated at $8 R_\odot$, while the star has $2 R_\odot$. What actually is plotted is λF_λ, where F_λ is the flux density received at the earth from a source in the Taurus star formation region, 140 pc away. In the infrared, the spectrum shows that $\lambda F_\lambda \propto \lambda^{-4/3}$, which is the classical (steady-state) disk spectrum. The spectrum falls off at short wavelengths because there is an inner temperature cutoff in the disk.

This type of theoretical spectrum does not fit most observed sources, which show a fairly wide variety of slopes and scatter in the $(\log \lambda, \log \lambda F_\lambda)$ plane. Figure 4.9 also shows the combined SED's of 7 classical T Tauri stars [8], along with another

4.3 Basic Theory of Disk Evolution

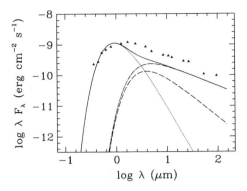

Fig. 4.9 The curves give idealized spectral energy distributions in a star-disk system. *Dotted curve*: Black-body spectrum of a star with $T_{\text{eff}} = 4{,}000$ K. *Dot-dashed curve*: Spectrum of a disk whose surface temperature decreases with $R^{-0.6}$ and which radiates as a black body at each annulus. *Long dashes*: spectrum of a disk whose surface temperature decreases with $R^{-3/4}$ and which radiates as a black body at each annulus. *Solid curve*: sum of the stellar spectrum and that of the disk with $T \propto R^{-0.6}$. *Triangles*: Averaged observed spectral energy distributions of seven typical T Tauri stars [8]. The theoretical curves are not meant to match the data; the important point to note is the slope of the curves in the IR (10–100 μm)

theoretical spectrum calculated with $T_{\text{eff}} \propto R^{-0.6}$, which provides a reasonably good fit with the observed slope of about $-2/3$. The relation between the infrared slope of a disk spectrum in this plane, s, and the surface temperature distribution of the disk, $T_{\text{eff}} \propto R^{-q}$ is

$$-s = 4 - \frac{2}{q}.$$

Figure 1.13 shows the spectrum of T Tauri, an example of a flat-spectrum source. The fit to that part of the spectrum is obtained with a radial distribution of surface temperature (in the disk) of $T_{\text{eff}} \propto R^{-0.515}$. This solution is approximately consistent with a flared disk which has more radiation at long wavelengths than does a "flat, black disk" spectrum.

A modified disk model [111] includes calculation of the flaring and the temperature distribution of a passive disk under the assumption that a warm, optically thin layer of dust rests above the disk photosphere. This layer, above the disk midplane by a few gas scale heights, re-radiates some of the starlight outward and also heats the disk below. The dust layer has a temperature T_{dust}, which is higher than the black-body temperature of the disk, because the dust absorbs optical radiation from the star efficiently, but radiates it less efficiently in the infrared. This effect results from the opacity curve as a function of wavelength shown in Fig. 3.9. Half of the radiation absorbed by the grains is re-radiated outward, and the other half heats the interior, which is at a cooler temperature T_i, vertically isothermal at a given R.

In these models the surface density distribution is assumed to be proportional to $R^{-3/2}$. The specific parameters assume a stellar mass of $0.5\,M_\odot$, a stellar radius of $2.5\,R_\odot$, and a stellar T_{eff} of 4,000 K, all typical of T Tauri stars. The results show

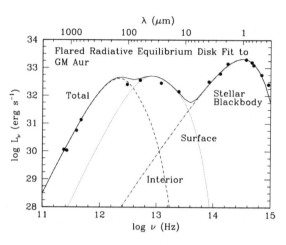

Fig. 4.10 The observed spectral energy distribution of GM Aur is given by *filled circles*. Curves refer to a theoretical model [111, 112]. *Dash-dot line*: contribution of the star. *Dotted line*: contribution of the warm dust layer. *Dashed line*: contribution of the cool disk interior. *Solid line*: the sum of the three contributions to the spectrum. Reproduced, by permission of the AAS, from [111]. © The American Astronomical Society

that the disk flares with $H/R = 0.04(R/1\mathrm{AU})^{2/7}$ inside 84 AU. The warm dust layer has temperature $T_{\mathrm{dust}} \approx 550(R/1\mathrm{AU})^{-2/5}$, and the interior disk temperature is $T_i \approx 150(R/1\mathrm{AU})^{-3/7}$ out to 84 AU, and constant at 21 K between 84 and 209 AU. An improved model [113] also shows that T_{dust} and T_i decrease approximately as $R^{-0.5}$. This model has two important consequences. First, the superheated but optically thin dust layer predicts that spectral features of the dust should appear in emission. In fact most T Tauri stars exhibit silicate, and in a few cases ice, features in emission. Second, the fact that the temperatures in the irradiated, flared, disk decrease less rapidly with distance than do the surface temperatures generated by internal viscous evolution ($T_{\mathrm{eff}} \propto R^{-3/4}$) shows that for the purpose of comparison with observations, stellar irradiation dominates internal heating beyond some radius. Even if accretion onto the star is occurring, the surface properties of disks can be calculated as if they were passively irradiated, longward of a wavelength of about 10 μm [113].

Figure 4.10 shows the comparison of the result of a theoretical face-on disk with the observed spectrum of GM Aur. The stellar photosphere dominates the short wavelengths, the dust layer dominates at intermediate wavelengths ("surface"), while the disk interior dominates at long wavelengths. The fit is good, but an inner hole in the disk, out to roughly 5 AU, is required in the model to fit the low emission at around 10 μm. The outer edge of the disk model is at 390 AU. A similar spectrum in which the known inclination of 60° is taken into account [112] requires essentially the same parameters. Improved observations of the spectral energy distribution with the Spitzer IRS instrument in the 5–40 μm spectral range led to a revised estimate

4.3 Basic Theory of Disk Evolution

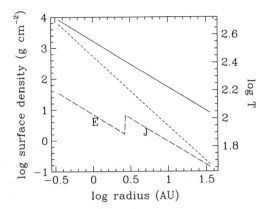

Fig. 4.11 The minimum mass solar nebula [210]. *Solid line:* gas surface density (left-hand scale) as a function of distance to the star. *Long dashed line:* solid surface density (left-hand scale). *Short dashed line:* temperature (right-hand scale). The positions of present Earth and Jupiter are indicated

of the inner edge of the optically thick disk of about 24 AU [101]. The presence of a central hole is required for a number of other young objects.

Numerous observations of disks, at various wavelengths, in combination with radiation transfer models, have been used to deduce that the grain sizes in disks are somewhat larger than interstellar grains, indicating grain growth, the first step toward planet formation.

The classical disk model known as the "minimum mass solar nebula" (MMSN) is shown in Fig. 4.11, as presented by Hayashi [210]. It is defined over the region from 0.35 to 36 AU. The solid surface density distribution is derived from estimates of the present solid masses of the planets. The surface density for each planet is then obtained by spreading out that mass from a point half-way to the next innermost planet to half-way to the next outermost planet. The resulting approximate fit is

$$\Sigma_{\text{solid}} = 7.1 (R/1\text{AU})^{-3/2} \text{ g cm}^{-2} \text{ for } 0.35\text{AU} < R < 2.7 \text{ AU}$$
$$\Sigma_{\text{solid}} = 30 (R/1\text{AU})^{-3/2} \text{ g cm}^{-2} \text{ for } 2.7\text{AU} < R < 36 \text{ AU}. \quad (4.77)$$

The gas surface density is then determined by adding sufficient gas, mainly hydrogen and helium, to each planet to bring it to solar composition:

$$\Sigma_{\text{gas}} = 1700 (R/1\text{AU})^{-3/2} \text{ g cm}^{-2} \text{ for } 0.35\text{AU} < R < 36 \text{ AU}. \quad (4.78)$$

The temperature in the Hayashi minimum-mass disk is derived under the assumption that the solid particles, which provide most of the opacity, have settled to the midplane of the disk, leaving most of it transparent to solar radiation. At each distance, the energy per unit time absorbed by a small solid particle is $L_* \pi a^2/(4\pi R^2)$ where a is the radius of the particle. Assuming that it reradiates over its entire surface, the radiated energy per second is $4\pi a^2 \sigma_B T_{\text{dust}}^4$. Equating these energies gives the temperature distribution in the disk (actually the dust temperature)

$$T = 280 \left(\frac{L}{L_\odot}\right)^{0.25} (R/1\text{AU})^{-0.5}. \qquad (4.79)$$

The jump in the surface density of solid material (lower curve) is caused by the condensation of ice at temperatures less than 170 K. This jump is known as the *snow line* and is located at 2.7 AU in the minimum mass nebula, but during actual disk evolution its location changes with time. The solid surface density at 5 AU is only $3\,\text{g}\,\text{cm}^{-2}$, a number which is used as a reference point for calculations of the formation of Jupiter by core accretion; in this model a core of heavy elements forms first by accretion of the solid particles and, once the core has reached a few Earth masses, gas is captured from the disk. Hayashi's estimate of the total mass of the disk is $0.013\,M_\odot$, but uncertainties in the solid masses of the planets lead to a possible range in that number from 0.01 to $0.07\,M_\odot$ [540]. In any case, the MMSN should be considered only as a rough approximation, because the distribution of surface density in typical evolving disk models is not a simple power law.

4.4 General Results of Disk Evolution

A simplified example of disk evolution is shown in Fig. 1.22 [323]. The one-dimensional disk evolution equation (4.56) is solved with an assumed viscosity that simulates gravitational instability:

$$\nu \approx Q^{-2} H^2 \Omega \quad \text{if } Q < 1$$
$$\nu = 0 \quad \text{otherwise} \qquad (4.80)$$

where the standard Toomre $Q = c_s \Omega/(\pi G \Sigma)$ and H is the scale height of the disk. This formulation, although very approximate, has the advantage that ν depends only on Σ and R and constants, so that equation (4.56) can be solved directly. The general result is that the surface density of the inner part of the disk decreases with time as material accretes onto the star, and the angular momentum that has to be lost by that material is transferred to matter in the outer part of the disk, which expands.

Another consideration regarding gravitational instability is the cooling time. Once the instability develops into a spiral arm pattern, shock dissipation occurs and the disk tends to heat up, resulting in an increasing Q and a tendency to return to stability. Only if the heat generated by shocks can be radiated efficiently, can the amplitude of the perturbations continue to grow and develop into fragmentation. If the cooling time is less than half an orbital period, the disk, if gravitationally unstable, can fragment [182]. Otherwise the disk evolves to a steady state with

4.4 General Results of Disk Evolution

a limited amplitude in spiral waves, and it transfers angular momentum without fragmentation.

The cooling time is the energy content per unit surface area (erg/cm^2) divided by the energy radiated per unit time per unit area. Assuming an isothermal vertical structure,

$$t_{\rm cool} = \frac{E\,\Sigma}{2\sigma_B T^4} \tag{4.81}$$

where $E = 2.5 R_g T/\mu$, the internal energy per unit mass of molecular hydrogen. Taking $\mu = 2.3$, one obtains

$$t_{\rm cool} \approx 200 \text{ yr } (\Sigma/10^3)(T/50)^{-3}. \tag{4.82}$$

Half of an orbital period is

$$t_{\rm orbit}/2 = \frac{\pi R^{3/2}}{\sqrt{GM_*}} \tag{4.83}$$

and if one equates this time with the cooling time one finds that the radius outside of which fragmentation could occur in a disk around a 1 M$_\odot$ star is 50–100 AU, depending on how Σ and T drop off with R. However, for the correct treatment of the cooling time, detailed numerical simulations with radiation transfer are required, taking into account the details of the opacity in the disk. The possibility of fragmentation will be discussed again in Chap. 5.

Once the disk has become gravitationally stable, the physical processes that must be considered include hydrodynamics, the effects of magnetic field, radiative transfer, which depends mainly on the properties of the dust, chemistry, which includes gas-phase reactions as well as reactions on grain surfaces, and the degree of coupling between the evolution of the gas and the evolution of the solids. The mechanism that drives disk evolution is not fully understood, but some possibilities include hydrodynamic instability, driven by the Keplerian shear in the disk [141], baroclinic instability, driven by a sufficiently steep outward radial temperature gradient ($T \propto R^{-1}$ or steeper) plus either rapid cooling by radiation or rapid thermal dissipation [254, 415], and the magnetorotational instability [25], the last being the most likely one.

The long-term evolution of a disk is best carried out in the (1 + 1)D approximation, in which the radial evolution equation (4.56) is solved in conjunction with a set of vertical structure models, calculated at each radius and at each time. The natural time scale is thus the viscous time ($t_{\rm visc} \approx [\Delta R]^2/\nu$) where ΔR is an appropriate distance interval, and the evolution can easily be followed for more than 1 Myr. However, for a more detailed examination of the basic physical processes, 2-D [in the (R,ϕ) plane] or 3-D simulations are required. These calculations run on hydrodynamic (roughly orbital) time scales, and the maximum time that can be covered is much less than the full evolution time of $>10^6$ yr.

An example of an evolution in one space dimension based on the standard "alpha" model with driving mechanism not specified [137] is shown in Fig. 4.12. The radial diffusion equation is solved with the full vertical structure equations and

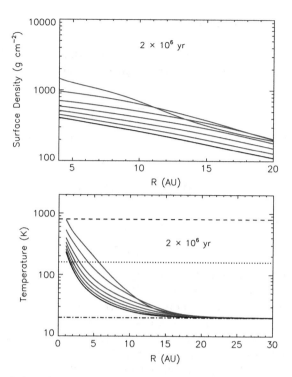

Fig. 4.12 The evolution of the gas in a disk with an assumed viscosity parameter $\alpha_{\rm visc} = 0.002$. *Top panel*: Gas surface density as a function of distance to the star, at various times. *Lower panel*: same for gas temperature. The time 2×10^6 years refers to the heavy solid line (lowest curve). Other times (top to bottom) are 3×10^4, 1.0×10^5, 3×10^5, 6×10^5, 1.0×10^6, 1.5×10^6 years. Reprinted, with permission from Elsevier, from [137]. © 2008 Elsevier, Inc.

$\alpha_{\rm visc} = 2 \times 10^{-3}$. The model includes a calculation of the evolution of the surface density of the solid component, which consists of ices of water, methane, ammonia, CO, and minor species. The plot shows the total surface density (which is dominated by the gas), and the temperature as a function of radius at various times. The initial disk mass is $0.12\,M_\odot$, the initial radius is 30 AU, and the initial gas surface density is proportional to $r^{-3/2}$. The general results of the calculations can be summarized as follows: (1) The evolution of an active disk with initial mass ≈ 0.1–$0.2\,M_\odot$ around a solar-mass star, with size comparable to that of the planetary system, takes place on a time scale of a few times 10^6 yr, for a value of $\alpha_{\rm visc}$ in the range 10^{-2}–10^{-3}, with lower $\alpha_{\rm visc}$ giving longer times. (2) After an initial adjustment period, the evolution is fairly independent of initial conditions. (3) Mass, surface density, temperature, and scale height at a given radius all decline with time, except in the very outer regions, which are expanding. (4) As the surface density decreases in the main part of the disk, the evolutionary time lengthens as a result of a decrease in the efficiency of angular momentum transport. The value of $\dot{M}$ onto the star declines, in agreement with observations (Fig. 4.8). (5) The inner regions of the disk tend

4.4 General Results of Disk Evolution

to evolve to a quasi-steady state, where $\dot{M}$ is independent of the distance to the central star. (6) In regions interior to 1 AU the temperature distribution is relatively flat, but farther out it declines as R^{-1}. (7) The initial disk is gravitationally unstable outside 30 AU, but the Q value increases with time, and for most of the evolution the disk is gravitationally stable. It is generally true that active disks of 0.1 $M_\odot$ or less are gravitationally stable inside ≈ 20 AU. (8) On the temperature plot, the dashed line at 800 K gives the minimum temperature for silicate crystallization. The dotted line at 160 K gives the sublimation temperature of H_2O. Note how this point (the "snow line"), moves inwards with time. The dash-dot line at 20 K gives the assumed ambient temperature in the surroundings of the disk.

In the absence of gravitational instability, the main alternative for driving the evolution of a disk is the magnetorotational instability (MRI), which amplifies an initially weak magnetic field and develops into turbulence [492]. A linear stability analysis [25], shows that a Keplerian flow is linearly unstable in the presence of a weak magnetic field. The Rayleigh criterion (angular momentum increasing outward) normally stabilizes a disk, but it does not guarantee stability in the presence of the field. Two- and three-dimensional numerical simulations [207, 208] extend into the non-linear regime and show growth rates in the linear regime which match those from the analytical calculations.

The MRI operates under fairly general circumstances, in particular, for the case of a Keplerian differentially rotating thin disk, magnetized by a uniform vertical field B_Z. Consider two fluid elements, located at the same distance R from the rotation axis of the disk, but separated vertically by a small amount and located on the same vertical field line. The instability is generated by a small radial displacement. Let element 1 move to $R - \delta R$ and element 2 move to $R + \delta R$. The field line is stretched and develops tension, and also it develops an azimuthal component, because element 2 is moving less rapidly in the ϕ direction than element 1. Element 1 is braked by the tension and loses angular momentum, while element 2 is speeded up, that is, the field acts in the direction to bring about co-rotation. As a result of the angular momentum exchange, element 1 moves inward even farther, and element 2 moves outward even farther, causing further stretching of the field in the R-direction, along with amplification.

The requirements for the instability are (1) the magnetic field must be coupled to the matter, (2) the angular velocity of the disk must decrease outwards, (3) there must be an initial poloidal field component present, and (4) the energy density in the field must be less than the thermal energy density. In connection with point (1), if Ohmic dissipation is present, the first term on the right-hand side of the magnetic induction equation (2.24) must be greater than the second, in other words the magnetic Reynolds number (2.26) must exceed unity. Thus the requirement can be written [521]

$$R_{em} = \frac{vL}{\eta_e} \approx \frac{V_A^2}{\eta_e \Omega} > 1 \qquad (4.84)$$

where η_e is the electrical resistivity, V_A is the Alfvén velocity, and V_A/Ω is the magnetic length scale.

The result of the instability is amplification of the field and generation of MHD turbulence, which results in outward transport of angular momentum [26], allowing an accretion flow of mass inwards. The growth time of the instability is fast, only one orbital period. One can approximate an effective viscosity by $\nu \approx V_A^2/\Omega$, and from numerical experiments the turbulence generates an effective $\alpha_{\text{visc}} \approx 10^{-2} - 10^{-1}$. Figure 4.13 shows a simple numerical example in two space dimensions of the distortion of the initially vertical field lines (solid lines). At time $t = 2$ material near $Z = 0$ has lost angular momentum and is moving inward, and material near $Z = \pm 0.4$ has gained angular momentum and is moving outward. The dashed lines are contours of equal angular momentum.

Magnetic field coupling is expected in the inner regions of the disk, inside 0.1–1 AU, where the temperature is 1,400 K or higher, as a result of a very low degree of thermal ionization. In the outer regions, perhaps beyond 10 AU, external cosmic rays can provide sufficient ionization. The cosmic rays are stopped once they have penetrated through a layer of order $\Sigma = 100\,\text{g cm}^{-2}$ [525], which accounts for the entire disk thickness in the outer regions. However in most disks there will be a magnetically inactive region near the midplane in the 1 AU to 10 AU range, where the source of viscosity is problematical. One possibility [181] is that there is still an ionized skin layer at these radii, through which accretion could progress, but that the regions near the midplane are inactive and are not accreting. It is also possible that another source of viscosity exists in the so-called "dead zone", or that turbulence generated in the surface layers by the MRI could result in transport of enough charge into the interior layers to keep those regions ionized enough to generate the MRI [521].

Suppose the MRI does not operate in a particular region of a disk. Is there any non-magnetic alternative? This question is still being actively investigated. The original proposed mechanism for angular momentum transport in disks [321] was turbulent viscosity arising from convective instability. It is clear that convection is present in disk models, driven in the vertical direction by the temperature dependence of the dust opacity, which decreases upwards. The main question is whether the convection leads to outward angular momentum transport. A critical quantity in this respect is the *Rayleigh number*, a measure of the ratio of buoyancy effects to the product of viscous effects and effects of thermal diffusivity.

$$Ra = -\frac{N^2 L^4}{\chi \nu} \qquad (4.85)$$

where χ is the thermal diffusivity [$k/(\rho c_p)$], where k is the coefficient of thermal conductivity and c_p is the specific heat at constant pressure), ν is the kinematic viscosity, L is a characteristic length, taken here to be the vertical disk thickness, and N^2 is the Brunt–Väisälä frequency, which can be written, for an ideal gas, as:

$$N^2 = -\frac{1}{\gamma \rho}\frac{\partial P}{\partial Z}\frac{\partial}{\partial Z}\ln\left(\frac{P}{\rho^\gamma}\right). \qquad (4.86)$$

4.4 General Results of Disk Evolution

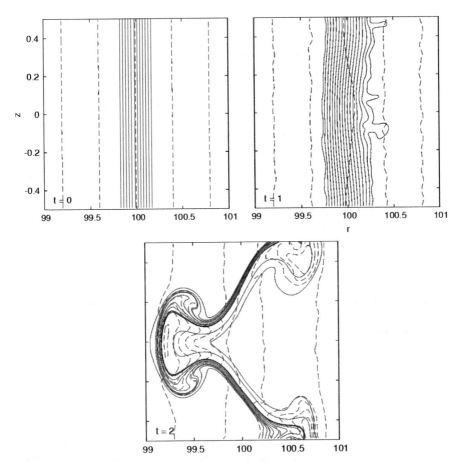

Fig. 4.13 The magnetorotational instability: a simple case in two space dimensions. The numerical simulation shown here is similar to that published by [207]. The computational domain of the axisymmetric simulation is a small section of the (R, Z) plane of a Keplerian accretion disk. The vertical component of gravity is neglected, and initial pressure and density are constant across the domain. At $t = 0$ part of the domain is filled with a uniform vertical field. Simultaneously, the temperature is adjusted such that the total pressure across the domain remains constant. *Solid curves*: field lines. *Dashed curves*: contours of constant angular momentum, with angular momentum increasing to the right at $t = 0$. At time $t = 2$ material near $Z = 0$ has lost angular momentum and is moving inward (to the left), while material near $Z = \pm 0.4$ has gained angular momentum and is moving outward (to the right). In three space dimensions the flow is more complicated but the general effect is same. Units of time and distance are arbitrary. Reproduced by permission from [66]. © 2007 Taylor and Francis Group LLC

Here Z represents the vertical direction in the disk, and γ is the adiabatic index $(\partial \ln P / \partial \ln \rho)_s$ where the derivative is taken at constant entropy. Note that if N^2 is negative, the region is unstable to convection. Studies with relatively low values of the Rayleigh number [441] show that angular momentum transport is inwards and does not allow the disk to accrete onto the star. However if the number is greater than about 10^6, the angular momentum transport can be outwards [314]. It is quite possible, but not yet completely determined, that the required conditions, that is, low viscosity and low thermal diffusivity, can exist in some regions of disks.

A second possible mechanism arises simply from the differential rotation in the disk. The argument that this drives turbulence is based mainly on laboratory experiments [433] in which fluid between two concentric rotating cylinders, with angular velocity decreasing outwards, is found to generate turbulence if the Reynolds number (R_e) is high enough. In general,

$$R_e = \frac{vL}{\nu} \qquad (4.87)$$

where ν is the kinematic viscosity, v is a characteristic velocity, and L is a characteristic length. In this case the Reynolds number is given by

$$R_e = \frac{R\Delta\Omega\Delta R}{\nu}$$

where ΔR is the distance between the cylinders, R is the average radius, and $\Delta\Omega$ is the difference in angular velocity between the two cylinders. However it is not at all clear whether such experimental results apply to accretion disks. The Rayleigh criterion states that a rotating fluid is linearly stable to axisymmetric perturbations, in the limit of zero viscosity, if

$$\frac{d(R^2\Omega)^2}{dr} > 0,$$

that is, the angular momentum per unit mass increases outwards. A Keplerian disk clearly satisfies this criterion. It is under investigation whether, nevertheless, a finite-amplitude perturbation or a non-axisymmetric mode could produce an instability [141].

A third hydrodynamic effect which is of interest is the generation of vortices in the disk. The density waves that they produce can result in outward angular momentum transport. A process that can produce vortices is the so-called *baroclinic instability*. Without going into the details of the complicated physics involved, it can arise if the density gradient and the pressure gradient in a disk are not aligned. Detailed studies [315, 415] show that the following conditions are needed: (1) there must be an entropy gradient with the entropy decreasing outward in the radial direction, which is equivalent to saying that the disk is unstable to convection in the radial direction according to the Schwarzschild criterion (8.8), or, in other words, N^2 (4.86) is negative. (2) The disk must have a radiative cooling time, or a thermal diffusion time that is relatively short, comparable to a few orbital periods. (3) The initial perturbation must be finite; that is, the instability is non-linear. Some numerical simulations show that once vortices form they can survive for hundreds of orbits; however the details as to the conditions for survival or their effectiveness in angular momentum transport are still not determined.

4.5 Disk Dispersal

Numerous mechanisms, most still under investigation, have been proposed to explain the fact that observational evidence for the presence of disks around newly-formed stars disappears once the stars reach an age of 1–10 Myr. Note that radiation from dust, rather than gas, provides most of this observational evidence. Important information on the nature of the clearing process is provided by the so-called *transition disks*. In general, a transition object has observed properties intermediate between those of classical T Tauri stars and weak-lined T Tauri stars. The usual observed characteristic is a deficiency in near-infrared radiation in comparison with normal CTTS, but normal CTTS radiation at 10 μm and longer wavelengths. The preferred interpretation of these observations is that there is an inner hole in the disk, where the dust emission is sharply reduced from that of the rest of the disk. These disks may be ones which are caught in the process of clearing. One piece of evidence to support this interpretation is that transition objects have lower accretion rates [383] than standard disks of the same mass, indicating that they are at a relatively late stage of their evolution. If so, the statistics of the number of such disks in comparison with the number of classical T Tauri stars gives an estimate of the time scale of the clearing process. This time scale is relatively short, about $1 - 2 \times 10^5$ years, as the number of transition disks is relatively small. A prime example of a transition disk is that around TW Hydrae, with an inner hole, inferred from both infrared and radio observations, of about 4 AU. Another example is GM Aurigae, where observations in the mm spectral region resolve the inner regions and show a deficit of material interior to 20 AU [237], in general agreement with the conclusions based on the spectral energy distribution discussed above.

We now summarize various mechanisms that are probably relevant to the clearing of disk holes and the eventual dispersal of a disk.

- **Grain growth**: During the evolution of the disk, whether it is turbulent or not, the small dust grains collide, stick, and grow to meter size. At the same time, once they become large enough, the particles settle to the midplane of the disk. This complicated subject will not be treated in detail, but see [138] and [541]. After a few thousand orbital periods, significant grain growth and settling will have occurred, that is, a few thousand years at 1 AU and a few times 10^5 years at 30 AU. As the settling and accretion proceed, the infrared opacity from the grains decreases considerably. Roughly if a is the grain radius, the cross section for absorption or scattering by a grain is proportional to a^2 while the mass of a grain is proportional to a^3. Thus the cross section per unit mass (opacity) decreases with grain growth. Assuming the emissivity $j_\lambda \approx \kappa_\lambda B_\lambda$, where B_λ is the Planck function and κ_λ is the opacity at wavelength λ, eventually, even in the near IR, the disk will no longer radiate appreciably and will be very difficult to observe. The disappearance of IR excess around young stars of ages 5×10^6–10^7 yr could be attributed to the growth process. This process is also one possible explanation for observed inner "holes" in disks. Thus the disk gives the appearance of having disappeared, even though no mass has been lost. A problem with this scenario is that grain growth times in the inner disk (1 AU or less) are far shorter than disk lifetimes, which means that the small grains, which provide most of the emission, must either survive or be replenished, perhaps by collisions that result in fragmentation rather than growth.

- **Planet Formation**: However, the gas would still remain, and there is a chance some of it could be incorporated into giant protoplanets. However, the total mass of the four giant planets in our solar system is only about 10% that of the minimum mass solar nebula, and giant planets are unlikely to form outside 30 AU. In the terrestrial planet zone, the time scale for planet formation up to Earth mass is much longer than observed disk lifetimes, and practically no gas is incorporated into these planets. Thus, planet formation is not an important mechanism for disk dispersal, although giant planets can open up gaps in the disk.

- **Accretion**: Accretion onto the star is observed, mainly through the ultraviolet excess, implying that gas is present in the disk. Furthermore, there is a strong correlation between the occurrence of accretion and the presence of near infrared excess in the star. Thus, indirectly there is a suggestion that loss of gas from the disk occurs on comparable time scales to the loss of small dust grains. The time scale for viscous accretion, assuming the standard "alpha" model, is $t_{vis} \approx (\Delta R)^2/\nu$. Approximating $H/R = c_s/(\Omega R)$ and assuming $H/R = $ constant $= 0.05$ then

$$t_{vis} \approx 5750 \text{ yr} \left(\frac{0.01}{\alpha_{visc}}\right) \left(\frac{R}{1 \text{AU}}\right)^{3/2}. \qquad (4.88)$$

Out to 10 AU, for the standard $\alpha_{visc} = 10^{-2}$ these times are less than 2×10^5 yr but are over 5 Myr at 100 AU. Thus the time scale depends strongly on distance, suggesting that the inner disk will clear rapidly, and will be replenished from the outer disk only on much longer time scales. This picture is inconsistent with the observed result that disks clear on similar time scales at all distances. The viscous accretion model also indicates that the accretion rate declines with time, suggesting that the transition time to an optically thin disk is long. However the observations regarding transition disks indicate the opposite.

- **Stellar winds and outflows**: Young stars with accreting disks also generate bipolar outflows, with mass loss rates roughly 0.05–0.1 times the disk accretion rates [201], thus at age 10^6 yr $\dot{M} \approx 10^{-9} M_\odot$ yr^{-1}. Ejection speeds are 100–200 km s^{-1}. The outflows do result in mass loss from the disk, but again on time scales too long to be compatible with observations. The young stars may also generate more nearly symmetric stellar winds. A rough estimate of the time scale for the stripping of the surface layers of a disk by a spherically symmetric wind [233] is over 10^7 yr for typical disk and wind parameters and is only weakly dependent on distance. In fact even the stellar wind may be collimated in the polar direction by magnetic fields, and the separate outflow of material from the disk generated by the photoevaporation process (see below) deflects the stellar wind, so that this process is unlikely to be important.

4.5 Disk Dispersal

- **Photoevaporation by nearby hot stars**: In the Orion Trapezium region there is observational evidence that material is flowing away from disks, as a result of the UV radiation from OB stars. One can distinguish between FUV radiation (6–13.6 eV) and EUV radiation (13.6–100 eV). EUV photons from the star ionize particles on the surface of the disk and heat the ionized region to about 10^4 K. The sound speed in this region is $c_s \approx 10$ km s^{-1}. If the sound speed exceeds the Keplerian orbital speed of the disk, the particles are unbound, and they flow away in a wind. In the case of FUV photons, which would dominate in the case of a nearby B star, the disk upper layer is heated to temperatures in the range 100–5,000 K, depending on the details of the heating and cooling processes. The gravitational radius, where the sound speed exceeds the escape speed, is given by

$$r_g = \frac{GM_* \mu m_p}{kT} \approx 100 \text{ AU} \left(\frac{1000}{T}\right)\left(\frac{M_*}{M_\odot}\right) \quad (4.89)$$

where μ is the mean molecular weight of the gas and m_p is the proton mass. Evidently, for the case of EUV radiation, $r_g \approx 10$ AU, however more detailed calculations of the evaporative flow [6] show that the effective escape radius is about $0.15 r_g$, a result which applies to both EUV and FUV irradiation. External photoevaporation tends to preferentially remove the outer disk, where the surface area is largest. Under the Orion Trapezium conditions, where OB stars are present and the density of stars in the cluster is high (5×10^4 stars per pc^3 in the central regions) photoevaporation can remove the outer parts of disks (outside 20 AU) in less than 3 Myr [233]. However although most stars do form in clusters, they do not undergo such extreme conditions; thus for the typical low-mass star, this mechanism is less effective than self-evaporation.

- **Photoevaporation from disk's central star**: Young stars emit significant UV and X-radiation; the UV photons are the most effective at evaporating the disk, but their observed flux is variable and somewhat uncertain. The effective escape radius is again about 0.15 r_g. The UV photons are produced in accretion shocks on the stellar surface, but probably more importantly, by active chromospheres in young stars. The details of hydrodynamic flow calculations are complicated, and observations of actual evaporative flows are not available, however the probable order of magnitude of the effect is given by the following relation [145]

$$\dot{M}_{\text{EUV}} \sim 4 \times 10^{-10} \left(\frac{\Phi_{\text{EUV}}}{10^{41} \text{s}^{-1}}\right)^{0.5} \left(\frac{M_*}{M_\odot}\right)^{0.5} \text{ M}_\odot \text{ yr}^{-1} \quad (4.90)$$

where Φ_{EUV} is the flux in photons per second produced by the central star, typically 10^{41} for a low-mass star. This relation suggests that (1) photoevaporation of a disk can happen very rapidly (10^5 yr) for a massive star, which has a very high UV flux, but (2) in the case of a 1 $M_\odot$ star the process is unimportant compared with the typical $\dot{M}_{\text{visc}}$ for viscous accretion until relatively late in the lifetime of the disk, >6 Myr. In the latter case, $\dot{M}$ for the evaporative flow is largest at the effective radius, is very small inside the effective radius, and declines as $R^{-1/2}$

outside the effective radius. The evolution of a disk subject to photoevaporation and viscous evolution then runs as follows: First, once $\dot{M}$ from photoevaporation exceeds $\dot{M}_{\rm visc}$ at the effective radius, a gap opens in the disk at that point. Second, the disk interior to that point (say ≈ 1 AU) rapidly (10^5 yr) drains onto the star and is not replenished, leaving an inner hole. Third, once the hole has been cleared the UV radiation from the star can impinge rapidly directly onto the outer disk, and the outer disk is cleared, from inside out, on the short time scale also of about 10^5 yr [12]. This result neatly explains the observed fact that outer disks, as observed in the submillimeter, disappear on about the same time scale as inner disks, as observed in the near infrared. It is also consistent with the fact that relatively few "transition" disks are observed [465].

The conclusion one could draw is that in theory the combined effects of viscous evolution and photoevaporation should completely remove disks around low-mass stars on a timescale of 10^7 yr. The fact that the observed main dispersal phase of a disk is rapid ($\sim 10^5$ yr) agrees with theoretical calculations.

4.6 Winds and Outflows

Class 0, Class I, and Class II sources are all, or almost all, associated with some kind of outflow. The phenomenon extends over the entire range of masses, from brown dwarfs to massive protostars. The outflows are important with regard to (1) clearing away of the infalling envelope and terminating the inflow, (2) determining the range of stellar masses and the star formation efficiency, (3) contributing to the maintenance of turbulence in the molecular cloud, supporting it against collapse, and (4) removal of angular momentum of disk material before it lands on the central star, perhaps explaining why T Tauri stars rotate slowly.

There are various types of observed flows. The bipolar molecular outflows are associated with all Class 0 and Class I sources and are observed on a extended scale (0.1 pc to several pc), usually through transitions in the CO molecule in the millimeter spectral region. A characteristic velocity on the large scales is 20 km s^{-1}. The optical jets are observed in ionized gas on smaller scales near the central stars of the outflows. The optical jets are also bipolar, but they are often associated with a blue shift in a forbidden emission line. The red-shifted component is often not visible, being hidden by circumstellar material, including the disk. The jets have velocities in the range 200–300 km s^{-1}. The optical jets are also associated with neutral flows, observed in a few sources in the 21 cm radiation of HI.

Closely related to the jets are the optical emission knots near young stars, first observed by Herbig and Haro in the early 50's [200,217]. Measurement of the proper motion of knots [127] showed they were moving away from the young star. The knots emit in Hα and in forbidden lines such as [SII]. An example is shown in Fig. 1.11. HH1 and HH2 are moving in opposite directions at velocities of about 350 km s^{-1}, with the main motion in the plane of the sky. The emission regions

4.6 Winds and Outflows

are thought to be caused by stellar jet material impacting relatively dense, slowly moving knots of material ahead of it and producing bow shocks. A given young object can have several different knots moving away from it. The kinematic ages (length scale divided by velocity) suggest that discrete ejection events occurred, separated by a few hundred to a thousand years, or at least a non-steady outflow. Derived velocities are 200–400 km s^{-1}.

All of these phenomena, which are most often bipolar and collimated rather than spherically symmetric, are known generally as *outflows*. They are practically always associated with disks, and the source of the outflow, according to most theories, is in the disk rather than in the central star. The outflows generally decrease in intensity and frequency of occurrence from Class 0 sources to Class II sources.

In contrast, main sequence stars, particularly those younger than the Sun, have *stellar winds*, powered somehow from within the star and accelerated in the very outer layers of the star, that are quasi-spherical. The weak-lined T Tauri stars, generally Class III, do not have an associated bipolar outflow, but it is very likely that they have the more nearly spherically symmetric stellar winds, because they show evidence of surface activity, like that of the Sun but more intense, for example from X-ray emission.

The bipolar molecular outflows are associated with objects that are still experiencing inflow of gas. The outflows exhibit a red-shifted and a blue-shifted lobe, when observed in the CO molecule, with the young stellar object (usually) found in between. In some cases, quadrupolar flows are observed, from sources that are either closely associated or in a binary. The classic example of a bipolar outflow is that of L1551 IRS5, the first one discovered [472]. A schematic diagram of that source is shown in Fig. 1.9. The general structure of the flow is thought to be a partially evacuated shell-like structure, with higher-velocity gas inside, but with a relatively slow-moving shell, observed in the CO molecule, at the outside. The upper left and lower right regions, respectively, have red-shifted and blue-shifted line profiles. Herbig–Haro knots are observed in the interior. The flows are perpendicular to the plane of the accretion disk, shown in the figure as the solid contours of CS emission. On a smaller scale, the protostellar source is in fact resolved into a 40 AU binary (Fig. 1.10).

In the CO outflows, typical flow velocities are a few to 100 km s^{-1} (15 km s^{-1} in the figure), and length scales are 0.1 pc to several pc. The typical "kinematical age", obtained from the observed scale divided by the observed velocity, is about $2-5 \times 10^4$ yr. However the total lifetime of a given flow is thought to be about 2×10^5 yr. The mass of the associated gas is $0.1-100\,M_\odot$, which is probably mostly gas from the molecular cloud swept up by the high-velocity flow from the center. The actual velocities are deduced to be parallel to the axis, with higher velocities radially farther from the source, which could be the result of momentum conservation and a decrease in density away from the source.

There is a class of flows known as "highly collimated", with the fastest gas being the most highly collimated and the closest to the axis. Figure 4.14 shows the well-collimated flow of HH211 [198]; the concentration of high-velocity gas towards the axis is evident. The central source is a class 0 object, the scale of the

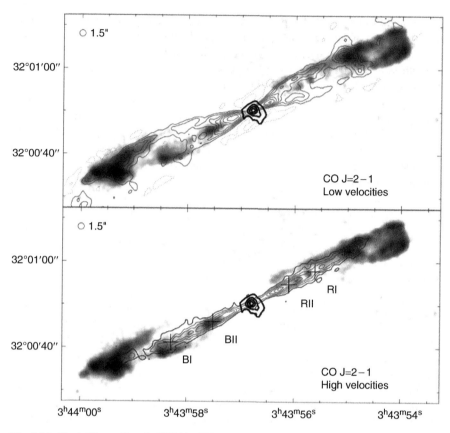

Fig. 4.14 The jetlike outflow in HH211. *Thin contours:* Contours of equal CO emission. *Thick contours:* Contours of equal emission in continuum radiation at 1.3 mm, representing dust emission from the central protostar. "Low" and "high" velocity correspond to absolute values of velocity less than or greater than, respectively, 10 km s^{-1} with respect to the system's velocity. The points marked RI and RII lie on the red-shifted lobe; those marked BI and BII, on the blue-shifted lobe. Credit: F. Gueth and S. Guilloteau: *Astron. Astrophys.* **343**, 571 (1999). By permission. © European Southern Observatory

outflow is 0.1 parsec, and the maximum velocity is about 40 km s^{-1} with respect to the system's velocity. What we see in the figure is the CO flow, not the optical jet near the star. The grey-scale patches at the head of the outflow are shock-excited molecular hydrogen. In general, the outflows from Class 0 sources are more strongly collimated than those from Class I sources.

The ratio of the mechanical luminosity ($0.5\dot{M}_{\text{wind}} v_{\text{wind}}^2$) to the luminosity of the source (L_{bol}) is typically 0.004 for high-mass sources and 0.03 for low-mass sources [562]. However total energy in the system is not conserved because some is radiated, so the luminosity ratio will vary with time for a given source. The flows are, however momentum-conserving. Figure 4.15 plots the momentum flux

4.6 Winds and Outflows

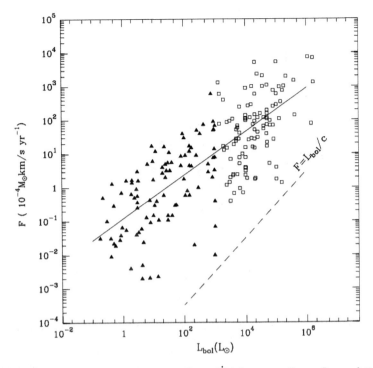

Fig. 4.15 The rate of transfer of momentum ($F = \dot{M}v$) in protostellar outflows, plotted as a function of the bolometric luminosity of the source. *Solid line:* least-square fit. *Dashed line:* the momentum transfer that could be delivered by radiation pressure. *Open symbols:* high-mass sources. *Filled symbols:* low-mass sources. The boundary between high-mass and low-mass is taken to be $L_{bol} = 10^3 \, L_\odot$, corresponding to a source mass of about $3 \, M_\odot$. Reproduced, with permission, from Y. Wu et al.: *Astron. Astrophys.* **426**, 503 (2004). © European Southern Observatory

($\dot{M}_{wind} v_{wind}$) in the outflow against the luminosity of the source. These quantities are quite well correlated. Also shown is the momentum flux that can be delivered by radiation pressure from the central source, demonstrating that radiation pressure is insufficient (by at least two orders of magnitude) to drive these flows. Since L_{bol} is mainly provided by accretion of matter from the disk to the central stellar core, this figure strongly suggests a connection between the strength of the accretion flow and the strength of the outflow.

Figure 4.16 shows a plot of outflow momentum flux as a function of the mass of the infalling envelope of the protostar [82]. The open circles are Class 0 objects (envelope mass larger than core mass) and the filled circles are Class I objects. Apparently by the time an outflow has reached the Class I stage it has lost a good deal of its collimation and outflow power, with less material at high velocity. The decrease in outflow strength with age is evident.

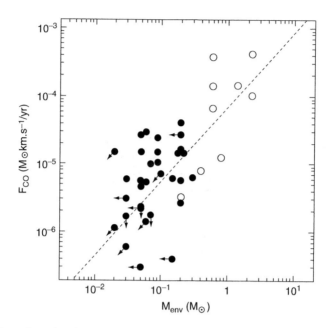

Fig. 4.16 Rate of transfer of momentum in protostellar outflows as a function of envelope mass. *Arrows:* upper limits. *Filled circles:* Class I objects. *Open circles:* Class 0 objects. The evolution of protostars is in the general direction of decreasing envelope mass. Credit: Bontemps et al.: *Astron. Astrophys.* **311**, 858 (1996), reproduced with permission. © European Southern Observatory

The energy and momentum from the optically observed jets seem to be sufficient to drive the molecular flows. The origin of the jets appears to be close to the central star but it is not known observationally precisely how close. Observations of the jets themselves in optical forbidden emission lines and in the radio continuum extend to within tens of AU from the central source, not close enough to probe the driving region of the flow. However it is strongly suspected that the flow originates in the disk near the star. Existing observations suggest that outflow velocities are highest near the rotation axis of the star, as in the case of HH211, and decrease as a function of distance from this axis. Future improvements in instrumentation will allow even better resolution of this region.

Outflows are also present in classical T Tauri stars and here they are closely linked to the presence of disks. The diagnostics of the outflows include the "P Cygni" profiles observed in some objects in Hα, CaII, and Na D. The otherwise symmetric emission line profile is cut by a deep absorption on the blueward side of the line, corresponding to absorption by material flowing toward the observer. Another diagnostic is the blueshifted line profile in forbidden lines of OI and SII. The corresponding red-shifted emission from the wind flowing away from the observer is presumably obscured, in part by the disk [153]. These outflow diagnostics are always associated with disk diagnostics, in particular infrared excess. However the weak-lined T Tauri stars show no IR or UV excesses characteristic of

4.6 Winds and Outflows

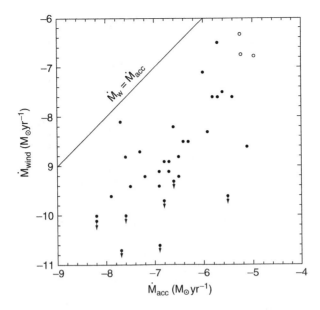

Fig. 4.17 Rate of disk accretion onto T Tauri stars plotted against rate of mass loss in the outflow. Reproduced, by permission of the AAS, from [201]. © The American Astronomical Society

disks, and they also lack the strong Hα and forbidden line emission that characterize outflows. Their Hα profiles are weak, narrow, and have no strong blue-shifted absorption. The forbidden [OI] is absent, as is infrared excess. However in other respects the WTTS and CTTS are similar, such as in strength of X-ray emission.

A more precise correlation between the rate of accretion of a disk onto the star ($\dot{M}_{acc}$) and the mass loss rate in the outflow ($\dot{M}_{out}$) was first determined by [201] and is shown in Fig. 4.17. In this case the $\dot{M}_{acc}$ is determined from the UV excess, which is a diagnostic of the flow rate of matter onto the star along magnetic field lines. The luminosities in several high-velocity forbidden lines are used to estimate $\dot{M}_{out}$. The proportionality constant between $\dot{M}_{out}$ and $\dot{M}_{acc}$ is usually quoted to be 0.05–0.1. The ratio of these two quantities (f_w) can also be roughly estimated for even younger objects [434] by use of Fig. 4.15. Assuming that L_{bol} is primarily accretion luminosity, which is proportional to $\dot{M}_{acc}$, while the momentum transfer rate in the wind is proportional to $\dot{M}_{out}$, one can derive $f_w \approx 0.1$ for both Class 0 and Class I objects.

The theory of the generation of the bipolar flows and optical jets, as well as that of the coupling between these flows and the slow rotation of T Tauri stars, is still under development and is discussed in detail by [427] and [454]. Magnetic coupling in a disk or star or both is involved in all models, and the various models can be classified according to whether the field is (1) external and passing through the disk [522], (2) generated in the star but interacting with the disk [462], (3) generated in the disk [519], or (4) primordial [336]. The flows can also be classified on the basis

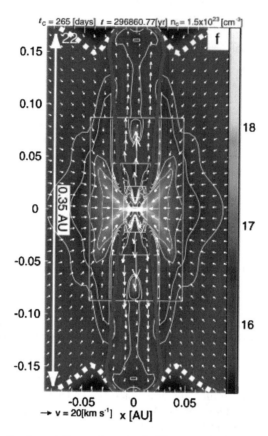

Fig. 4.18 Cross section through the rotation axis of a 3D magnetohydrodynamic simulation of the collapse of a protostar with a magnetic field. The early stages of a bipolar outflow are seen, as upward- and downward-moving arrows near $x = 0$. The x-direction gives the distance from the rotation axis; the vertical axis indicates the distance above or below the equatorial plane. *Arrows*: velocity vectors with a maximum of $20\,\mathrm{km\,s^{-1}}$. *Colors*: particle density in cm^{-3}. The quantities n_c, t_c, and t are, respectively, maximum density in the stellar core (at $x = 0, y = 0$), the time after the formation of the high-density stellar core, and the time since the beginning of collapse. Reproduced, by permission of the AAS, from [336]. © The American Astronomical Society

of the outflow type, either a flow emanating over a considerable range of radii in the disk [58, 402], or originating close to the star near the point where the disk angular velocity equals the stellar angular velocity (X-wind [462]), or originating in the star itself [345].

An example of the effect of the primordial field is shown in Fig. 4.18. The entire collapse of a protostar from an initial molecular cloud core, in 3D including an initial uniform field and rotation, is calculated with non-ideal magnetohydrodynamics. The initial condition is a Bonnor–Ebert sphere at a density of $10^4\,\mathrm{cm}^{-3}$. The collapse is followed through an increase of 19 orders of magnitude in density, to the time when the stellar core has formed. After 300,000 years a disk is formed in the center on a

4.6 Winds and Outflows

scale of only 0.1 AU, and an outward-flowing bipolar jet is created which reaches maximum velocity of about 30 km s^{-1} and is collimated toward the axis. The upflow in this case is driven primarily by a magnetic pressure gradient. This result may represent the very earliest phase of the generation of a bipolar flow.

Other models for outflows are based on assumed disk structures and magnetic field configurations at later stages of evolution, corresponding to Class II objects. One of the important ideas in this work concerns the accretion flow of material from disk to star along magnetic field lines above or below the equatorial plane. One might have expected that the accretion of mass and angular momentum from the disk would spin up the star, while in fact T Tauri stars are rotating quite slowly. However, it was pointed out [265] that if the star contained a kilogauss field that truncated the disk out to a few stellar radii, then the stellar magnetic field lines passing through the disk outside of the corotation distance would tend to transfer angular momentum from star to disk, slowing down the star, and that a steady state of relatively slow rotation could be reached where the spinup torque on the star is balanced by the spindown torque. This model was extended [462] to show that the excess angular momentum that is stored in the disk is then removed from the system in the outflow, thus contributing to the solution of the angular momentum problem.

Figure 4.19 shows very schematically the Shu "X-wind" model. The origin of the magnetic field is in the star. The disk is truncated by the field at a few stellar radii. The truncation radius is also very close to the co-rotation radius where the angular velocity of the stellar rotation equals that of the orbiting disk. The stellar dipole field has B $\approx$ 1 kG, as deduced from observations of several T Tauri stars; the assumption of a dipole field is not essential to the success of the model [365]. Just outside the truncation radius, the accretion flow pinches the dipole field lines inwards, toward the star. At the co-rotation point the accretion flow from the disk divides, part of it following magnetic field lines onto the surface of the star (the "funnel flow"). The magnetic field configuration in the inner region is such that most of the angular momentum of the accreted material is transmitted back to the disk near the co-rotation point, so that the star is actually spun down. At the equilibrium state it is rotating relatively slowly as observed.

Just outside of the co-rotation point, there is a potential gradient, forcing material outward. The mechanism here is different from that shown in Fig. 4.18. Charged particles are tied to the uniformly rotating magnetic field lines and move outward. These magnetic field lines are bent inward, making an angle of roughly 60° or less with the surface of the disk. The centrifugal force increases with increasing distance from the rotation axis, so particles continue to be accelerated in the X-wind, transferring angular momentum off the disk. In this manner, matter can be accelerated to escape speed. As long as the flow velocity is sub-Alfvénic, the magnetic field dominates the flow, the acceleration continues, and the angular momentum in the flow increases. Once the flow reaches the Alfvén surface where the flow velocity exceeds the Alfvén velocity, the rotating magnetic field no longer controls it, but the toroidal (ϕ-component) of the field becomes important, and the inward force generated by it ("hoop stress") collimates the flow.

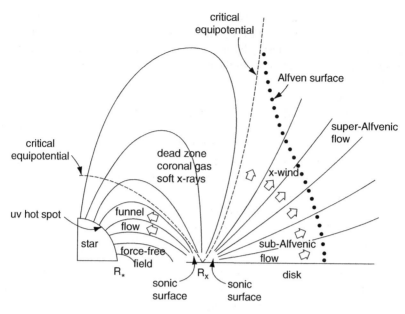

Fig. 4.19 Schematic diagram of an accreting T Tauri star and outflow generation. R_x is the point where star and disk are co-rotating. The Alfvén surface marks the points where the outflow velocity in the X-wind begins to exceed the Alfvén velocity. The "dead zone" consists of subsonically moving material of low density that does not participate in either the funnel flow or the X-wind; note that it is not the same as the "dead zone" that occurs in the interior of disks when the magnetic field is not coupled to the matter. A force-free magnetic field implies that the ratio of gas pressure to magnetic pressure is very small. The critical equipotential is that surface where the effective potential, containing both gravitational and centrifugal contributions, is zero in a frame of reference that rotates with the stellar angular velocity. Outside the outer branch of this curve, material is subject to the magnetocentrifugal effect and flows outward. Below the inner branch, material can flow inwards to the star. Reproduced, by permission of the AAS, from [462]. © The American Astronomical Society

Another possible mechanism for generating outflows is known as the "disk wind", as the flow is generated not near a single disk radius but over a range of radii, extending out to at least a few AU. This mechanism was originated [57] to explain jets generated near black holes. A magnetic field passing through the disk is required; it could be a primordial field or it could be generated through dynamo action in the disk. The origin of the outflow from various radii in the disk is consistent with the observed decrease in velocity away from the rotation axis of the outflow, as it can be shown that the magnitude of the outflow velocity is closely related to the Keplerian velocity in the disk. The basic mechanism for driving the flow is the same magnetocentrifugal effect that operates in the X-wind model. Numerous numerical simulations [427] have discussed the detailed physics appropriate to the generation of outflows, and have shown that this process is feasible. It has not yet been determined whether disk winds or X-winds dominate the generation of bipolar flows; both mechanisms could in fact operate, and an axial component, similar to that shown in Fig. 4.18, could coexist as well.

4.7 Summary of Disk Evolution

A disk formed from the collapse of a cloud core with angular momentum in the observed range would be expected to evolve as follows:

1. After initial collapse, the disk will be gravitationally unstable and will evolve under the effect of gravitational torques generated by spiral waves. The time scale will be short, comparable to 100 orbital periods, so that a considerable amount of mass can be accreted by the star in a few thousand years. FU Ori outbursts may be generated toward the end of this stage.
2. Viscous evolution starts, generated by processes that are not fully explained. Further accretion onto the star occurs on a time scale of 10^6 yr, until the disk accretion rate onto the star is reduced to small values. Then the various dissipation processes, probably dominated by photoevaporation, remove much of the remaining mass. Gas and dust disappear from the disk on about the same time scale.
3. At all stages of the evolution of the disk, the disk angular momentum and the local magnetic field interact to generate a fast bipolar jet, originating within a few AU of the central star. On larger scales the fast (200 km/s) jet entrains and accelerates local molecular cloud material to form bipolar CO flows. Observations suggest a mass outflow rate on the order of 10% of the accretion rate of disk material onto the star.
4. During the evolution of the disk, dust grains coagulate into larger particles and settle to the midplane. As the mean particle size increases, the infrared opacity decreases, and correspondingly, the observed infrared excesses decline.
5. The solid material begins to coagulate into planets. The residual small dust grains tend to be blown out by radiation pressure or forced into the central star by the Poynting–Robertson effect once the gas is gone, but the dust can be regenerated by collisions between the larger "rocks" that still remain in orbit. This second-generation dust is likely to be the material observed in the very low mass remnant "debris disks" around young main-sequence stars such as Beta Pictoris.

4.8 Problems

1. (a) A rotating cloud core of mass M, initial uniform angular velocity Ω, and initial sound speed c_s, starts out near equilibrium with thermal and gravitational energies in balance (the initial rotational energy is small).

It collapses, with conservation of angular momentum, to an equilibrium Keplerian disk. Show that its outer radius at the final stage is

$$R_{ct} = \text{constant} \frac{\Omega^2 G^3 M^3}{c_s^8}. \tag{4.91}$$

(b) Calculate R_{ct} for an initial cloud of 1 $M_\odot$ with $\Omega = 10^{-14}$ s^{-1}, with uniform density and $T = 10$ K. What is the specific angular momentum j at R_{ct}? What is the average j for the whole cloud? Compare with j for planetary orbits in the solar system.

2. The envelope of a Class 0 protostar is observed at a wavelength of 1 mm and has a flux density of 1 Jansky. The distance is 100 pc. What is its mass? Compare results with the two different opacity curves in Fig. 3.9.

3. (a) Assume that the radiative diffusion equation holds for the vertical direction in a disk at a given distance R from the star. If $F = \sigma T_{\text{eff}}^4$, where F is the radiative flux and T_{eff} is the surface temperature at R, show that the midplane temperature T_c at R is given by

$$T_c^4 \approx \frac{3}{4} \tau T_{\text{eff}}^4 \tag{4.92}$$

where the vertical optical depth $\tau = \kappa_c \Sigma/2$, Σ is the surface density, and κ_c is the midplane opacity. Assume τ is large.

(b) Combine this equation with the equation for vertical thermal balance

$$2\sigma T_{\text{eff}}^4 = \frac{9}{4} \Sigma \nu \Omega^2 \tag{4.93}$$

where Ω is the orbital frequency. Find T_c as a function of Σ, Ω, and constants. Use the "alpha" viscosity $\nu = \alpha_{\text{visc}} c_s H$, where H is the scale height and c_s is the isothermal sound speed. You can use the thin disk approximation $H = c_s/\Omega$. Assume the opacity κ_c is constant = 1. Note that a typical value of α_{visc} is 0.01.

(c) Now calculate the central density, assuming an ideal gas with molecular weight $\mu = 2$.

(d) Calculate the cooling time of the disk as a function of Ω. What do you conclude about the possibility of fragmentation in this disk?

4. A uniformly rotating uniform sphere of radius 1×10^{17} cm and a mass of 1 $M_\odot$ has $\Omega = 2 \times 10^{-14}$ rad/s. It collapses to a system consisting of a T Tauri star of 0.98 $M_\odot$ and a Keplerian disk of 0.02 $M_\odot$, conserving total angular momentum. The star is uniformly rotating with a surface rotational velocity of 30 km s^{-1} and a radius of 4 $R_\odot$. What is the outer radius of the disk? Assume that the surface density distribution in the disk goes as $R^{-1/2}$, where R is the distance

4.8 Problems

from the star. Compare with the sizes of observed disks. What do you conclude (see Fig. 4.1)?

5. Use the information from Problem 3 to calculate a simple numerical disk model from 1 AU to 20 AU for a central star of 1 $M_\odot$. Assume that the disk surface density scales as R^{-1} and that the total mass between 1 AU and 10 AU is 0.03 $M_\odot$. Calculate $\Sigma(R)$, $T_c(R)$, $\rho_c(R)$, and $T_{\text{eff}}(R)$. Is the disk gravitationally unstable anywhere?

6. This problem illustrates a moderate version of the "beaming" effect of radiation along the polar axis of a flattened, rotating cloud.

    (a) Calculate numerically the density distribution along the polar axis and very close to the equatorial plane in a rotating cloud, according to (4.3). Assume the outer cloud radius is 10^{17} cm, the rotation rate is $\Omega = 7 \times 10^{-15}$ s^{-1} and the interior mass is 0.75 $M_\odot$. The accretion rate is $10^{-5}\,M_\odot$ yr^{-1}.

    (b) Calculate the optical depths in the polar and equatorial directions, integrating inward to the dust destruction front at 1 AU. The mean opacity can be taken to be 0.2 cm^2 g^{-1}. Explain the result.

    (c) Repeat the calculation for an interior mass of 0.25 $M_\odot$.

Chapter 5
Massive Star Formation

We will define a high-mass star as one with a final mass of greater than $10 M_\odot$. On the main sequence this group would include stars of spectral type O, B0, and B1. Although stars in this mass range are few in number compared with low-mass stars, they are extremely important with regard to galactic chemical evolution and the physics of the interstellar medium. Stars in this mass range (actually down to about $8 M_\odot$) evolve to become Type II supernovae which produce and eject a significant amount of the heavy-element material in the Galaxy. The heavy elements act as cooling agents in the interstellar gas, promoting the formation of later generations of stars. The UV radiation from the massive stars and the concomitant generation of HII regions, along with strong stellar winds, and eventually supernovae remnants stir, energize, and heat the ISM, producing fascinating observable structures. The resulting dynamical effects are thought, in some cases, to induce the formation of a new generation of stars in the surrounding medium (Chap. 2). On the other hand, the resulting ionization and heating can erode nearby molecular cloud material and suppress subsequent star formation in the neighborhood. The brightness of HII regions allows them to be a useful tool in estimating the star formation rate in distant galaxies. The UV radiation from massive stars can have a significant effect on the evaporation of protostellar disks around lower-mass stars in the same cluster.

Although the basic picture of collapse of a rotating molecular cloud core applies to both high-mass and low-mass star formation, there are a number of significant differences between the two cases. For example, (1) the accretion time of the envelope $(M/\dot{M})$ for high mass stars is longer than the gravitational contraction time (1.8) of the equilibrium stellar core. For example, $t_{KH} = 10^4$ yr for $50 M_\odot$ while the accretion time is closer to 10^5 yr, so that the core can reach the main sequence while accretion is going on. On the other hand, the low-mass stars, with $t_{KH} = 3 \times 10^7$ yr and an accretion time of 3×10^5 yr for $1 M_\odot$, finish their accretion while the cores are still in the gravitational contraction phase, well before the main sequence, with surface temperatures around 4,000 K. (2) In massive star formation, when the stellar core approaches the main sequence, the radiative acceleration in the infalling envelope, arising from radiation interacting with the dust, becomes more important than gravity. In the low-mass case, gravity always dominates in the infalling envelope. (3) It was earlier suggested [459] that massive stars form in magnetically supercritical molecular cloud cores, and that low-mass stars form in subcritical cores by ambipolar diffusion. Although magnetic fields are undoubtedly important in both mass ranges, there is no strong observational evidence in support of this suggested bifurcation; while massive stars undoubtedly do form under supercritical conditions, it is entirely possible for low-mass stars to form under such conditions as well. Rather it is becoming apparent from observations [52, 418] that massive stars form in a highly turbulent environment from high-mass, high-density ($n \approx 10^6$ cm^{-3}) cores, while low-mass stars form from relatively quiescent lower-mass cores of density $n \approx 10^5$ cm^{-3}. (4) The protostellar evolution time is not much different [353] between high-mass stars ($\approx 10^5$ yr) and low-mass stars ($\approx 3 \times 10^5$ yr); thus $\dot{M}$ for accretion of high mass stars must be significantly greater than that for low-mass stars. This requirement is apparently consistent with the formation of high-mass stars in very turbulent high-density regions; this initial condition naturally provides a high mass accretion rate.

In this chapter, we point out the theoretical and observational difficulties in studying high-mass star formation, consider the probable initial conditions in molecular clouds, summarize the suggested theoretical solutions to the problem of high-mass star formation, and address the question: are there theoretical and observational upper limits to the mass of a star?

5.1 Information from Observations

Although high-mass stars in the process of formation are very bright, there are difficulties in observing them because they are relatively few in number, they are found typically only at large distances from the Sun, leading to limited spatial resolution, and they are located in regions that are highly obscured in optical radiation. In addition they are almost always found in clusters, so at their large distances it is difficult to resolve individual objects in the cluster. But additional clues as to their formation may be obtained from the properties of already-formed massive stars. A few important major features that have been determined from the existing massive stars as well as from the suspected massive star-forming regions are discussed in this section.

5.1.1 The Present Massive Star Population

Massive stars apparently form primarily in clusters, along with numerous low-mass stars. The present locations of massive stars can be roughly divided into four groups. First, many postformation stars are observed in gravitationally bound OB clusters, good examples of which include the Orion Nebula cluster, the dense cluster associated with the galactic HII region NGC 3603, and the R136 cluster in the 30

5.1 Information from Observations

Doradus star formation region in the Large Magellanic Cloud (LMC). The latter is the most massive cluster of the three and contains on the order of 100 O stars packed within a radius of a few parsecs. Second, some massive stars are found in OB associations, for example Scorpius OB2 and Orion OB1. In this case the stars are 1–10 pc apart, and the association is gravitationally unbound. The relationship between the bound OB clusters and the associations is not clear, but one possible scenario [272] is that a bound cluster forms with a low star formation efficiency, after which a considerable amount of remaining gas is expelled. A large fraction of the stars becomes unbound and forms an expanding association, while a smaller fraction, those near the center of the initial cluster, is able to remain as a bound cluster, smaller than the original one. The third category is the so-called "runaway" OB stars, which are individual stars in the field with velocities of 40 km/s or more, which presumably were ejected from a cluster or association. Finally, fourth, there are a few OB stars in the field that cannot be traced back to an origin in a cluster or association and could have formed independently; this group consists of <10% of massive stars.

The fraction of massive stars in binary or multiple systems, both spectroscopic and visual, is higher than in solar-type stars. This quantity is difficult to measure, because of the scarcity and large distances of the massive stars; furthermore their extreme brightness makes it difficult to detect faint companions. Estimates of the fraction of systems that are binary or multiple range up to 80%, although this fraction is highly variable from cluster to cluster. The fraction seems to be similar in OB clusters and OB associations [575] but somewhat lower for field stars [341]. Triple systems and higher multiples are common among massive stars as are "twins", binaries with practically equal-mass components. A particularly well-defined (although small) sample of massive stars that has been used to determine the multiple star fraction is the young Orion Nebula Cluster [421]. Out of 13 stars observed, in the mass range 3.5–45 $M_\odot$, eight turned out to have at least one companion, giving a lower limit of 61% for the multiple star fraction, uncorrected for possible undetected companions. This fraction is significantly higher than that for solar-mass stars [147], 44%, also uncorrected. Note however that the total number of companions among the 13 primaries is 14: there are three triple systems, one quintuple system and four binary systems among the eight stars with companions. The fraction of systems that are triple or higher is also greater than that for solar-mass stars.

The initial mass function, on the average, has a slope quite similar to that of stars in the 1–5 $M_\odot$ range, that is, a Salpeter-like slope of d log (N)/d log (m) = -1.35, although the slope can vary from cluster to cluster because of small-number statistics. The slope is thought to be about the same in OB clusters and OB associations. There is no "feature" or change in slope of the IMF near the transition between the high-and low-mass stars, indicating that the differences in physics between high-mass and low-mass star formation do not affect the origin of the IMF. However, particularly for massive stars the determination of the IMF from observations is uncertain and involves some assumptions. In general, the following ingredients are needed to determine the IMF from stars presently on the main sequence: the luminosity function (since masses are usually not directly determined), the mass-luminosity relation, main-sequence lifetimes, and the correction for evolved stars. The latter correction is important for massive stars, since the present-day mass function is not the same as the initial mass function, because of the short time scale of stellar evolution. To make the correction an assumption regarding the history of the galactic star formation rate is needed. A summary of the actual procedure is given in the Appendix to this chapter.

The question arises whether there is an upper mass limit for stars. The most massive stars whose masses have been accurately measured, through analysis of the orbit of a double-lined spectroscopic eclipsing binary, are the two components of WR 20a, with masses of 83 ± 5 and 82 ± 5 $M_\odot$ [75, 429]. The orbital period is 3.7 days, the eccentricity is zero, and T_{eff} for the primary is about 42,000 K. The orbital separation is only 0.25 AU, about 6 stellar radii. Both components are Wolf-Rayet stars, indicating mass loss, so the original masses could have been somewhat higher. But stars have formed with masses higher than 80 $M_\odot$. A double-lined spectroscopic eclipsing binary, NGC 3603-A1, has been analyzed to obtain masses of 116 ± 31 and 89 ± 16 $M_\odot$ [450].

The upper limit can also be estimated indirectly through observations of the upper end of the main sequence in various clusters. The most massive clusters observed are R136 in the LMC and the Arches cluster near the Galactic center. Various authors [575] have shown that there is an apparent upper limit of about 150 $M_\odot$. However, extrapolation of the Salpeter-slope IMF in these massive clusters indicates that stars with up to about 750 and 500 $M_\odot$ in the two clusters, respectively, would have been present if they had formed [172, 542].

The rather well-accepted upper limit of about 150 $M_\odot$ may have been broken, however, by observations of the highly luminous star R136a1 in the LMC [124]. The analysis was again indirect, based upon the observed luminosity and spectrum, as compared with theoretical stellar evolutionary models and model atmospheres, combined with an estimate of the mass loss rate of this Wolf-Rayet star. The same technique was applied to NGC 3603-A1, and the mass obtained agreed with the dynamically determined mass (above). The resulting estimate for the initial mass of R136a1 is 320^{+100}_{-40} $M_\odot$.

It is not clear whether the upper limit, which is uncertain because of mass loss from the most massive stars, is physical, or statistical, resulting from the extreme rarity of the most massive stars. The possible physical mechanisms that could limit the mass include (1) pulsational instability generated by nuclear reactions in the core of the star [309], (2) limits on accretion during star formation arising from radiation effects on the infalling gas (Sect. 5.2), and (3) rapid mass loss from the surface of newly-formed massive stars owing to radiation pressure in the absorption lines near the stellar photosphere [284]. In cases (2) and (3) the mass limit would be expected to increase as the metal content of the region decreases. It is not clear which of these processes is the most important.

5.1.2 Formation Sites

An observed example of a region of massive star formation is shown in Fig. 5.1. Seven specific features are pointed out. Feature 1 is a young compact cluster dominated by a triple system of massive stars. The cluster has formed in the head of a massive dust pillar. It is suggested that here is an example of triggered star formation: outflows from the massive cluster R136 (not on frame) could have shaped the pillar and induced the collapse of part of it to form this secondary cluster. The new stars have blown off the top of the pillar with their HII regions and winds, and thus the cluster is visible in both the infrared and visible pictures.

Features 2 and 3 are also young stars or clusters still embedded in dusty regions; thus they represent a somewhat earlier stage of evolution than does Feature 1. They are very bright in the infrared image, but very faint in the visible. Feature 4 is a very red star that has formed within a compact dust cloud.

Feature 5 is another young triple-star system with a surrounding cluster of fainter stars. Features 6 and 7 are glowing patches which are interpreted, at least tentatively,

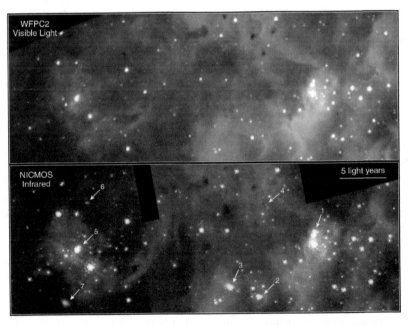

Fig. 5.1 Active region of massive star formation near the star cluster R136 in the Large Magellanic Cloud. The *top panel* shows a visual image taken with Hubble's Wide Field and Planetary Camera 2. The *lower panel* shows the same region in the near infrared, taken with Hubble's Near Infrared Camera and Multi-Object Spectrograph. Although this region is 55 kpc away, details of some individual objects are discernible (see text). WFPC2 image credit: NASA, John Trauger (Jet Propulsion Laboratory), James Westphal (California Institute of Technology). NICMOS image credit: NASA, Nolan Walborn (STScI), Rodolfo Barba (La Plata Observatory). Courtesy Space Telescope Science Institute. (5 light years = 1.5357 pc)

Fig. 5.2 The star-forming region RCW108, about 1,200 pc away in the Milky Way. The image is a composite of X-ray data from the Chandra X-ray Observatory (*blue points*) and an infrared image obtained with the Spitzer Space Telescope. Most of the X-ray sources are part of a young cluster NGC 6193. It is thought that the formation of the cluster of massive stars seen in the infrared image was triggered by effects of massive stars outside the dusty region, in NGC 6193. At the same time, the hot, massive stars in 6193 may be eroding the dust clouds around the forming cluster through their high-energy radiation. Image credit: NASA/JPL-Caltech/CXO/CfA. Courtesy Spitzer Science Center

to be caused by bipolar jets from Feature 5 impacting onto surrounding dust clouds. They are symmetrically located on opposite sides of Feature 5 and possibly originate in a disk around one of the objects in the triple-star system. This image at least suggests that bipolar flows are produced by massive stars, possibly involving similar mechanisms to those found for low-mass stars. It has been shown observationally that massive stars forming closer by, in our Galaxy, do generate outflows [455] (see below).

Another example of observed star formation, closer by in the Milky Way, is shown in Fig. 5.2. The young cluster NGC 6193, of which many of the X-ray points are members, contains massive O stars which are producing an HII region; the brightest O stars are off the image to the left. The edge of this region is eating its way into the dusty cloud shown in the infrared image, and gradually eroding it. At the same time this region is thought to be a good example of induced star formation [558]. A dense region of newly formed massive stars, known as RCW108-IR, is seen in the IR image.

The initial conditions for massive star formation seem to require molecular cloud cores with surface density $\Sigma = M/(\pi R^2)$ higher than that for low-mass stars [354]. For example, $100\,M_\odot$ within a radius of 0.1 pc gives $\Sigma \approx 0.7\,\mathrm{g\,cm^{-2}}$, while

 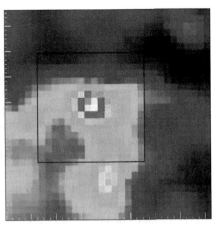

Fig. 5.3 An example of an Infrared Dark Cloud (G29.55+00.18) observed with Spitzer at 8 μm (*left*) and Herschel at 350 μm (*right*). The mass of the dark cloud is estimated to be ≈ 500 $M_\odot$, its radius 1.7 pc, its central density 3×10^4 cm^{-3}, and its temperature 16 K. It appears in absorption against the bright IR background at 8 μm, and in emission at 350 μm. Credit: D. Stamatellos et al.: *MNRAS* **409**, 12 (2010). Reproduced with permission from John Wiley and Sons. © 2010 Royal Astronomical Society

1 $M_\odot$ within a radius of 0.05 pc (a typical low-mass core) gives $\Sigma \approx 0.027$ g cm^{-2}. The high-Σ conditions are believed to exist in the so-called Infrared Dark Clouds (IRDC's), which are regions of molecular clouds that are dense enough so that they have several magnitudes of dust extinction in the infrared, around 10 μm. They are observed as absorption patches against the diffuse background Galactic emission from hot dust at those wavelengths. Observations [414] from the Infrared Space Observatory and [154] from the Midcourse Space Experiment at mid-infrared wavelengths have revealed a large number of such objects [466]. They also are observed, from dust emission in the millimeter continuum [428], to have dense, compact cores. These cores could represent the formation sites of massive stars or even clusters. Figure 5.3 shows observed images of one of these clouds at two different wavelegths [485].

At least some of the massive molecular cloud cores found in infrared dark clouds could represent the initial stage of star formation. The observational characteristics of such cores would be emission from dust at mm and submm wavelengths, molecular lines observed in the mm, and no evidence, from strong mid-infrared radiation, which would indicate that a collapsed protostar had already formed. Typical physical characteristics include: masses of 100 to a few thousand $M_\odot$, mean densities of 10^5 cm^{-3}, sizes of 0.25–0.5 pc, and temperatures around 15 K. Observations are not sensitive enough to rule out low-mass protostars in these objects, but they probably have not yet formed high-mass protostars [183]. Mean values of linewidths were found [479] in a sample of suspected high-mass starless cores to be 1.6 km s^{-1}, indicating the presence of turbulence.

As examples of the next phase of evolution after the high-mass starless core phase, one would expect to find some infrared dark clouds with evidence of protostar formation as indicated observationally, for example by the presence of emission at 8 μm (along with the dust emission in the mm and submm). Many surveys of such objects have been undertaken [52]. Their overall characteristics are similar to those of starless cores (preceding paragraph) except the temperatures are slightly higher (22 K) and the turbulent linewidths somewhat greater (2.1 km s^{-1}.) They are often associated with water, OH, and methanol (CH_3OH) masers which are thought to originate either in molecular outflows or disks. Water masers are found also around low-mass protostars, but OH and CH_3OH only around high-mass protostars. It is not clear what physical process is associated with the maser production, but it could be shock activity. Massive collimated molecular outflows have in fact been observed for a number of massive protostars, and their general properties are consistent with those around low-mass protostars [52]. The observations indicative of outflows, as for low-mass protostars, are red-shifted and blue-shifted emission lines in, for example, the CO molecule. However, no collimated outflows have been observed for high-mass protostars exceeding 30 $M_\odot$. There are cases of observations of high-mass protostars with wide-angle CO outflows; an example is one of the infrared sources in the massive star-forming complex W49A, located 11.4 kpc from the Sun. This source is estimated [470] to be a protostar of 45 $M_\odot$. It is suggested that this object has been accreting from a disk, which powers the outflow as in the case of low-mass stars. The inner part of the disk is in the process of being cleared out, possibly explaining the wide angle of the outflow.

An example of the spectral energy distribution of a high-mass protostar [366, 367] is shown in Fig. 5.4. The peak of the SED is at about 100 μm. Its precise evolutionary stage is difficult to determine, but it is not yet at the stage where it has a detectable HII region, and very possibly its core has not reached the main sequence. The object is considered to be a Class 0 protostar because (1) its stellar core mass ($\approx 10 M_\odot$) is much less than the mass of the collapsing envelope (a few hundred $M_\odot$), and (2) the ratio of submillimeter to bolometric luminosity is about 3×10^{-3}. Also there is a molecular outflow with properties similar to but scaled up from those of low-mass Class 0 objects. The high accretion rate onto this object ($\sim 10^{-3} M_\odot$ yr^{-1}) is probably sufficient to suppress the formation of an HII region.

The presence of the outflows strongly suggests that disks are also present in the high-mass protostars, although they have proved difficult to detect. The highest-mass star around which a Keplerian disk has been definitely detected is in the early B spectral range, with a mass of about 7 $M_\odot$ [105]. The luminosity of this object (IRAS 20126+4104) suggests that it is a protostar accreting at about 2×10^{-3} $M_\odot$ yr^{-1}. In some cases, for suspected high-mass protostars, velocity gradients have been detected (e.g. [167]) perpendicular to the direction of the outflow. However the velocity structure is not that of a Keplerian disk. In this case the structure could either be an inflow of rotating gas, possibly the precursor of a Keplerian disk, or a massive, self-gravitating disk whose orbital velocity as a function of distance from the star is not Keplerian. Thus the observations at least suggest that high-mass protostars go through a disk phase, although the disk may be short-lived.

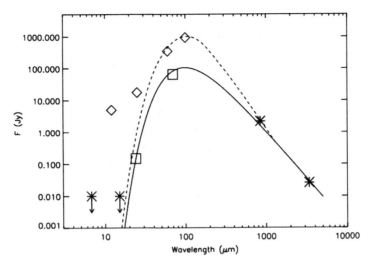

Fig. 5.4 The spectral energy distribution of a suspected high-mass protostar IRAS 23385+6053 whose stellar core has not yet reached the main sequence. *Diamonds:* observations at the four IRAS satellite wavelengths. *Open squares:* mid-infrared data from the Spitzer satellite. *Asterisks:* a combination of observations with the James Clerk Maxwell Telescope and the Owens Valley Radio Observatory in the mm and submm region, and upper limits in the infrared at 6.75 μm and 15 μm obtained from the ISO satellite. There are also upper limits at 2 cm and 6 cm, from the VLA, indicating that if an ultracompact HII region is present at all, it must be very faint. The 12 μm and 25 μm IRAS fluxes arise from sources outside the actual protostar. The *dashed line* is a fit to the data based on an assumed spherical envelope model with power-law distributions of density and temperature, including the 60 μm and 100 μm IRAS points. The *solid line* is a fit using the more reliable Spitzer data instead of the IRAS data. Credit: Molinari et al.: *Astron. Astrophys.* **487**, 1119 (2008). Reproduced with permission. © European Southern Observatory

Once a massive protostar has gained enough mass so that its core has evolved to the main sequence, the core, while still accreting, starts to emit ionizing radiation. In the earlier phases, there is still high-density gas accreting onto the core, and the ionizing flux can ionize only a small volume. The observational evidence for this phase is thought to be cm radiation, along with the mid-infrared emission and mm emission characteristic of the earlier phases. The cm radiation is produced by free–free emission from the ionized gas. The ionized region at this phase is known as a *hypercompact* HII region, defined as having a size less than 0.01 pc and a density of $\sim 10^6$ electrons cm^{-3}. It is probably produced by a single O or B star that is still accreting, with the ionized region comprising only a small fraction of the protostellar volume. The high-pressure ionized region is prevented from expanding by the dynamic pressure of the infalling material, and the ionized particles are still gravitationally bound to the star [250, 536]. The presence of broad radio recombination lines emitted in at least some of these regions indicates the presence of accretion, rotation, and outflow activity.

At some stage, most of the accretion onto the central star will occur via the disk, and the density and inflow rate in other directions, in particular the polar direction, will decrease. Also, as the star accretes more mass, its ionizing flux increases, and the HII region will be able to expand into almost the entire protostellar volume, excluding the disk. At this stage the HII region is called an *ultracompact* HII region. It is distinguished from the hypercompact HII region by the fact that the former is larger, with size of ≈ 0.1 pc, and has lower density (10^4 electrons/cm^3). Also, the line widths in the radio are smaller. The ultracompact HII regions are also detected through their radiation at cm wavelengths [560]. Even at this stage, all of the optical, UV and near IR radiation produced by the protostellar core is absorbed by the infalling dust, and the object is detected primarily at mid-infrared wavelengths.

The size of an ultracompact HII region is comparable to the original molecular cloud core which spawned the massive star. The relevant size scales are the Strömgren radius, which characterizes the size of an ionized region, and the gravitational radius which is defined by

$$r_g = \frac{GM_*}{c_i^2} \tag{5.1}$$

where c_i is the sound speed in the ionized gas, about 10 km s^{-1}. For a 50 M$_\odot$ star this radius is about 400 AU or 6×10^{15} cm. It marks the boundary for an ionized region, inside of which gas can continue to rain down upon the protostar, and outside of which accretion is suppressed. By comparison, the Strömgren radius, which defines a volume in which the rate of ionization by stellar photons is balanced by the rate of recombination of electrons onto hydrogen atoms

$$r_S = \left(\frac{3N_i}{4\pi \alpha_R n_e^2}\right)^{1/3} \tag{5.2}$$

where N_i is the number of ionizing photons per second, n_e is the electron density in cm^{-3}, and α_R is the recombination coefficient of electrons onto hydrogen atoms to levels $n = 2$ or above (recombinations to the ground state, $h\nu > 13.6$ eV, immediately result in a nearby ionization so do not affect the overall balance; $\alpha_R = 3 \times 10^{-10} T_{\text{HII}}^{-3/4} = 3.5 \times 10^{-13}$ cm^3 s^{-1} for $T_{\text{HII}} = 8,100$ K). For a 30 M$_\odot$ star, $N_i \approx 10^{49}$ s^{-1}, and the typical density in a dense molecular cloud region is 10^4 cm^{-3}, giving $r_S = 0.13$ pc, assuming uniform density for simplicity. Thus, since $r_S \gg r_g$, the ultracompact HII region is not gravitationally bound to a single protostar and is beginning to ionize the molecular cloud material in the neighborhood. At this stage the HII region is expected to be photoionizing the disk [232] and evaporating it, resulting in escape of the material from the protostar.

The feedback as a result of the HII region may have ultimately some influence on the final mass of the star, but a more important feedback effect is that of radiation pressure on the dust, which we discuss in the next section. Even if the HII region is present, accretion can still take place through a disk. It is more likely that the final mass is determined by the mass of the initial molecular cloud core, modulo an efficiency factor that depends more on the properties of the mass outflow than on those of the HII region. However once the HII region expands beyond the region that is gravitationally bound to the star, it begins to have a more global effect. The HII region of a single star or a combination of HII regions from several massive stars produce a high-pressure bubble in the surrounding molecular cloud material as a result of the high temperature in the ionized gas, typically 8,000 K. The material begins to expand, and when the ionized region reaches the edge of the cloud, rapid mass loss from the cloud occurs. With sufficient ionizing flux, the star-forming region in the molecular cloud can be destroyed.

5.2 The Problem of Radiation Pressure

Consider now a single massive core, a subunit of a massive molecular clump, out of which a massive star can form. The question of cluster formation will be discussed below. Although there is a wide spread in observed properties, for definiteness let us say that the core has $100\,M_\odot$ and a radius of 0.1 pc, therefore a mean density of 1.4×10^{-18} g cm^{-3}. It is initially supported against gravitational collapse by turbulent motions, possibly combined with magnetic effects. The thermal Jeans mass under the core conditions ($T \approx 15$ K) is less than $1\,M_\odot$. Nevertheless the observed virial parameter $\alpha_{\rm vir} = 5\sigma_x^2 R/(GM)$, is of order unity, where σ_x is the one-dimensional velocity dispersion and R and M are, respectively, the radius and mass of the core. For the assumed core properties, the mean turbulent speed $v_{\rm turb}$ turns out to be about 1.6 km/s, consistent with observations.

The cloud is destabilized once the turbulence starts to decay, and collapse sets in. In a thermally dominated core the implied mass accretion rate is given by (3.15), $\dot{M} \approx c_s^3/G$ which, for a low-mass core at $T = 10$ K gives $\dot{M} \approx 2 \times 10^{-6}\,M_\odot$ yr^{-1}. For the case of an initially turbulence- dominated core, the turbulent speed is added in quadrature to the sound speed,

$$\dot{M} = \frac{(c_s^2 + v_{\rm turb}^2)^{3/2}}{G} \tag{5.3}$$

and $\dot{M} \approx 10^{-3}\,M_\odot$ yr^{-1} for the same T. Also, if one takes the total mass divided by the mean free-fall time (5.6×10^4 yr) one obtains the same approximate $\dot{M}$. However the actual collapse is not a true free fall, and these estimates for $\dot{M}$ should be reduced by at least a factor 2, giving about 2×10^5 year to form a $100\,M_\odot$ star. In fact the final mass would be reduced by an unknown factor because of the effect of outflows. The question is whether the final mass of perhaps $50\,M_\odot$ can ever be reached.

As the collapse proceeds, thermal effects eventually dominate, the Jeans mass is low, and one might expect the core to fragment into several pieces; this issue will be further discussed below. For the moment, assume that most of the mass will be accreted onto a single stellar core, which can build up to high mass (although some mass will be lost in the outflow). For simplicity, consider first the case of spherical symmetry.

The stages of protostellar evolution, once collapse sets in, are at first very similar to those for low-mass stars. The gas collapses isothermally for six orders of magnitude increase in central density and becomes highly centrally condensed. The adiabatic collapse phase then ensues for the matter near the center. It heats up and the collapse slows down until the central temperature reaches 2,000 K; then a second collapse begins in the center, induced by the dissociation of H_2. Once the dissociation has completed and the central temperature reaches 20,000 K the stellar core begins to form in hydrostatic equilibrium, initially with a mass of only a few thousandths of a solar mass.

The accretion phase begins, but it differs from that for low-mass stars in that the equilibrium core, once it exceeds about $10 M_\odot$, contracts on the same time scale as the accretion takes place. The point at which the core arrives on the main sequence depends on the accretion rate. In simple spherical models [406] at constant accretion rates, stars arrived at the main sequence with masses of 8, 10, and $15 M_\odot$, at accretion rates of 10^{-5}, 3×10^{-5}, and $10^{-4} M_\odot$ yr^{-1}, respectively. For a variable accretion rate [354] in the range $10^{-4} - 10^{-3} M_\odot$ yr^{-1} the initial main-sequence mass was about $20 M_\odot$. The difference in mass as a function of accretion rate depends on the ratio of the characteristic accretion time to the thermal adjustment time of the star. For high accretion rates, the star is not able to radiate away its excess thermal energy and contract to the radius it would have in the absence of accretion. The higher the accretion rate, the larger the radius of the star remains in comparison with the thermally adjusted radius. An additional effect comes when deuterium starts burning in a shell outside the core of the star; this effect results in a temporary increase in radius as a result of the deposition of nuclear energy which cannot be radiated rapidly. Once the star becomes massive enough, its thermal adjustment time shortens to the point where it can contract to the main sequence, and subsequent evolution moves it up along the main sequence. The object becomes very luminous, the radiation from the core surface shifts toward the UV region, ionizing photons are produced, and the question of radiative acceleration of the dusty infalling gas must be considered.

The opacity of the dust in the infalling envelope is much higher in the optical and UV than in the infrared. The opacity as a result of dust is far greater than that of the gas, but all species of dust have essentially evaporated at temperatures above 1,500 K; this temperature defines the *dust destruction front* which, for example, lies at about 8 AU for a star with $L/L_\odot = 1,000$ [557] and scales with $L^{1/2}$. To evaluate the effect of radiation in slowing down the infall, we consider a point with distance r from the star slightly larger than that of the dust destruction front. We wish to compare the force of gravity with that from the radiation pressure gradient. If the following condition is satisfied, infall of a given layer can occur:

$$\frac{GM_*}{r^2} > -(1/\rho)(dP_r/dr) = \bar{\kappa} F/c \qquad (5.4)$$

5.2 The Problem of Radiation Pressure

from basic radiation transfer theory, where $\bar{\kappa}$ is the opacity averaged over frequency

$$\bar{\kappa} = \frac{1}{F} \int_0^\infty \kappa_\nu F_\nu d\nu, \tag{5.5}$$

a flux-weighted mean. Here ρ is the density, P_r is the radiation pressure, $F(r)$ is the radiative flux from the star integrated over all frequencies, and F_ν is the flux at frequency ν. The luminosity $L_* = 4\pi r^2 F$, where the asterisk subscript refers to stellar properties. Setting the ratio of forces to 1, we find a critical value of

$$L_*/M_* = 4\pi c G/\bar{\kappa} = 1,600 L_\odot/M_\odot \tag{5.6}$$

above which radiation pressure dominates gravity and material can no longer accrete. The value of $\bar{\kappa}$ has been chosen to be the appropriate one[1] at the temperatures in the dusty region, where, between $T = 1,000$ K and $T = 300$ K, the mean opacity is roughly constant at 8 cm² g⁻¹ [420]. This value of L_*/M_* corresponds to a main-sequence star of about 15 $M_\odot$, above which further accretion tends to be shut off. Note that in the spherical case the accretion luminosity $(GM_*\dot{M}/R_*)$, where R_* is the main-sequence radius, has to be included as well. However, if, as is likely, much of the accreting material falls onto a disk at radii much larger than that of a main-sequence star, then the accretion luminosity is less than the main-sequence stellar luminosity.

Note that the same argument gives the Eddington limit, where, in the absence of dust, the opacity at the stellar surface is given by electron scattering, $\bar{\kappa} = 0.2(1 + X)$, where X is the hydrogen mass fraction. The critical value of L_*/M_* is then 36,500 $L_\odot/M_\odot$, corresponding to a mass of over 200 $M_\odot$, comparable to the highest mass known. Or, if the star is accreting, the spherical accretion luminosity $GM\dot{M}/R$ cannot exceed the Eddington limit, giving a limiting accretion rate $\dot{M} \approx 10^{-2} M_\odot$ yr⁻¹ for 100 $M_\odot$.

In most of the infalling dust envelope, the radiation is thermalized and takes on the temperature of the local gas. A given layer can continue to collapse onto the star as long as gravity exceeds the radiation pressure gradient. However an additional problem occurs at the dust destruction front itself, where the optical/UV radiation from the star directly impinges on the dust. The mean opacity is higher than 8 cm² g⁻¹ by about a factor of 3, and in the UV part of the spectrum it reaches 200 cm² g⁻¹. Thus one might expect the limiting mass to be much less than 15 $M_\odot$. This problem was investigated by Wolfire and Cassinelli [557] who made detailed (spherical) models of infalling envelopes around main-sequence stars, including detailed grain properties. They compared the rate of momentum transfer by the material of the infalling envelope, $\dot{M}v$, where v is the infall velocity, to the rate of momentum input to the dust from the stellar radiation field L_*/c, where c is the velocity of light. The condition for continued infall at the dust destruction front is

---

[1] The flux-weighted mean opacity is approximated by the Planck mean (5.13) in this case.

$$\dot{M} > \frac{L_*}{c(2GM_*/r_d)^{1/2}} \tag{5.7}$$

where it is assumed that the envelope is at free fall at the dust sublimation radius r_d. Evaluating this expression for a 30 $M_\odot$ star on the main sequence ($L_* \approx 10^5 L_\odot$), $\dot{M}$ has to be larger than $8 \times 10^{-5} M_\odot$ yr^{-1}. For 100 $M_\odot$ the requirement is $10^{-3} M_\odot$ yr^{-1}. The initial conditions in a turbulent core mentioned at the beginning of this section satisfy these conditions.

However condition (5.7) applies only at the dust destruction front. In the outer regions of the infalling envelope, say at a radius of 100 r_d, where the free-fall velocity is much lower and in fact the infall velocity is less than that of free fall, the required $\dot{M}$, from (5.7), to allow a 100 $M_\odot$ star to form goes up to more than $10^{-2} M_\odot$ yr^{-1}, and for a 30 $M_\odot$ star up to more than $10^{-3} M_\odot$ yr^{-1}. Thus the expected infall $\dot{M}$, based on conditions in the initial turbulent core, is insufficient to allow the collapse of the outer envelope for a central mass of above about 30 $M_\odot$. In that outer envelope, where the radiation has been converted into the infrared part of the spectrum, the force condition (5.6) still applies, based as previously mentioned, on a mean opacity in the infrared of 8 cm^2 g^{-1}.

Various suggestions have been made regarding overcoming the radiation-pressure problem for massive stars.

- Non-spherical geometry, with most of the mass arriving at the star through a disk and most of the luminosity (and associated radiation pressure) going in the direction of the poles where, also, there may be an outflow cavity
- Collisions and mergers of intermediate-mass stars in the dense core of a young cluster
- Non-standard grain properties in massive star formation regions, for example, reduction in grain abundance by a factor of 4 or a considerable increases in the typical grain size over interstellar values [557]
- A Rayleigh-Taylor instability, visible only in three-dimensional simulations, which allows material to break through the expanding radiation bubble and continue to accrete onto the star.

The first point was examined in a full two-dimensional hydrodynamic collapse calculation, with rotation and radiative transfer included [566]. The initial conditions were similar to that mentioned at the beginning of this section; for initial masses of 30, 60 and 120 $M_\odot$ the radii were, respectively, 0.05, 0.1, and 0.2 pc. The initial temperature was 20 K, giving a ratio of thermal to gravitational energy of 0.05 in each case. Turbulent effects were not included but the initial condition is consistent with the turbulent core models [354]; effectively the calculation assumed that the presumed initial turbulence had already decayed. The clouds rotated with $\Omega = 5 \times 10^{-13}$ s^{-1} in each case. For 60 $M_\odot$, for example, the mean density is 1.0×10^{-18} g cm^{-3}, $\rho \propto r^{-2}$, the ratio of rotational energy to the absolute value of gravitational energy is 0.09, the free-fall time is 6.5×10^4 yr, and $\dot{M} = 10^{-3} M_\odot$ yr^{-1}.

5.2 The Problem of Radiation Pressure

The equations solved are given in Chap. 4 as (4.9), (4.12), (4.17), and (4.29) through (4.32), except that the energy equation is expanded to include separate equations for the radiation energy and the gas energy (see below, Sect. 5.3). The viscosity parameter $\alpha_{\rm visc}$ is set to a constant, for treatment of the evolution of the disk. Radiation transfer, treated by the method of flux-limited diffusion (Chap. 4), includes extinction from both absorption and scattering. The grain components that are included in the opacity calculation are amorphous carbon, silicates, and ice-coated silicates. The calculation includes the formation of a disk around the central protostar, as well as its subsequent photoevaporation by the UV radiation from the central object. Results are compared for two different assumptions regarding the radiative transfer: first, a "grey" (frequency-independent Rosseland mean opacity) calculation, and, second, a full frequency-dependent calculation.

The star at the center is not resolved; it is contained in a "sink" zone of radius 40 AU, into which mass can fall but from which no mass can escape. The luminosity is determined from stellar evolutionary tracks (Chap. 8) and includes contributions from contraction, nuclear burning, and accretion. The optical/UV radiation properties of the central object are calculated, and this radiation heats the infalling envelope. Fragmentation cannot be treated in an axisymmetric 2-D calculation. A disk similar to those calculated by [566], before the effects of radiation pressure take hold, is shown in Fig. 5.5.

For the two $60\,M_\odot$ cases, the frequency-dependent case reaches a higher total mass, $34\,M_\odot$, as compared to $20\,M_\odot$ in the frequency-independent case, before radiative acceleration disperses the infalling material. The inflow is first halted and reversed in the polar direction, allowing accretion to continue in the disk region in the equatorial plane. In the presence of the disk, the radiation from the central object becomes highly non-isotropic. This "flashlight effect" results in much higher radiative flux in the polar direction, where the density is lower than in the equatorial plane, and the average radiative acceleration is reduced in the disk as compared with the spherical case. As the final mass is approached, the infall has been reversed everywhere even though most of the radiation, especially the hard UV radiation, is directed outwards in the polar direction. The disk is quickly dissipated. The difference between the frequency-independent (grey) cases and the frequency-dependent cases arises mainly because the opacity in the UV due to dust is much higher than the opacity in the IR. Thus the high-frequency radiation, which is the most effective at accelerating grains outwards, is most strongly collimated toward the poles, which constitute the lowest-density channel of escape of the radiation. Thus this radiation is able to accelerate a smaller amount of material outwards than in the grey case, where the radiation at the mean frequency is less concentrated toward the pole.

The results show that an initial molecular core mass of $30\,M_\odot$ produces a final star of mass $31.6\,M_\odot$ in the frequency-dependent case (inflow across the outer boundary is allowed) and $19.1\,M_\odot$ in the grey case. An initial core mass of $120\,M_\odot$ produces a final star of mass $42.9\,M_\odot$ in the frequency-dependent case and $22.9\,M_\odot$ in the grey case. Clearly the correct treatment of radiation transfer has a significant effect on the final mass. But the result may be dependent on the initial conditions,

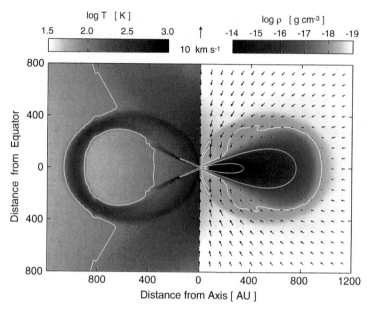

Fig. 5.5 Example of a calculation in two space dimensions [575] of the formation of a star of $10\,M_\odot$, including the effects of rotation. The plot is in the (R, Z) plane, where Z is the distance above or below the equatorial plane, and R is the distance from the rotation axis. The time is 65,000 yr after the start of the collapse of the core of the molecular cloud. The stellar core is unresolved at the center (0,0) and has a mass of $7.0\,M_\odot$, the disk has $2.8\,M_\odot$, and the infalling envelope has $0.2\,M_\odot$. The *white contours* on the *right-hand panel* correspond to densities of log $\rho = -18, -16, -14, -12$. The *white contours* on the *left-hand panel* correspond to temperatures of log $T = 2.0, 2.5, 3.0$. The *arrows* are velocity vectors, with length proportional to speed. *Dark bands* in the temperature plot indicate shock compression. Reproduced by permission. © Annual Reviews 2007

which used a constant number of thermal Jeans masses for all cases, meaning the higher masses were initially larger with lower densities and longer free fall times. An increase in $\dot{M}$ for the $120\,M_\odot$ case could have allowed its mass to grow even larger. An equivalent calculation without rotation was not performed, so it is not clear to what extent the non-spherical geometry aided star formation. However it is clear that the original mass limit of $15\,M_\odot$ in the spherical case has been significantly increased. This result is in fact an upper limit, because outflows were not included. However the outflows, although they result in mass loss, also clear out a cavity in the polar direction which would intensify the beaming effect that preferentially directs radiation toward the poles [280].

The second mechanism, stellar mergers, was considered in a numerical simulation [77]. Observationally, massive stars seem to form in the central regions of rich stellar clusters. To get a reasonable time scale for collision, however, comparable to the lifetime of a massive star, the stellar density has to be on the order of 10^8 stars per pc^3, if collisions between single stars are considered, or 10^6 stars per pc^3 if mergers

5.2 The Problem of Radiation Pressure

of pre-existing binaries are considered [78]. Even in the center of the Orion cluster the stellar density ($\sim 10^4$ stars per pc^3) is nowhere near that high, and in the dense cluster W3 IRS5 [358] the five observed OB protostars within a radius of 0.015 pc give a stellar density of $\sim 5 \times 10^5$ stars per pc^3 plus an unknown contribution from lower-mass stars. The merger scenario would require that clusters produce a short-lived high-density core, which later reexpands as a result of the ionizing photons and outflows of the high-mass stars formed there.

The numerical simulation, which is discussed here as an example of cluster evolution, is a combination of N-body and SPH methods. There are initially 1,000 stars in a bound cluster with a Gaussian density distribution and a mean density of about 200 stars per pc^3. Additionally there is gas with 91% of the total mass. All of the stars are of equal mass (0.5 M$_\odot$ in the numerical example). The initial cluster radius is 1 pc, and stars and gas have zero velocity. Initially there are 1,000 thermal Jeans masses in the cloud. A million particles are used. An SPH particle is assumed to have accreted onto a star if it is gravitationally bound and if it is within 20 AU. Two stars are assumed to have merged if they are gravitationally bound and within 2 AU of each other. The calculation was followed until 7% of the gas accreted onto stars. In fact a high-density core is produced, with $\sim 10^8$ M$_\odot$ pc^{-3}, with massive stars forming there and resulting in mergers. Radiative feedback from the massive stars was not included, which would tend to decrease the accretion rate of gas onto stars. The final mean stellar mass is 0.8 M$_\odot$ and the most massive star has 50 M$_\odot$. During the simulation, 19 mergers occur, 5 of which help to build up the most massive star. A number of binaries are produced by 3-body capture. The mergers are generally produced by hardening of close binaries. The term "hardening" refers to loss of angular momentum from the orbit of the binary, and consequent shrinking of the orbit. The loss of angular momentum occurs partly from interactions of a third star with a pre-existing binary, partly from accretion by the binary of low-angular-momentum gas, and partly from transfer of angular momentum from the orbit to a circumbinary disk. In the simulation being described, the actual merger processes took place as a result of a close encounter of the binary with a third star, which carried away some angular momentum. The most massive star, which is a binary, obtained half of its mass by gas accretion and the other half by mergers. The mass function at the end looks strikingly like the IMF of Salpeter ($dN/dm \propto m^{-2.35}$).

The specific outcome may have been affected by the initial conditions and the liberal criterion for a merger to occur, but the authors probably are correct in pointing out that the formation of a massive star must be considered not in isolation, but in connection with the formation of an entire cluster. Also, a model of this type is reasonably consistent with the high percentage of massive stars that exist in binaries. Although the binary merger process produces, at least temporarily, a single star, it is easy for the object to capture a new star into orbit, in the dense environment in the cluster center. A problem is, however, that mergers would tend to suppress the formation of disks and associated outflows.

An observational result, however, seems to cast considerable doubt upon the merger scenario. Observations of $v \sin i$, the spectroscopic rotational velocity of young stars in the range 0.2–50 M$_\odot$[556], show that the ratio of the rotational velocity to the equatorial breakup velocity (centripetal acceleration equal to gravity at the equator) is nearly constant with mass and has a median value of ≈ 0.15. There is no sign of a discontinuity over this entire mass range. Furthermore the specific angular momentum of rotation is a smooth function of mass, again, with no evidence of a discontinuity. This observation suggests that there is a unique star formation mechanism, at least up to $50\,M_\odot$: if mergers played a role at high mass one would expect rotational velocities above the average in that mass range. Thus, while the occurrence of mergers is not ruled out, they probably play a minor role in massive star formation.

The third point, involving the suggestion that grain properties are somehow different in high-mass star forming regions as compared with the interstellar medium in general, has little if any observational support. As mentioned above, it would take a significant reduction in the dust-to-gas ratio to allow the formation of a $60\,M_\odot$ star. However it is conceivable that in the process of collapse, the grain properties could be modified by coagulation: at a given total mass, the mean opacity of large grains is significantly less than that of small grains. A calculation involving 30 grain species [496] showed that coagulation and settling effects became important in the disk once it had formed, and shattering effects become important in shocked regions, but the effects were negligible in the infalling material outside the disk. Thus at best the grain effects could reduce the radiative acceleration in the disk.

The last point, three-dimensional effects, is discussed in the next section.

5.3 Full 3D with Radiation Transfer

The key to solving the massive star problem is to perform full three-dimensional numerical simulations with radiation transfer. Here the physics becomes very complicated, and the amount of mass growth that can be followed becomes somewhat limited. A further problem that must be overcome is that inevitably very dense localized regions form, which severely limit the time step (Chap. 4). In SPH, the particles in such a region, once a certain density criterion is met, are combined into a single "sink" particle which is treated according to N-body dynamics. An analogous method has been developed for Eulerian grid-based numerical simulations [279]. The main questions that 3-D simulations could answer that could not be answered in 2-D are the following: (1) will the 100 solar mass collapsing cloud actually build up into a massive star, or will it fragment into small pieces? and (2) are there 3-D effects that permit the accretion to high masses, in spite of the problem of radiation pressure?

First we extend the set of equations given in Chap. 4 to three space dimensions, including radiative transfer in the flux-limited diffusion approximation, and including the dynamics of both the gas and the sink particles, which in fact are unresolved protostars. In two space dimensions the sink particle is fixed at the center and is more accurately called a sink zone; it is allowed to accrete mass. In 2D the center is the only likely location where a high-density region might form. In three

5.3 Full 3D with Radiation Transfer

space dimensions a high-density concentration may occur anywhere in the grid, and the sink particles are followed as Lagrangian elements as they move through the grid and accrete mass. The set is further extended over that in Chap. 4 to include separate equations for the energy of the gas and the energy of the radiation. The following equations [275, 277] are written in a cartesian coordinate system and neglect quadratic and higher-order terms in v/c:

$$\frac{\partial \rho}{\partial t} + \nabla \cdot (\rho \mathbf{v}) = -\Sigma_i \dot{M}_i W(\mathbf{x} - \mathbf{x_i}) \tag{5.8}$$

$$\frac{\partial (\rho \mathbf{v})}{\partial t} + \nabla \cdot (\rho \mathbf{v} \, \mathbf{v}) = -\rho \nabla \Phi - \nabla P - \lambda \nabla u_{\text{rad}} - \Sigma_i \dot{\mathbf{p}}_i W(\mathbf{x} - \mathbf{x_i}) \tag{5.9}$$

$$\frac{\partial (\rho E)}{\partial t} + \nabla \cdot (\rho E \mathbf{v}) = -P \nabla \cdot \mathbf{v} - \rho \mathbf{v} \cdot \nabla \Phi - \kappa_P \rho (4\pi B - c u_{\text{rad}})$$
$$+ \lambda \left(2\frac{\kappa_P}{\kappa_R} - 1 \right) \mathbf{v} \cdot \nabla u_{\text{rad}} - \Sigma_i \dot{\epsilon}_i W(\mathbf{x} - \mathbf{x_i}) \tag{5.10}$$

$$\frac{\partial u_{\text{rad}}}{\partial t} + \nabla \cdot \left(\frac{3 - R_2}{2} u_{\text{rad}} \mathbf{v} \right) = \nabla \cdot \left(\frac{c \lambda}{\kappa_R \rho} \nabla u_{\text{rad}} \right) + \kappa_P \rho (4\pi B - c u_{\text{rad}})$$
$$- \lambda \left(2\frac{\kappa_P}{\kappa_R} - 1 \right) \mathbf{v} \cdot \nabla u_{\text{rad}} + \Sigma_i L_i W(\mathbf{x} - \mathbf{x_i}) \tag{5.11}$$

$$\nabla^2 \Phi = 4\pi G \left[\rho + \Sigma_i M_i \delta(\mathbf{x} - \mathbf{x_i}) \right]. \tag{5.12}$$

Here $\mathbf{x_i}$, $\dot{M}_i$, $\mathbf{p_i}$, and M_i are respectively the position, mass accretion rate, momentum, and mass of the ith sink particle. The quantities ρ, $\mathbf{v}$, E, and u_{rad} are respectively the gas density, the gas velocity, the internal plus macroscopic kinetic energy, per unit mass, of the gas, and the energy per unit volume in the radiation field, in the rest frame of the computational grid. [Note that in (4.16) the quantity E is just the internal energy per unit mass of the gas, not including the kinetic energy].

In (5.8), (5.9), and (5.10), the last term represents, respectively, the rate of transfer of mass, momentum, and energy, per unit volume, from the gas to the particles by accretion. Thus $\dot{\epsilon}_i$ is the rate of transfer of total energy, thermal plus kinetic. The summation Σ_i is over the i sink particles that have been created. In (5.11) the last term represents the energy input to the radiation field arising from the luminosities L_i of the protostellar sink particles. This term in an important feedback mechanism. Each particle is followed as it accretes gas, and the appropriate luminosity is calculated according to evolutionary models of isolated protostars. The function $W(\mathbf{x} - \mathbf{x_i})$ is a weighting function, with dimensions of inverse volume, which defines the region in the gas flow which is affected by accretion onto particle i, which is taken to be a few grid cells in the vicinity.

In (5.8) through (5.11), the second term on the left-hand side represents the advection of the relevant variable arising from the flow through the fixed grid. In (5.12), Φ is the gravitational potential and δ is the Dirac delta function, chosen to represent the spike in density at $\mathbf{x_i}$. The Rosseland mean opacity κ_R is given by

(8.40), and the Planck mean κ_P is

$$\kappa_P = \frac{1}{B} \int \kappa_\nu B_\nu(T) d\nu \tag{5.13}$$

where B is the Planck function $B_\nu(T)$ integrated over all frequencies. Both of these mean opacities are independent of the local radiation field and can be tabulated as a function of density, temperature and composition. In (5.10) and (5.11) the terms involving $(4\pi B - c u_{\text{rad}})$ represent the exchange of energy between matter and radiation: the term $\kappa_P \rho 4\pi B$ is the radiation emitted by the gas, while the term involving $c u_{\text{rad}}$ is the radiation absorbed. The quantities κ_P and κ_R are evaluated in the reference frame comoving with the gas, while all other quantities are evaluated in the fixed rest frame of the grid.

The first term on the RHS of (5.10) represents the rate of work done by the gas through expansion or contraction, and the second represents the effect on the energy generated by the external force of gravity. The first term on the RHS of (5.11) represents the energy transferred by radiation (see 4.9). The quantities relevant to flux-limited diffusion [316] are

$$\lambda = \frac{1}{R_{\text{lp}}} \left(\coth R_{\text{lp}} - \frac{1}{R_{\text{lp}}} \right) \tag{5.14}$$

$$R_{\text{lp}} = \frac{|\nabla u_{\text{rad}}|}{\kappa_R \rho u_{\text{rad}}} \tag{5.15}$$

$$R_2 = \lambda + \lambda^2 R_{\text{lp}}^2 \tag{5.16}$$

where λ is the flux limiter and R_2 is the Eddington factor.[2] An alternate approximation for λ is given in Chap. 4. In (5.9) the term $-\lambda \nabla u_{\text{rad}}$ is the radiation force $\kappa_R \rho \mathbf{F}/c$, where $\mathbf{F}$ is the radiative flux; (this term is analogous to the ∇P term for the gas). The $\nabla \Phi$ term represents the gravitational force. In (5.10) the term involving $\lambda \mathbf{v} \nabla u_{\text{rad}}$ is the work done by the radiation field on the gas, and in (5.11) the same term with a minus sign represents the work done by the gas on the radiation field. The second term on the LHS of this equation represents the advection of radiation.

The individual protostars, once formed, move through the gas according to

$$\frac{d\mathbf{x}_i}{dt} = \frac{\mathbf{p}_i}{M_i} \quad \text{and} \quad \frac{d\mathbf{p}_i}{dt} = -M_i \nabla \Phi + \dot{\mathbf{p}}_i \tag{5.17}$$

where, in the momentum equation, the first term on the right-hand side represents the gravitational effect and the second the accretion effect. A crucial quantity is

---

[2]In general, the Eddington factor, either at a given frequency or integrated over all frequencies, is the ratio of the radiation pressure to the radiation energy density. It usually varies between the values of 1/3 (for a nearly isotropic radiation field) and 1 (for a highly beamed, non-isotropic, radiation field).

5.3 Full 3D with Radiation Transfer

then $\dot{M}_i$, the gas accretion rate onto the sink particles, which can be roughly approximated by the Bondi-Hoyle-Lyttleton rate. If a point mass moves through a medium with density ρ_∞ at a velocity v_∞, it accretes at the Hoyle-Lyttleton rate [235]

$$\dot{M} = \frac{4\pi G^2 M^2 \rho_\infty}{v_\infty^3} \tag{5.18}$$

where the density and velocity are defined at a distance reasonably far from the point mass. Bondi [76] then extended this formula to include gas pressure within a medium with sound speed c_∞. The revised formula, usually known as the Bondi-Hoyle accretion rate, is

$$\dot{M} = \frac{4\pi G^2 M^2 \rho_\infty}{(v_\infty^2 + c_\infty^2)^{3/2}}. \tag{5.19}$$

However, as pointed out by [152] there are numerous simplifying assumptions in this formula, as compared to the actual accretion flow in the protostar. For example, the gas in the accretion flow is assumed to have negligible mass in comparison with the point mass. In actual numerical simulations the accretion rate onto the particles is adjusted to local flow conditions, using (5.19) as a guide [279].

Numerical simulations of the formation of a massive star from a turbulent core have been reported, but this difficult computational problem is still work in progress. An example is a full three-dimensional collapse calculation with grey radiative transfer, using mean opacities κ_R and κ_P and radiative feedback from the cores that have formed, with an adaptive-mesh grid code [276]. The initial condition is a turbulently supported core of radius 0.1 pc and 100 $M_\odot$, with a centrally condensed density distribution. The turbulence is allowed to decay. Some fragmentation occurs, but much less than what happens if the run is approximated to be purely isothermal with no feedback. It is mainly the increase in temperature of the infalling gas, generated by the luminosity of the main core, that limits fragmentation. The results show that most of the mass goes into one dominant star with most of the mass being added by accretion of gas rather than by mergers. It gains mass at 10^{-3}–$10^{-4}\, M_\odot$ yr^{-1}. A disk forms around the massive star, and gravitational instability results, producing rapid accretion onto the star as well as disk fragmentation. Most of the fragmentation in fact occurs in the disk, but the fragment masses remain low, and some of them merge with the central star. An example of the disk fragmentation is shown in Fig. 5.6. This occurs about 20,000 yr after the beginning of the simulation. However the central star, which is accepting mass from the disk at a rate of $10^{-4}\, M_\odot$ yr^{-1} at this time, continues to accrete, and relatively little mass goes into the fragment.

In two different runs up to 2×10^4 yr (only about 1/3 of the initial mean-density free-fall time) the most massive star accreted 5.4 and 8.9 $M_\odot$, respectively, so the effect of stellar UV radiation on the further buildup of the star was not considered. The two runs differ only in the statistics of the initial spectrum of velocity perturbations, which correspond to those expected in supersonic turbulence.

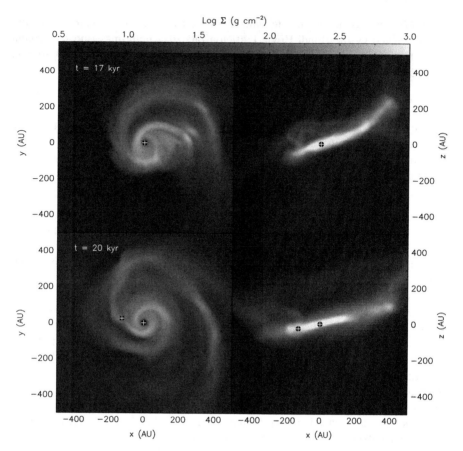

Fig. 5.6 Numerical 3D simulation of the formation of a massive star from a turbulent core. Stars are indicated by *crosses*. The *upper panel*, at a time of 17 kyr, shows the column density, integrated along lines of sight face-on to the disk as a function of position in the (x, y) plane (*left*) and perpendicular to the disk axis in the (x, z) plane (*right*). Distances on the axes are given in AU. The lower panel shows the same at 20 kyr. The generation of spiral arms as a result of gravitational instability is evident. The fragment that has formed in the disk at the later time has a mass of only 0.2 $M_\odot$, in comparison to the central protostar, which has a mass of 5.4 $M_\odot$. Reproduced by permission of the AAS from [276]. © 2007 The American Astronomical Society

The overall conclusion is that massive stars form from massive cores, and that mergers play only a minor role.

Further calculations from a similar initial condition allowed the calculation to be carried to a higher total mass [277]. The angular momentum, instead of arising from the random turbulent velocity field, was assumed to be generated by an overall uniform rotation with rotational energy 2% that of the absolute value of the gravitational energy. At 57 kyr the configuration consisted of a binary with masses 41.5 $M_\odot$ and and 29.2 $M_\odot$ with an estimated semimajor axis of 1,280 AU (Fig. 5.7).

5.3 Full 3D with Radiation Transfer

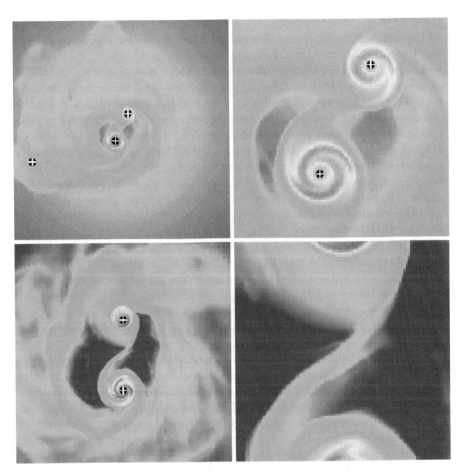

Fig. 5.7 Numerical 3D simulation of the formation of a massive star system from a uniformly rotating core of 100 $M_\odot$ [277]. Stars are indicated by *crosses*. The *upper panels* correspond to a time 41.7 kyr, the lower, to 55.9 kyr. The logarithmic color scale indicates the column density in a plane perpendicular to the rotation axis, with a range from 10^{-1} (*dark blue*) to 10^3 (*orange*) $\mathrm{g\,cm^{-2}}$. The box size is 966 AU in the *right-hand panels* and 3,864 AU in the *left-hand panels*. The generation of a massive binary is evident; note circumstellar disks as well as circumbinary material. The orbit expands with time and reaches a semimajor axis of 1,280 AU at the end of the simulation. From Krumholz et al.: *Science* **323**, 754 (2009). Reprinted by permission of the AAAS. © The American Association for the Advancement of Science

Young O-type stars in fact are often contained in systems with similar properties. The companion again was formed by disk fragmentation.

Once the star reached about 17 $M_\odot$, radiation effects became important and gas was driven outward in the polar direction. Along with this outflow a "radiation bubble" developed, both above and below the equatorial plane, in which radiation pressure dominated gas pressure. At least two different effects allowed continuing accretion onto the star. Infalling gas hit the boundary of the bubble and was shocked.

However it was able to travel along the edge of the bubble until it hit the disk, which was shielded from the radiation and allowed continued accretion onto the star. Eventually, when the radiation pressure became strong enough, a second effect took over. The bubble itself became unstable and non-axisymmetric, developing a clumpy structure. Most of the mass became relatively dense in the clumps, and because of the high optical depth there, the radiation tended to flow around the clumps through the low-density material between them. The radiation can ionize and heat the low-density material, and the resulting expansion of these regions can further compress the higher-density clumps. The net force on the clumps themselves resulted in inward motion even though, on the average over the surface, the radiation pressure gradient dominated the gravitational force. This effect is a form of the classical Rayleigh-Taylor instability, which involves a heavy gas on top of a light gas in a gravitational field: dense fingers of the heavy fluid move downwards through the light fluid. In the massive star case, the radiation-filled cavity plays the role of the light fluid. Material continues to infall onto the disk, and buildup of mass is expected to continue after the 57 kyr cutoff, up to an undetermined final mass. The final mass of $41.5\,M_\odot$ is considerably higher than the maximum mass obtained in 2-D simulations ($22.9\,M_\odot$) with similar assumptions regarding the radiation transfer but somewhat different initial conditions [566]. Again, most of the mass growth occurred through gas accreting through the disk, not by collisions with small fragments.

At least three additional effects, not included in the above calculation, could tend to allow further accretion onto the star despite radiation pressure. First, the calculation used grey radiation transfer rather than frequency-dependent transfer, in view of computer time limitations. In 2-D, as mentioned above [566], frequency dependence allowed accretion to a higher mass than did grey radiation transfer. The same trend is likely in 3-D. Second, outflows were not considered. While they do limit the amount of mass that finally accretes onto the star(s) to roughly 1/3 that of the initial molecular cloud core, they reduce the effect of radiation pressure over much of the solid angle away from the pole by allowing much of the radiation to escape along the polar outflow cavity, thus accentuating the "flashlight effect". Third, if magnetic fields had been included, the photon bubble could have developed another instability [520]. The instability occurs if magnetic pressure and radiation pressure are larger than gas pressure and the bubble is optically thick to its own thermal photons. Shocks develop in the infalling gas, allowing dense regions of gas to form and to continue to move inward, while allowing the photons to escape through the low-density regions between the shocks.

The overall conclusion is that although the problem has not been completely solved, there are reasonable physical mechanisms that can overcome the radiation pressure barrier to the formation of a massive star. However, the two examples discussed above show that the details of the outcome are sensitive to even small changes in the initial conditions, for which there are many possibilities. A further calculation [275] shows how the outcome depends on the assumed surface density of the molecular cloud core. As above, the core contains $100\,M_\odot$ and includes a random turbulent velocity field such that the kinetic energy is comparable to the gravitational energy. Surface densities of 0.1, 1, and 10 g cm^{-2} are considered. For the lowest value, radiation feedback effects are relatively unimportant, and fragmentation produces a cluster of low-mass stars. For the highest value, most of the mass that ends up in stars is contained in a single massive star. Fragmentation is suppressed because of the heating of the gas by the radiation resulting from accretion onto the first protostellar cores. In the high-surface-density case, the higher density results in higher $\dot{M}$ and therefore higher accretion luminosity than in the low-surface-density case. Furthermore, the high-density gas has a higher optical depth and is able to trap the radiation more efficiently than in the low-density case. This result reinforces the conclusion that massive stars are most likely to form from molecular cloud cores of high surface density.

5.4 Massive Star Formation: Competitive Accretion or Monolithic Collapse?

There has been an ongoing debate regarding the mechanism for the formation of massive stars, which involves primarily the initial condition for the formation. In the models described in the previous section, the basic premise was that a single molecular cloud core of ∼0.1 pc size would result in the formation of a single, or perhaps a binary, massive star, provided that the surface density is high enough. The cloud core has presumably formed by condensation from a much larger unit, a molecular cloud clump with a mass of thousands of $M_\odot$, but once the core has formed, the resulting collapse is little influenced by the remainder of the material in the clump, which is overall stable to collapse under gravity. Thus low-mass stars would form from individual low-mass cores, and high-mass stars from high-mass cores. The entire mass of the core would not necessarily go into the final star; there is an efficiency factor involved as a result primarily of outflows during the formation process, but that factor is likely to be fairly uniform over a set of cores. This model is often referred to as *monolithic collapse*.

There are several observational consequences of this picture. First, the initial mass function of stars is essentially already determined by the process of fragmentation of the cloud clump to form the cores. Thus the observed core mass function must be very similar to the observed IMF. There is considerable observational evidence that this is in fact the case. For example, in the mass range 0.5–3 $M_\odot$ [370] and in the range 1.7–25 $M_\odot$ [53] the core mass spectrum $dN/dm \propto m^{-2.5}$ (with substantial uncertainty at the massive end), very close to the stellar mass spectrum. Second, the fact that the cores in the model are non-interacting strongly suggests that the efficiency of star formation in a given clump should be relatively low, that is, only a small fraction of the clump actually evolves to cores that are unstable to gravitational collapse, over a dynamical time. The comparison with observations is accomplished through the use of (2.56), the efficiency of star formation per free-fall time $\epsilon_{\rm ff}$ [283]. The results, based on structures in the density range $10^2 - 10^5$ cm^{-3}, show an efficiency of only a few percent per dynamical time, not too different from the overall value in larger-scale molecular clouds, about 1% per free-fall time [576]. Third, in young systems, the spatial distribution and velocity distribution of young stars and cloud cores should be similar, again reasonably consistent with observations [274].

The second model, competitive accretion, is based on the premise that star formation is controlled not by collapse of individual cores, but by the overall collapse of a much larger region, containing initially gas with several thousand $M_\odot$. The fragmentation into a cluster of stars, containing high-mass and low-mass members, takes place after the overall collapse has started. Individual fragments form at low mass and compete for the accretion of the remaining gas; also there can be interactions among the various fragments. One might expect a massive star to form in the center, since as the cloud collapses and develops a gravitational potential well, a considerable amount of gas can be funnelled toward the center. In fact massive stars are found to be preferentially located near the centers of clusters, but it is not entirely clear that they formed there. They could have settled to the center after formation as a result of dynamical interactions with other stars. Thus massive stars do form, but the majority are low-mass, because the typical thermal Jeans mass at 10 K in such a cloud is about 1 $M_\odot$. In this picture, then, massive stars must form in clusters, as is generally observed.

An example of a calculation involving competitive accretion [37] involves an SPH calculation with sink particles, starting with a cloud of 500 $M_\odot$ and a radius of 0.4 pc, mean density of about 10^5 cm^{-3} and a temperature of 10 K. Although the typical star-formation clump actually is somewhat larger than this, computational requirements limit the size of the region that can be practically computed, and the basic features of a competitive accretion model should be represented. With a total of 3×10^7 particles, masses down to the brown dwarf range can be resolved. Radiative feedback from the stars that have formed is not included, nor are magnetic fields. The initial cloud is supported against collapse by an assumed turbulent velocity field, with turbulent energy approximately equal to the absolute value of gravitational energy, and with random turbulent velocities consistent with Larson's linewidth – size relation. As the evolution starts, shock dissipation results in a loss of turbulent energy, and collapse begins. The initial fragmentation of the cloud into low-mass cores is followed by the buildup of the mass of these cores primarily as a result of competitive accretion. The cores themselves would exist only for a very transient period before developing protostars in their interiors; thus the "starless core" phase is brief, and by the time a fragment has built up to high mass, it is very likely to contain an infrared source.

The end result of the calculation was the production of 459 stars and 795 brown dwarfs, with an overall mean mass of 0.15 $M_\odot$. Thus the overall star formation efficiency was high, about 20% in 1.5 initial free-fall times. Several subclusters formed, which then merged into one centrally condensed cluster. In the centers of the subclusters the formation of massive stars could begin. In the actual simulation the most massive star to form had about 7 $M_\odot$, but it was continuing to accrete. (A similar simulation [79] with a higher total cloud mass but fewer SPH particles produced a star of 27 $M_\odot$). The IMF is in good agreement with the observed one

5.4 Massive Star Formation: Competitive Accretion or Monolithic Collapse?

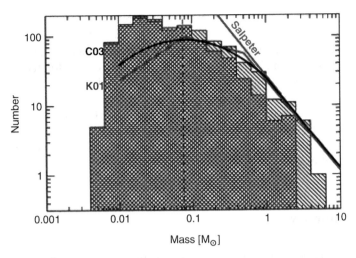

Fig. 5.8 The initial mass function produced in a numerical simulation of cluster formation with an initial total mass of 500 $M_\odot$. The single-hatched region includes all 1,254 stars and brown dwarfs that were formed, while the double-hatched region includes only those objects which have stopped accreting. The line labelled "Salpeter" gives the Salpeter IMF [443], that labelled K01 gives the Kroupa IMF [270], and that labelled C03 gives the Chabrier IMF [107]. Credit: M. R. Bate: *MNRAS* **392**, 590 (2009). By permission of John Wiley and Sons. © Royal Astronomical Society

down to about 0.1 $M_\odot$ (Fig. 5.8) but it overproduces brown dwarfs. Numerous comparisons with the observed properties of binary and multiple systems are possible; they are treated in Chap. 6. The main mechanism for termination of accretion onto a given fragment is dynamical interaction with another fragment, ejecting it into a region where little gas density remains. Thus brown dwarfs could maintain their low masses by ejection from an unstable multiple system, soon after formation [431].

The main criticism of the competitive accretion model is that it is very efficient. If the three-dimensional calculations had been continued for longer times, most of the mass in the initial clouds would have ended up in stars. This problem is most likely a result of the neglected physical effects – radiative feedback and magnetic fields – both of which tend to suppress fragmentation. The higher temperatures near stars caused by radiative feedback tend to suppress disk fragmentation. In fact, simulations [38, 424] with magnetic fields and radiative feedback included, but with a much smaller initial cloud mass because of computational limitations, at least partially resolve these problems. The mean stellar mass increases and overall star formation efficiency per free-fall time drops to below 10%, in better agreement with observations. A further simulation of cluster formation [526] using SPH with an initial volume of about 1 pc^3 and a mass of 760 $M_\odot$ includes heating from the accretion luminosity and the contraction luminosity of the protostars that form. Compared with an isothermal simulation with the same initial conditions, the heating results in far fewer fragments and an increase in the average mass, clearly favoring the formation of massive stars. Another result of the calculation is that the accretion rate onto a given protostar is highly variable in time.

Thus the advantages of the competitive accretion model are, first, that it considers the massive star formation problem in the context of cluster formation, a picture which is consistent with observations. Second, it is able to predict an initial mass function, also consistent with observations. Third, it can be used to explain the observed properties of binary and multiple systems. However the computational demands for this model, considering the large number of stars that eventually are produced, are markedly greater than for the monolithic collapse model, to attain the same level of physical input and numerical resolution.

5.5 Summary

It is clear that high-mass star formation differs from low-mass star formation, but in general there is not a sharp distinction: in many respects there is a smooth transition between the two cases, as regards, for example, the importance of radiation pressure in the infalling protostellar envelope and the ratio of time scales of accretion and quasi-static contraction to the main sequence. There is no obvious change in slope of the initial mass function in the high mass region as compared to the $1-10\,M_\odot$ region, although the IMF at high masses is difficult to determine. High-mass stars form in turbulent cores, but again there could be a smooth transition between weakly turbulent low-mass cores and highly turbulent high-mass cores. There is no clear observational evidence on this point.

Bipolar outflows are observed among high-mass protostars, and their properties are similar but scaled-up versions of those of low-mass stars, although at the very high-mass end they appear to be less sharply collimated. Their formation mechanism, as in the low-mass case, is likely to be magnetic interaction near the star-disk interface, with radiation pressure playing a lesser role. Disks are thus inferred to be present among high-mass stars, although the observational evidence is scant, and they are certainly produced in numerical simulations.

There does appear to be a smooth transition in the binary frequency as a function of mass, with high-mass stars formed preferentially in binaries or higher-order multiples. In the monolithic collapse picture of massive star formation, the binary formation mechanism is fragmentation in a massive, gravitationally unstable protostellar disk. In the competitive accretion scenario, the preferred formation mechanism is three-body capture in a young cluster-formation interactive environment, although some disk fragmentation is possible as well. The question of binary formation is considered in more detail in the next chapter.

In the monolithic collapse picture, supersonic turbulence in molecular clouds, in a region that is overall stable to gravitational collapse, produces occasional high-density fluctuations which can form molecular cloud cores and then collapse as individual stars. Thus fragmentation occurs first, then collapse. There is little interaction between cores once they have formed. Observations show that the mass function of the cores is very similar to that of the observed IMF. Thus the mass function of the cores determines the mass function of the stars, taking into account an efficiency factor, most likely caused by outflows, of roughly 50%. Thus massive stars form from massive cores. Radiation feedback, as well as magnetic fields, prevent the core from fragmenting into many small pieces.

In the competitive accretion picture, a region of a molecular cloud, containing perhaps a few thousand $M_\odot$, becomes overall unstable to collapse, overcoming the effects of turbulent pressure. Fragmentation into smaller masses then occurs after the overall collapse starts. The fragmentation starts with a number of low-mass cores, which compete for the remaining gas. The fragments can interact strongly once they are formed. The central regions are favored to accrete more gas than the outlying regions, and massive stars tend to form there. Mergers of fragments can occur, and the spiralling together of the massive components of a close binary system could be a way of producing a very massive star. The end result is the formation of a cluster, with massive stars preferentially at the center. Numerical simulations of this process also show that the mass function is in very good agreement with the IMF. Whether this process or the alternative one dominates is yet to be worked out, but the structure of observed molecular clouds could account for both possibilities.

A significant feature of high-mass stars is their production of ionizing photons, which, above a threshhold level, can ionize and eventually disrupt the surrounding molecular cloud material, suppressing further star formation in the region. However this property does not depend on any particular star-formation mechanism.

There is in fact an internal barrier to formation of very massive stars, caused by the high luminosity of the stellar core once it reaches the main sequence at 10–20 $M_\odot$, while still accreting. The collapse of dusty infalling regions can, in principle, be reversed by the radiative force exerted by the stellar photons. A combination of effects can allow inflow to continue; they include (1) a high accretion rate ($\sim 10^{-3}\, M_\odot\, yr^{-1}$) which is a consequence of the relatively high density and consequent short free-fall time in the initially turbulence-supported molecular cloud core, (2) disk formation and bipolar outflows, which channel the radiation in the polar direction and allow accretion through the disk, and (3) hydrodynamic and hydromagnetic instabilities, evident only in three-dimensional simulations, in which "fingers" of high-density infalling material can penetrate through the high-radiation-pressure environment and accrete onto the stellar core. Nevertheless there is an upper limit to the mass of a star, and, although it is not clear what causes the limit, the effects of radiation pressure, outflows, and the formation of HII regions may all be significant.

5.6 Appendix to Chap. 5: Determination of the IMF

The basic quantity to be determined is the initial mass function, usually given the symbol $\xi(\log m)$, which is the number of stars formed over the history of the galaxy in a given volume of space, per unit logarithmic mass interval. It is usually expressed as number of stars per unit volume. However Scalo [447] defines it per unit area, that is, the number of stars per pc² in the Galactic disk. To get the volume density in this case one must divide by the scale height of the Galactic disk for the particular population of objects that is being counted. The IMF includes stars which have evolved and are no longer visible. It can be normalized in various ways, for example it could be defined as the fractional number of stars in the given mass interval in the given volume. It is also useful to define the slope of the IMF,

$$\Gamma(m) = -\frac{\partial \log \xi(\log m)}{\partial \log m}\Big|_m, \qquad (5.20)$$

which is independent of mass if the mass function is a power law, and which for the Salpeter IMF gives $\Gamma = 1.35$.

A summary of the actual procedure, based on field stars in the nearby region of the Galaxy, follows [107, 271, 447]. It should be emphasized that the procedure is complicated enough so that numerous uncertainties are introduced, particularly for massive stars. It is assumed that $\xi(\log m)$ is time-independent and a smooth function of mass. The fundamental observed quantity is the luminosity function, $\phi(M_v)$, which is the total number of stars observed at absolute visual magnitude M_v per unit volume, per unit magnitude interval. Note that distances are needed for all stars included in the count. The main-sequence stars in the sample are then used to convert this distribution into a mass function $\phi_{ms}(\log m)$, from the main-sequence mass-luminosity relation. This mass function is the number of main-sequence stars per unit volume, at mass m per unit logarithmic mass interval. It is obtained from the relation

$$\phi_{ms}(\log m) = \phi(M_v)\left[\frac{dM_v}{d\log m}\right] g_{ms}(M_v) \qquad (5.21)$$

where g_{ms} is the fraction of stars at a given M_v that are on the main sequence, that is, the actual observations are corrected for stars that have evolved away from the main sequence. The derivative in brackets is obtained from the well-known main-sequence mass-luminosity relation obtained from stellar models and calibrated through use of those stars, in binaries, whose masses can be determined directly. The relation that should be used in the brackets is the *average* luminosity of a main-sequence star of a given mass during its main-sequence lifetime.

The function $\phi_{ms}(\log m)$ is known as the Present-Day Mass Function (PDMF). However, particularly for massive stars, that is not the same as the IMF, which refers to the frequency distribution of stellar masses at birth in the region considered, integrated over the lifetime of the Galactic disk. However the massive stars evolve in a time short compared with the lifetime of the galactic disk, so a correction must be made for stellar evolution, taking into account the main-sequence lifetime of a star of given mass as well as the history of star formation over the lifetime of the disk. For stars less than about $0.9\,M_\odot$, whose main-sequence lifetimes are greater than the age of the disk, the PDMF is essentially the IMF.

5.6 Appendix to Chap. 5: Determination of the IMF

To make the correction, define a "creation function" $C(\log m, t)$ which is the number of stars born per unit volume in the disk during the time interval t to $t + dt$ in the mass range $\log m$ to $\log m + d \log m$, averaged over a suitably large volume around the Sun. This function is distinct from the "birthrate", to be defined below. Let the age of the galactic disk be $T \approx 1.0 \times 10^{10}$ yr. Then the total number of stars ever formed in the particular region of the Galaxy, per unit volume, is

$$N_{\text{tot}} = \int_{\log m_{\text{low}}}^{\log m_{\text{high}}} \int_0^T C(\log m, t) \, dt \, d \log m \qquad (5.22)$$

where m_{high} and m_{low} are respectively the upper and lower limits to stellar mass.

For stars with main-sequence lifetimes $\tau(m) < T$, the PDMF is

$$\phi_{\text{ms}}(\log m) = \int_{T-\tau(m)}^{T} C(\log m, t) dt \qquad (5.23)$$

because all stars born between $t = 0$ and $T - \tau(m)$ have evolved away from the main sequence. And the actual IMF is

$$\xi(\log m) = \int_0^T C(\log m, t) dt. \qquad (5.24)$$

Now assume that the creation function is a product of separate functions of mass and time, $C(\log m, t) = H(\log m) B(t)$ where $B(t)$, the birthrate function, is the total number of stars born per unit volume per unit time at time t. The quantity $H(\log m)$ is closely related to the IMF, as shown below. As mentioned above, $\xi(\log m)$ is time-independent but the birthrate can vary with time.

$$B(t) = \int_{\log m_{\text{low}}}^{\log m_{\text{high}}} C(\log m, t) d \log m. \qquad (5.25)$$

Thus the average birthrate, over the entire lifetime of the galactic disk, is

$$\langle B \rangle = \frac{1}{T} \int_0^T B(t) dt. \qquad (5.26)$$

Now define the relative birthrate $b(t)$ as the absolute birthrate in units of the average birthrate:

$$b(t) = \frac{B(t)}{\langle B \rangle}. \qquad (5.27)$$

Thus for example at the present time $b(T)$ is the birthrate now divided by the average birthrate in the past. The function $b(t)$ has the property

$$\int_0^T b(t)dt = T. \tag{5.28}$$

Now combine different expressions for N_{tot} and use the definition of $\xi(\log m)$

$$\begin{aligned} N_{\text{tot}} &= \int_{\log m_{\text{low}}}^{\log m_{\text{high}}} \int_0^T H(\log m) B(t) \, dt \, d \log m \\ &= \langle B \rangle T \int_{\log m_{\text{low}}}^{\log m_{\text{high}}} H(\log m) d \log m \\ &= \int_{\log m_{\text{low}}}^{\log m_{\text{high}}} \xi(\log m) d \log m \end{aligned} \tag{5.29}$$

so

$$\xi(\log m) = \langle B \rangle T H(\log m) \tag{5.30}$$

and the creation function can be expressed as

$$C(\log m, t) = \xi(\log m) \frac{b(t)}{T}. \tag{5.31}$$

Then the IMF is given by (5.23)

$$\xi(\log m) = \frac{\phi_{\text{ms}}(\log m) T}{\int_{T-\tau(m)}^T b(t) dt} \tag{5.32}$$

which again is the number of stars ever formed in the disk per pc^3 per unit $(\log m)$ interval. This expression holds if $\tau(m) < T$; otherwise $\xi(\log m) = \phi_{\text{ms}}(\log m)$.

Note in fact that the IMF is observed to be continuous across the boundary between low-mass stars and high-mass stars, so in itself it does not provide any evidence in favor of different star-formation mechanisms in the two groups.

For massive stars, $\tau(m) \ll T$, so

$$\xi(\log m) = \frac{\phi_{\text{ms}}(\log m) T}{\tau(m) b(T)}. \tag{5.33}$$

For a star with a main-sequence lifetime of only 3 Myr, the correction to $\phi_{ms}(\log m)$ to get $\xi(\log m)$ is very large, up to several thousand, assuming $b(T)$ is of order unity. In fact there are several arguments [447] which indicate that the past average birthrate did not differ from the present birthrate by more than a factor 2. Note also that the main-sequence lifetime of a star of $2\,M_\odot$ is about 10^9 yr, so that the approximation applies. Only for the mass range $1-2\,M_\odot$ does one need to know the detailed star formation history $b(t)$.

An alternate method for determining the IMF is through observations of star clusters. A cluster's age can be reasonably well determined, and typically the spread

5.6 Appendix to Chap. 5: Determination of the IMF

of ages in a given cluster is much less than the actual age. Thus the luminosity function below the main-sequence turnoff directly gives the IMF for that cluster, without the necessity of corrections for stellar lifetimes or a variable past history of the stellar birth function. The IMF so derived represents a particular time point in Galactic evolution; it is not integrated over time as is the case for the IMF derived for the field stars. However, the slopes in various mass ranges can be compared between different clusters and between the clusters and the field. It should be noted that the cluster formation process depends on local conditions and may not reflect the IMF on larger scales in the Galaxy. The main problem with most clusters is that the total number of stars, particularly massive stars, is relatively small, so there will be statistical fluctuations in the IMF. Also, clusters tend to be mass-segregated, with more massive stars at the center and low-mass stars in the outer regions. A true IMF must include all the low mass stars, for which it is difficult to establish membership. In fact some of lowest-mass members may have escaped. In relation to the massive star problem, those clusters with massive stars still in them are the youngest clusters, for which the spread in ages is comparable to the actual age. A correction must be made for stars that have evolved, as well as for a possible non-constant birthrate function.

An independent method for obtaining the IMF is to use very young clusters, most of whose stars are in the pre-main-sequence phase. Here the mass-luminosity relation cannot be used, and stellar masses are obtained from theoretical evolutionary tracks. This method is described in more detail in Chap. 8.

In spite of the difficulties involved in determinations of the IMF, its form in the galactic disk and in clusters, including globular clusters, is very similar. There are cluster-to-cluster variations in Γ, but on the average, the results are consistent with $\Gamma \approx 1.35$ above $1 M_\odot$. Below that mass, it is clear that there is a reduction in Γ, with a turnover to negative values at a characteristic mass of about $0.2 M_\odot$. In the brown dwarf regime the IMF is also uncertain, as indicated by the two different approximations to it given in Fig. 1.3. There is no mass-luminosity relation for brown dwarfs, because they never reach the main sequence. Thus ages need to be estimated to determine the IMF for brown dwarfs; given an age, the mass corresponding to a given luminosity can be obtained from theoretical evolutionary tracks. Brown dwarf IMF's can be determined in young clusters [108] where the ages are known, and it is clear that the number of brown dwarfs decreases with decreasing mass, down to $\approx 0.02 M_\odot$, and there is negligible difference between the cluster IMF and the field IMF for these objects. Taking into account Poisson noise, dynamical evolution of clusters, and the corrections needed for undetected binaries, there is no solid evidence for a variable IMF among the various systems in the Galaxy that have been observed [107, 270]. An exception could be the IMF near the Galactic center; there has been considerable discussion about a possible difference in the IMF there versus the rest of the Galaxy. Also, in the case of primordial star formation (Chap. 7), with zero metal abundance, it is very likely that the IMF differed from that in the present Galaxy, but it is not possible to obtain the relevant observations.

Note the numerous problems in determining the IMF:

- To get the luminosity function, distances must be known and reddening must be corrected for
- Stars move through the galactic disk during their lifetime
- The corrections to the PDMF to obtain the IMF for massive stars are large
- There is a limited number of observable massive stars
- The correction for unresolved binaries can be significant and has not been included in the above analysis [270]
- There is mass segregation in the Galactic plane
- In clusters there are problems with small-number statistics, especially for massive stars, and with the possible escape of low-mass stars
- In clusters, membership is a problem; however one can assume that all stars have about the same age and distance.

Figure 5.9 gives an example of the comparison between the IMF and the PDMF for stars above 1 $M_\odot$, as derived from observations for field stars in the galactic disk.

Star formation is known to occur in molecular clouds with supersonic turbulent velocities in which magnetic fields also play a role. Thus the IMF is determined by the complicated physics of such a region. A definitive theory of the IMF has yet to be worked out, yet important contributions toward that goal have been presented. The numerous theories that have been presented [81, 107, 160] include both analytical and numerical approaches, and many of them do provide satisfactory agreement with the observations. One of the many numerical examples is shown in Fig. 5.8 which shows that the simulation of fragmentation and cluster formation can produce a reasonable IMF. In this case competitive accretion is the main process that shapes

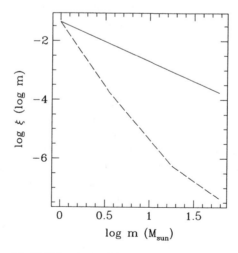

Fig. 5.9 Comparison of the IMF for single objects in the region of the Galaxy close to the Sun (*solid line*) and the present-day mass function (*dashed line*). Data from [447], converted to volume density and fit by [107]. The unit of ξ is $(\log m)^{-1}$ pc^{-3}

5.6 Appendix to Chap. 5: Determination of the IMF

the IMF. The overproduction of brown dwarfs in this simulation can no doubt be corrected with improved physical assumptions in the calculations.

On the analytical side, one approach [404] relies upon the statistical properties of turbulence. The main assumptions are (1) the velocity varies with length scale according to Larson's first finding (Chap. 2); (2) protostellar cores are formed by hydrodynamic or magnetohydrodynamic isothermal shocks, and the core size and density are determined by the thickness and density of the post-shock layer, (3) the number of shocks scales as the inverse cube of their size, and (4) a protostellar core collapses if its mass exceeds the Bonnor-Ebert mass. Without going into all the details, the result is that the IMF has a power-law slope close to the Salpeter value at masses above 1–2 $M_\odot$, and a turnover at around 0.25 $M_\odot$, depending on the mean column density, magnetic field, particle density, and temperature in the molecular cloud.

The turnover, and the sparse number of low-mass objects, can be understood on the basis of the Jeans mass in the molecular cloud. At molecular core temperatures of around 10 K, to get gravitational collapse for a low-mass object, say 0.05 $M_\odot$, the density of the protostellar core has to be relatively high compared with typical molecular cloud densities. In the theory, the highest-density cores which are unstable to collapse are produced by the strongest shocks generated by the supersonic flows. However the higher the density of the core that has to be produced, the less probable it is that the shock compression is strong enough to result in that density.

It should be noted that the monolithic collapse picture of star formation relies heavily on the properties of turbulence to explain the IMF, since in that picture the core mass function essentially determines the IMF. In the competitive accretion model, turbulence also plays a role, but not as large a one. The initial condition is a turbulent clump, so the initial protostellar cores that are formed will depend on the properties of that turbulence. However the later buildup of the IMF depends on the more complicated gravitational potential and hydrodynamics of the collapsing cloud. In this model, the prestellar core mass function has little relation to the final stellar IMF.

The theory involving turbulence gives the mass function of protostellar cores, not the final stars. The analytic results are tested by the use of three different magnetohydrodynamic computer codes in three space dimensions [405] and are confirmed. The results indicate that to get the correct power-law slope above 1 $M_\odot$ the magnetic field must be included; it is important in shaping the IMF even if globally the magnetic energy is less than the turbulent energy. Another numerical simulation with magnetic fields [317] of the production of self-gravitating protostellar cores also is in agreement with the analytical model [404] and the observed IMF.

An alternate approach to explain the characteristic mass where the IMF reaches a peak is to consider the thermal properties of the interstellar gas. Take the density range $n = 10^5 - 10^8$ particles per cm^{-3}, which we have previously (Chap. 3) approximated as being isothermal. In fact there is a slow dependence of T on n. Below some critical value of the particle density (n_c) the gas cools as it is compressed, based upon detailed heating and cooling calculations [302, 304]. If we define the index γ by the relation $P = K\rho^\gamma$, then for an ideal gas, the temperature $T \propto \rho^{\gamma-1}$. Cooling upon compression implies $\gamma < 1$. The result of the calculations gives $\gamma \approx 0.7$ in the region below n_c because the heating rate has a weaker dependence on density than the cooling rate. Turbulent compression causes substantial increase in density, reducing the Jeans mass and promoting fragmentation. However for $n > n_c$, γ changes to 1.1, resulting in a reduction of the tendency to fragment. Thus the Jeans mass at n_c is identified as the characteristic mass scale for fragmentation, and it should correspond to the peak mass in the IMF. The theoretical heating/cooling calculations give $n_c \approx 2.5 \times 10^5$ cm^{-3} [240] and the corresponding Jeans mass is about $0.3\,M_\odot$. However the value for n_c is uncertain, since it is difficult to determine it from observations. Also, the peak of the IMF depends on other factors besides n_c.

Another analytic version of the theory of the protostellar core IMF [214, 215], based upon statistical properties of turbulence, includes some additional effects such as turbulent support (in addition to thermal) of the protostellar cores and a non-isothermal equation of state, and finds good agreement with the observed IMF even without a magnetic field. However the low-mass end of the IMF of the cores is strongly dependent on the thermodynamical properties of the interstellar gas, in particular the value of γ, which are only approximately known. The results are found to be in agreement with numerical simulations of a non-isothermal gas without magnetic fields [240], given the proper choice of parameters in the simulations. In these simulations the turbulence is not allowed to decay but is continuously driven to maintain a constant turbulent Mach number. The value of n_c is taken to be a parameter, with values of γ of 0.7 and 1.1, below it and above it, respectively. The simulations give a characteristic protostellar mass of $0.5\,M_\odot$ for $n_c = 2 \times 10^5$, and about $0.2\,M_\odot$ for $n_c = 2 \times 10^6$. Again, feedback effects in the determination of the final stellar IMF are not considered.

The theoretical models in general predict the protostellar core mass function, or at best the protostellar mass function as represented by sink particles in numerical simulations. An example of the latter is shown in Fig. 5.8. The actual final stellar IMF is still a further step. However as mentioned previously, observationally the core mass function scales with the IMF for stars [15], and theoretically, outflows should result in an efficiency factor of 25–75% [346], very roughly in agreement with observations. At the low-mass end, the IMF is clearly continuous between the regimes of brown dwarfs and low-mass stars, and there are several other arguments [160] that indicate that brown dwarfs form by the same mechanism as stars, for example, their spatial distribution and disk fractions are very similar. Clearly further theoretical clarification of the origin of the IMF is a challenging problem for the future.

5.7 Problems

1. A massive star with solar metal abundance starts to form in a molecular cloud core with a radius of 0.1 pc, and mass of 100 $M_\odot$, a temperature of 10 K, and a mean turbulent velocity of 2 km/s. For the purposes of this problem, assume spherical symmetry.

    (a) What is the "thermal" Jeans mass? What is the "turbulent" Jeans mass? When the core starts to collapse, what will be the approximate mass accretion rate onto the "stellar" core, once it forms?
    (b) Once the stellar core has reached 20 $M_\odot$, its luminosity is 50,000 $L_\odot$. What is its contraction time to the main sequence in comparison to the free-fall time? Take a main-sequence radius of 5 $R_\odot$. What is the implication of this result?
    (c) By the time half of the mass has accreted onto the stellar core, it is radiating at 5×10^5 $L_\odot$. Compare the rate of photon momentum transfer into the cloud, L/c, with the rate of transfer of momentum from the collapse, $\dot{M}v$. Evaluate these quantities (1) at the dust destruction front, assuming free fall there, and (2) at the half-radius point in the infalling envelope, where the collapse is going at half the free-fall rate. Is the massive star likely to grow beyond 50 $M_\odot$?

2. Assume that a massive molecular cloud core can be represented approximately by a polytrope, $P = C_3 \rho^\gamma$ where P is the (mostly) turbulent pressure, ρ is the density, and C_3 is a constant [354]. Furthermore, assume that the pressure and density distributions are power laws in radius, with $P = C_1 r^{-m}$, and $\rho = C_2 r^{-n}$. Note that the mean turbulent speed $c_t = \sqrt{(P/\rho)}$ is in general not constant.

    (a) Show from the equations of hydrostatic equilibrium (Sect. 3.1) that $n = 2/(2-\gamma)$.
    (b) Assume that $n = 1.5$, roughly consistent with observations. Then, if the mass of the core is 100 $M_\odot$ and the radius 0.1 pc, calculate the density at the outer edge of the core, where the pressure matches that of the surrounding clump. What is the corresponding free-fall time t_ff?
    (c) Now assume that the polytrope starts to collapse and that the accretion rate onto the forming star in the center is $\dot{m} = m/t_\mathrm{ff}$ where m is the current mass of the star. Assume that a constant fraction 0.5 (with time) of the core mass M ends up in the star. Show that the stellar mass is given by

    $$m = m_F (t/t_{sf})^{4-3\gamma} \qquad (5.34)$$

    where m_F is the final stellar mass and t_{sf} is the total time. Then calculate $\dot{m}$ as a function of time, m_F, and t_{sf}.
    (d) For the case $n = 1.5$ you will see that $\dot{m}$ increases with time. Is this result physically reasonable? Which of the assumptions made is likely to have led

to this result? What parameter can be changed to result in a constant accretion rate?

(e) Show that $t_{sf} = (4 - 3\gamma)t_{fs}$ where t_{fs} is the free-fall time at the outer edge of the core. Then eliminate the time-dependence in part (c) and calculate $\dot{m}$ as a function of m and other parameters. How long does it take to form a star of $50\,M_\odot$?

3. Assume the IMF has the single-object form given in (1.7). A region in the solar neighborhood is observed to have 125 stars and brown dwarfs, within a volume of $1{,}000\,\mathrm{pc}^3$ with the lower mass limit at $0.02\,M_\odot$.

(a) What is the ratio of the number of brown dwarfs to the total number of objects? The boundary between brown dwarfs and stars is at $0.076\,M_\odot$.

(b) What is the mass of the most massive star in the region (to nearest $1\,M_\odot$)?

Chapter 6
Formation of Binary Systems

Recent observational studies of the properties of binary systems among young stars indicate that the majority of binaries are formed very early in the history of a star, perhaps during the protostellar collapse. Observational studies of these early phases also point to the fact that most stars are formed in binary or multiple systems. Major observational facts to be explained include the present-day overall binary frequency and how it varies with primary mass,[1] the non-negligible occurrence of multiple systems, and the distributions of period, eccentricity, and mass ratio among the individual binaries. Theoretical calculations of the collapse of rotating protostars during the isothermal phase indicate instability to fragmentation into multiple systems. This process in general produces systems with periods greater than a few 100 yr, although somewhat shorter periods are possible. Fragmentation during later, optically thick, phases of collapse tends to be suppressed by pressure effects. Therefore, major theoretical problems remain concerning the origin of close binaries. Fission of rapidly rotating stars, tidal capture, and three-body capture have been shown to be improbable mechanisms for formation of close binaries. Mechanisms currently under study include gravitational instabilities in disks, orbital interactions and disk-induced captures in fragmented multiple systems, hierarchical fragmentation, and orbital decay of long-period systems. Single stars, on the other hand, most likely result from escape from multiple systems during the early phases of evolution, but they could also form by the collapse of clouds of low angular momentum, coupled with angular momentum transport after disk formation.

Further issues include: what is the interaction between binaries and disks? How do binary orbits evolve after the binary has formed? Are some binaries disrupted? Why are close brown-dwarf companions rare among G–K main-sequence stars? How does the cluster environment affect binary formation and evolution? What is the influence of magnetic fields on the binary formation process? How are binary formation processes modified in the early universe?

---

[1]The primary star in a visual binary system is often operationally defined simply as the more luminous star of the pair; here we denote the more massive star in the system as the primary.

This chapter summarizes some relevant observational data [144], reviews a number of the suggested formation processes [505], and concentrates on recent theoretical results, primarily 3-D numerical hydrodynamical simulations [192].

6.1 Observational Data

Although no direct observation of the formation of a binary system has been made, systems at very early stages have been identified, for example IRAS 16293-2422, shown in Fig. 1.8 [534, 561]. The components A and B in the figure are suspected to be a binary protostar with separation 750 AU and a total mass of about 3 $M_\odot$. Another good example is the well-observed system L1551 IRS5, which appears as a binary with separation 40 AU and total mass about 1 $M_\odot$ (Fig. 1.10), when observed at high angular resolution with the VLA [440].

The formation process must explain the following observed properties of binary/multiple systems:

1. Approximately 60% of main-sequence solar-type stars have one or more companions [2, 147]. We are using the *multiplicity fraction* to define the frequency of binary/multiple systems:

$$\text{MF} = \frac{B+T+Q}{S+B+T+Q} \qquad (6.1)$$

where, for a given number of primary stars observed, S is the number of systems that are single, B the number that are binary, T the number that are triple, and Q the number that are quadruple. Thus, for example, if a given observed binary later turns out to be a hierarchical triple, the MF doesn't change. Note that an alternate form of this definition is the *companion star fraction* (CSF) where the numerator is replaced by $B + 2T + 3Q$. The primary star is the most massive in the system.

The standard sample of Duquennoy and Mayor [147] consists of all primaries between spectral types F7 and G9 within 22 pc of the Sun and visible from the Northern hemisphere (164 stars). Although surveys are not as complete for other spectral classes on the main sequence, results indicate that the frequency is highest for high mass stars (at least 70% [303]) and relatively low for the M dwarfs, in which only 30% of the systems are multiple [287]. The fact that the M dwarfs are by far the most numerous class of stars implies that in fact more than half of all observed main-sequence systems are single stars.

Extensive surveys have recently been made for the presence of binaries among pre-main-sequence stars [144, 262, 344, 349]. The frequency is in general greater than that of the F and G type stars on the main sequence. There seems to be an excess of about a factor 2 in the Taurus-Auriga star formation region in the mass range 0.5 – 2 $M_\odot$ and the separation range 19–1,900 AU. The lower-mass Taurus young stars have a smaller frequency, but still high compared to main-sequence stars in

6.1 Observational Data

the same mass range (0.1–0.5 M$_\odot$.) The shorter-period spectroscopic binaries have not been surveyed sufficiently completely to reach any firm conclusions about the multiplicity fraction, but there is a suggestion that they are also overabundant with respect to main-sequence stars [144]. Thus, considering the full range of binary separations and masses above about 0.3 M$_\odot$, the multiplicity fraction in Taurus-Aurigae is close to 100% [263]. The observed binary excess applies not only to the Taurus-Aurigae region but also to other nearby star-forming regions such as TW Hydrae and Rho Ophiuchi. The binary frequencies of the classical T Tauri stars and the weak-lined T Tauri stars are about the same. On the other hand, in the more distant and dense Orion Nebula cluster region the binary frequency among the Class II sources appears to be about the same as for the main sequence [349].

In recent years sufficient observational information has been obtained for protostars (Class I objects) in the Taurus region and in the Rho Ophiuchi region so that their binary properties can be ascertained. The binary frequency in those regions, in the projected separation range 110–1,400 AU [143], is very similar to that of the T Tauri stars, and both are almost a factor 2 higher than the main-sequence frequency for G dwarfs. Further surveys of Class I objects in several clouds [144] show a similarly high binary frequency. The overall conclusion is that stars are almost always born in multiple systems, preferentially binaries, and that the multiplicity declines with time, after a few times 10^5 yr. The mechanism is suspected to be disruption of loosely-bound systems – those preferentially with low-mass companions.

2. The periods of main-sequence G-type binaries form a smooth distribution in the range $0 < \log P$ (days) < 9 with a single maximum at about 180 yr, corresponding to a mean separation of 30 AU [147], as shown in Fig. 6.1. In the figure, the dotted line gives the actual observed data, and the solid line gives the derived frequency, corrected for incompleteness, with error bars. The period distribution for M stars is similar, but the peak is not accurately determined; it lies in the range 9–220 yr, corresponding to separations of 3–30 AU [173]. For very low mass stars and brown dwarfs the mean separations of about 4 AU are clearly smaller than those for the G stars [120]. The distribution of specific angular momenta of binary orbits of young stars is compared with that of the spin of molecular cloud cores in Fig. 6.2. Clearly the distributions overlap. In fact they may overlap more closely than shown, because the longer period binaries are excluded from the survey that is plotted, and the cloud cores with low rotation are not represented because of an observational lower limit at around $\log j = 20.5$, where the unit of j is cm^2 s^{-1}. Also, the spin angular momenta of cores may have been overestimated, as it was assumed they are uniform-density spheres. Still, the spectroscopic binaries,[2] and in general the shorter-period systems, have much lower orbital angular momentum than that indicated by the spin of the cloud cores (see Table 2.2 in Chap. 2). The figure supports the assumption that there

---

[2] A spectroscopic binary is a pair which can not be visually resolved but in which the presence of a companion can be deduced from a periodic Doppler shift in the spectral lines of one or both of the components. The orbital periods generally lie at the low end of the distribution.

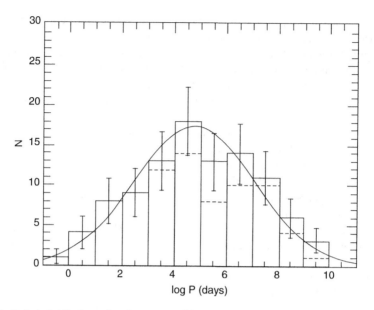

Fig. 6.1 Period distribution of main-sequence binary G dwarfs, corrected for incompleteness (*solid histogram*). The data are fit by a Gaussian-like curve. Credit: A. Duquennoy, M. Mayor: *Astron. Astrophys.* **248**, 485 (1991). Reproduced by permission. © European Southern Observatory

is an efficiency factor in the formation of a binary system from a given core: not all the angular momentum of the core ends up in the binary orbit, possibly as little as 10%. Furthermore, the figure shows (solid line) that the wide range of binary periods observed at the present time (Fig. 6.1) must have been set up at very early times. The figure also gives a clue as to why the binary fraction decreases between star-forming regions and the main sequence. The theoretical model shown takes into account stellar interactions and binary disruptions in a young cluster. The initial cluster, represented by the solid histogram, was assumed to contain 200 binary systems with a total mass of 128 $M_\odot$. The cluster was dynamically evolved until it completely dissolved, with the result that many of the longer-period orbits were disrupted and the multiplicity fraction decreased (short-dashed histogram).

3. The distribution of mass ratios ($q = m_2/m_1$ where $m_2 < m_1$ is the secondary mass) is a difficult function to determine, and the full range of primary masses and orbital periods has not been completely sampled. Accurate mass ratios in general are obtainable only for double-lined spectroscopic binaries,[3] which are relatively few in number. Thus error bars for this function are large. Figure 6.3 shows the mass ratios derived by [348] for a sample of 62 main-sequence spectroscopic binaries (43 of which are double-lined) with relatively short periods, all to the left of the

---

[3] Systems in which the spectral lines of both components are seen and exhibit Doppler shifts at the same period.

6.1 Observational Data

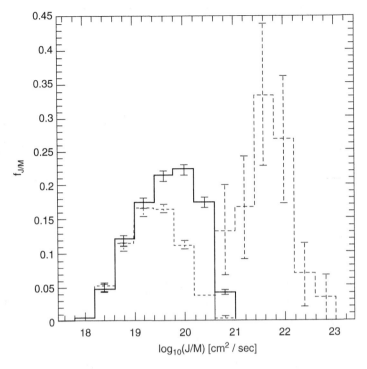

Fig. 6.2 Specific orbital angular momenta of binaries (J/M). The *dashed histogram* gives data from observations of the rotation of molecular cloud cores [191]. The *solid histogram* gives a model [269] of the initial binary population, in general agreement with observations. The *short-dashed histogram* gives a model of the distribution after interactions in a forming cluster, which cause some binary and multiple systems to break up and liberate single stars. This final distribution is in agreement with observations on the main sequence. Credit: P. Kroupa: *MNRAS* **277**, 1507 (1995). Reproduced by permission. © 1995 Royal Astronomical Society

peak in Fig. 6.1. Primary masses are in the range 0.6–0.85 $M_\odot$ and periods in the range 2–3,000 days. The lower part of the figure shows the results corrected for incompleteness; however the corrections are very uncertain for low q. Thus in the region where the results are reliable, $0.3 \leq q \leq 1$, the distribution is quite flat. For very low $q < 0.08$, other data show that for solar-type stars there is a definite shortage of companions with separations less than 5 AU; this phenomenon is known as the *brown-dwarf desert*. A larger sample with accurately determined q's in the range $q > 0.85$ shows the flat distribution but with a narrow peak, probably statistically significant, at $q > 0.95$ [327]. This peak is not resolved in Fig. 6.3. These systems with nearly identical masses are known as "twins".

For the longer-period visual binaries, the distribution of q is more difficult to measure and is somewhat uncertain in details, but it is clearly different from that of the short-period binaries. The distribution $N(q)$ is generally found to be rising slowly as q decreases. In [348] it is found that a period of 100 days separates the

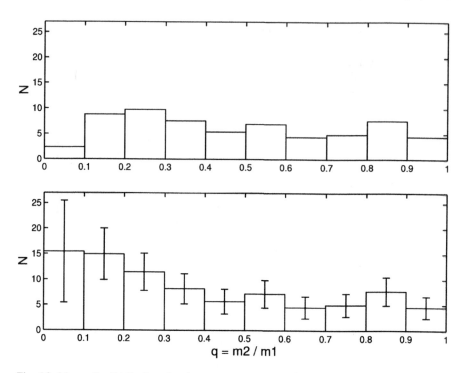

Fig. 6.3 Mass ratio distribution of main-sequence spectroscopic binaries with primary masses in the range 0.6–0.85 M$_\odot$. *Top:* the directly measured distribution; *bottom:* the distribution corrected for undetected binaries. Reproduced, by permission of the AAS, from [348]. © 2003 The American Astronomical Society

systems with a flat q distribution from those where the frequency increases toward lower q, but this borderline is uncertain. The present mass ratio distribution for all main-sequence G stars was analyzed by [147], over the full range of periods. It is reasonably consistent with an initial distribution in which both the primary and secondary masses are picked at random from the standard IMF [2, 147, 269, 409, 410]; this characteristic is thought to hold for periods longer than 100 yr. In the case of pre-main-sequence binaries the mass ratios are more difficult to measure, but indications are that the q distribution is similar to that of main-sequence stars [262].

4. Binaries probably formed with a wide range of eccentricities. Field main-sequence systems with P < 10 days have circular orbits, probably as a result of post-formation tidal evolution. At longer periods, practically all orbits are eccentric [147]. Figure 6.4 shows the eccentricity vs. orbital period for main-sequence M dwarfs and for main-sequence G and K dwarfs, both in the field and in open clusters [523]. Note that the distributions are very similar. The orbital eccentricity distribution of pre-main-sequence binaries is very similar, also with a tidally circularized population at very short periods. The maximum period for primarily

6.1 Observational Data

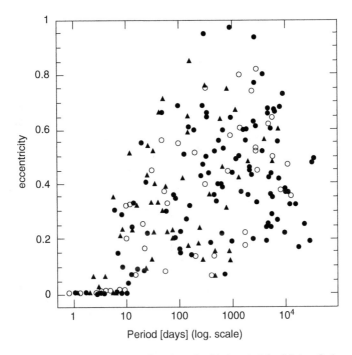

Fig. 6.4 Binary orbital eccentricity as a function of orbital period for M dwarfs (*open circles*), G and K dwarfs in the field (*filled circles*), and G and K dwarfs in open clusters (*triangles*). Credit: S. Udry, M. Mayor, J.-L. Halbwachs, F Arenou: in *Microlensing 2000: A new Era of Microlensing Astrophysics*, ed by J. Menzies, P. D. Sackett (Astron. Soc. of the Pacific, San Francisco, 2001). Reproduced by permission. © Astronomical Society of the Pacific

circular orbits is somewhat shorter than that shown in Fig. 6.4, consistent with the shorter time available for tidal circularization of the orbits. These distributions are close to "thermal", that is, all binding energies are equally probable, as given by the relation

$$dN = f_0 f(e) de = f_0 2e de \qquad (6.2)$$

where dN is the number of systems with eccentricity between e and $e + de$ and f_0 is a normalization constant.

5. Most binaries formed during the star formation phase or during protostellar collapse, as indicated by the fact that binary stars are detected among pre-main-sequence stars and protostars with ages as young as 10^5 yr and that pre-main-sequence and protostar binary frequencies are generally greater than the corresponding main-sequence frequencies.

6. Evidence from observations of a few pre-main-sequence systems suggests that the components of a given binary are coeval, at least to within an age difference of 10^6 yr [544]. Figure 6.5 shows the H-R diagram for four different systems, with evolutionary tracks from [408]. In the case of GG Tau, all four components fall in the age range 1–2 Myr. Also, the sum of the masses of the two massive (A) components,

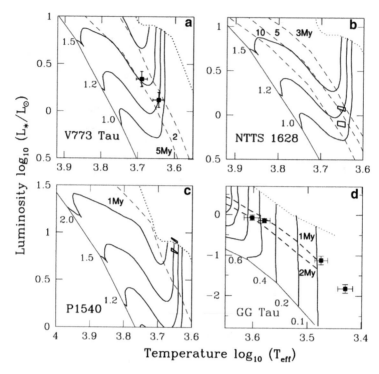

Fig. 6.5 Pre-main-sequence evolutionary tracks in the (log T_{eff}, log $L/L_\odot$) diagram, for solar-metallicity stars with masses, in $M_\odot$, as listed at the main-sequence end of the tracks [408]. The components of four different non-eclipsing spectroscopic binaries/multiples are indicated: V773 Tau, NTTS 162814-2427, P 1540, and the GG Tauri quadruple system. The *dashed lines* are isochrones, labelled with the time in Myr. The *dotted line* represents the birthline. The figure illustrates the fact that, taking into account the uncertainties, the components in each system were formed at about the same time. Reproduced, by permission of the AAS, from [408]. © 2001 The American Astronomical Society

deduced from the tracks, is about $1.3\,M_\odot$, in excellent agreement with that obtained from observations of the Keplerian velocity in the circumbinary disk surrounding these two objects [268, 439]. An image of this disk is shown in Fig. 6.6.

The example of GG Tauri clearly shows the importance of binary systems for calibrating pre-main-sequence tracks. For example the masses of low-mass objects, in the brown dwarf range or even down to the 10 Jupiter mass range, are often determined just from comparing their positions in the H-R diagram with evolutionary tracks. The calibration is done by measuring, if possible, the masses of pre-main-sequence binaries by methods based on dynamics. An example is the determination of the sum of the masses in a binary by using velocity measurements in a circumbinary disk as shown in Fig. 6.6. Clearly the method can also be used, if the disk can be resolved, for single stars with circumstellar disks. A second, potentially even more accurate method is to discover pre-main-sequence eclipsing

6.1 Observational Data

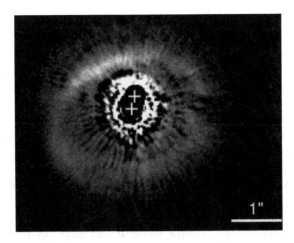

Fig. 6.6 Hubble Space Telescope image of the circumbinary disk around the Aa and Ab components of the quadruple system GG Tau. The images of the stars themselves have been subtracted, and their positions indicated by *crosses*. The image was taken in the visual region of the spectrum at a central wavelength of 555 nm. The two stars have a projected separation of 35 AU and they are of spectral types M0 and M2. The inner edge of the disk falls at about 150 AU from the center of mass of the binary. The disk mass is about 0.1 $M_\odot$. Reproduced by permission of the AAS from [268]. © 2005 The American Astronomical Society

double-lined spectroscopic binaries, of which a few are known. Measurements of the eclipse light curve and the spectroscopic radial velocity curve give the individual masses. Then the luminosity and surface temperature of the individual components have to be determined accurately so the components can be placed in the H-R diagram. Comparison with evolutionary tracks, as shown in Fig. 6.5, then gives the "track" masses, which can then be compared with the actual "dynamical" masses. This kind of comparison is considered in more detail in Chap. 8.

7. Hierarchical multiple systems are common; they have a wide range of properties, but an example would be a primary with a close secondary with a period of 10 days and a wide companion with a period of 30 yr. This type of triple system is dynamically stable if the ratio of the semimajor axes of the outer to the inner orbit is above a critical value, roughly 5, depending on the inner and outer mass ratios [155]. The overall frequency of triple systems is about a quarter that of binary systems. Most period ratios in such systems fall in the range $10–10^4$ [170]. A good example is the system T Tauri itself, whose originally discovered component, T Tauri N, is a 2 solar mass star in a 100 AU orbit about T Tauri S, which itself is a binary with a separation of about 13 AU (Sect. 8.3). A clue to the formation mechanism of such systems can be obtained by examining the statistics of the degree to which the angular momenta of the inner and outer orbits are aligned. If ϕ is the angle between the angular momenta, and if $\phi = 0$ corresponds to alignment, then the observed systems are typically not aligned [490]. However the angles ϕ are not randomly distributed; rather they average between $67°$ and $79°$.

6.2 Basic Formation Mechanisms

A number of different ideas have been put forward regarding the origin of binary systems. The original suggestion put forward by Laplace in 1796 (see [499]) was that binaries formed from separate stellar nuclei during star formation; the nuclei then somehow came into orbit. The physics behind this picture was never fully explained, so we may assume that the separate nuclei combined by capture. Capture into orbit of two independently formed unbound stars in the presence of a dissipative process was suggested in 1867 by Stoney (see [11]). Fission, meaning the breakup by dynamical instability of a rapidly rotating object in hydrostatic equilibrium, is attributed to Kelvin and Tait in 1883 (see [499]) and was strongly advocated by Jeans [241].

Fragmentation, on the other hand, refers to the breakup of a rotating protostar during the hydrodynamical collapse phase. It was originally proposed by Hoyle [234], who based the argument simply on the fact that an isothermal collapsing cloud, originally dynamically unstable on a large scale, becomes dynamically unstable on smaller and smaller scales because of the density increase during collapse. That is, as the density increases and the temperature remains constant, the Jeans mass decreases. However it was pointed out [326] that this process cannot continue indefinitely, because once the center of the cloud becomes optically thick, as it continues to collapse the temperature, and as a result the Jeans mass, starts to increase. Thus there is a lower limit to the mass of a fragment, given roughly by the minimum Jeans mass along the temperature-density evolution of the protostar (Fig. 3.6). This limit, which depends on the dust opacity, has been shown in various calculations to be about $0.01\,M_\odot$.

Later work has shown that rotation is a crucial element that stops the overall collapse of the cloud and allows individual fragments to separate out [62]. Thus, a molecular cloud evolves and develops slowly-rotating dense cores, which, after gravitational collapse has set in, spontaneously fragment into two or more pieces, perhaps induced by small density perturbations in the initial cloud core.

A variation of this process involves triggering of the fragmentation process by cloud–cloud collisions [426] with the angular momentum of the orbit arising from the initial condition of an off-center collision. The high expected values of semimajor axis and eccentricity could later be reduced by interaction of the two components with each other's disks. An example of a numerical calculation of such a collision is shown in Sect. 2.9 (Fig. 2.18) in which the outcome was a single star with a circumstellar disk [252]. However, only a slight change in the initial condition, an increase in the relative velocity of the two clouds from $1\,\text{km}\,\text{s}^{-1}$ to $2\,\text{km}\,\text{s}^{-1}$, results in the formation of two protostars with circumstellar disks, which later capture each other into a binary with a circumbinary disk. The two components of the binary have masses of about $1.5\,M_\odot$ each, and the separation is about 40 AU [253].

Finally, the disks that form around young stars can, under the right conditions, become gravitationally unstable and could possibly fragment [9]. Disk instability, if the Toomre Q is ≈ 1 and the disk is massive enough and cool enough, can produce stellar-mass secondaries or possibly giant planets. A basic requirement for disk fragmentation [182] is that the cooling time be comparable to or shorter than the orbital time. The following sections concentrate on the four basic processes of capture, fission, spontaneous fragmentation, and disk instability. However analysis of large numerical simulations of cluster formation (the probable dominant mode of star formation) from an initially turbulent cloud strongly indicates that there is not one single process by which binary formation occurs. Although fission has been practically ruled out, the complicated interactions between stars and gas during cluster formation result in binary formation by a combination of processes, including fragmentation, capture, and probably gravitational instability in disks [117].

6.3 Capture

Two independently formed stars can be captured into orbit if (1) a third body is present to take away the excess energy, if (2) the encounter is close enough so that tidal dissipation performs the same function, or if (3) a dissipative medium, such as residual gas in a young cluster or a circumstellar disk, is present. Considering processes (1) and (2), expected capture rates in the galactic disk or in even young dense clusters have been shown to be far too slow to explain the observed binary frequency [84]. The presence of residual gas, either ambient in a newly forming cluster or in the form of disks around young stars could change the picture. However stellar encounters with disks could also have the opposite effect of truncating or ejecting the disk, so capture occurs only in a limited range of circumstances [118]; this process may account for a small fraction of binary or multiple stars in a forming cluster [117].

Three-body captures produce wide separations, two-body tidal captures produce very close separations, and disk captures give separations comparable to the disk outer radius. The combination of such processes is unlikely to produce the observed smooth period distribution. The capture process also predicts (1) a wide range of eccentricities, as observed, (2) a wide range of mass ratios, essentially uncorrelated, as observed at least for wide binaries, (3) non-coevality of the components, for which there is no evidence except at the 10^6 yr level, and (4) non-alignment of the angular momentum in spin and orbit, for which there is evidence in some systems. Capture is unlikely under normal circumstances in the galaxy, so its role must be examined through detailed numerical simulations of cluster formation where fragmentation could produce protostars with typical separations close enough so that capture could occur. Indeed N-body simulations of the interaction of a number of point masses with arbitrary initial conditions do result in captures. The treatment of such systems as an N-body problem has been extensively investigated (e.g. [527]), and simulations have shown that the results are extremely sensitive to changes in the initial conditions and that a wide variety of orbital parameters and eccentricities is possible. However in actual cluster formation the gas plays a very important role, so the process must be simulated by full three-dimensional hydrodynamic calculations, e.g. [39]. Captures will play some role in the formation of binaries in such a simulation, but other processes are active as well, and a description of the results of such a simulation will be deferred until later in the chapter.

6.4 Fission

The physical process of fission is quite distinct from fragmentation, as it occurs in configurations which are assumed to be in hydrostatic equilibrium. As a star accumulates during protostar collapse, or as it contracts toward the main sequence after disk accretion has been completed, it tends to spin up, if angular momentum is conserved, and the ratio β of the rotational energy to the absolute value of the gravitational energy increases. When β obtains a critical value, the star becomes unstable to non-axisymmetric perturbations. It has been hypothesized that breakup into orbiting subcondensations then occurs. Because only a small amount of angular momentum can be stored in a star, this mechanism would produce close binaries.

Consider the contraction of an object of constant mass and constant angular momentum. The classical path to fission was based on the properties of uniform-density configurations [333]. The contracting sequence of axisymmetric Maclaurin spheroids, which are defined to be uniformly rotating and have constant density, becomes secularly unstable, in the presence of a dissipative mechanism, to deformation into triaxial objects (Jacobi ellipsoids), when $\beta = 0.138$. Evolution was then envisaged to proceed along the sequence of contracting Jacobi ellipsoids until a point of dynamical instability to a pear-shaped mode was encountered at $\beta = 0.163$. Fission was thought to result, but analytical methods were not sufficient to demonstrate this outcome. It is now thought that fission cannot occur by this route, because even if the point of secular instability at $\beta = 0.138$ is reached, the dissipative mechanism, which is viscosity in the case of a pre-main-sequence star, has a time scale which is long compared with the time scale for further contraction of the star.

An alternative path to fission that has been considered involves the continued axisymmetric contraction of the Maclaurin spheroids beyond the point of secular instability, until a point of dynamical instability to non-axisymmetric perturbations is reached at $\beta = 0.274$. It has been shown that analogous points of instability exist along sequences of polytropes,[4] that is, centrally condensed configurations [400]. However it has not been shown that even this instability results in fission, even though it does result in the development of a triaxial object. The situation is well summarized by Tohline [505].

---

[4] A polytrope is a hydrostatic structure in which the pressure is assumed to obey the relation $P = K\rho^{(n+1)/n}$, where K is a constant and n is the polytropic index

The fission scenario has several major problems [499]. First, T Tauri stars are observed to rotate very slowly. They have far too little angular momentum to reach a point of secular or dynamical instability. It would seem that fission could occur during the period of buildup of the stellar core by accretion from a disk and the associated increase in angular momentum. In fact the theory of disk accretion, mediated by a magnetic field [462], shows that the spin of the star is regulated by the disk and remains at relatively low values. Second, angular momentum considerations require the binary mass ratio to be $\sim 1:10$, in disagreement with observations. That is, if the result of fission were a system with comparable masses, and if the angular momentum of the original spin of the star were conserved, the two stars would overlap in space. Third, even if the critical β for dynamical instability were reached, several independent 3-dimensional numerical hydrodynamical calculations [148] show that the result is in fact deformation of the initially axisymmetric object into a triaxial configuration, but fission does not occur because of transfer of angular momentum out from the core by spiral waves, so that β is reduced below the critical value.

Figure 6.7 shows a calculation [550] starting from a rapidly rotating polytrope of index 0.8 with $\beta = 0.31$. Thus at the initial time the object should be unstable to deformation into an ellipsoidal-type structure and perhaps to fission, on a dynamical time scale. The initial stage (a) is axisymmetric and very flattened. The figure shows that on a time scale of about 10 central rotation periods the system develops spiral arms which exert torques on the central object, so that the eventual outcome is a more slowly rotating central object plus a ring. The same result occurs if the critical β is approached (through accretion) from below [85]; this calculation uses a realistic equation of state and includes radiative transfer. Even though fission does not occur promptly once $\beta > 0.274$, the ellipsoidal structure can still evolve through further contraction and possibly eventually become unstable to fission at a later time [308]. It is a difficult 3-D numerical problem to test this hypothesis, and attempts so far have not produced conclusive results [505].

6.5 Fragmentation

Various forms of fragmentation can be identified, including (1) fragmentation during the collapse of a low-mass core, producing a binary or a small multiple system, (2) Fragmentation of a higher-mass core, leading to formation of a small cluster, (3) fragmentation of a gravitationally unstable disk, and (4) fragmentation induced by a cloud–cloud collision. We first consider the collapse of a single, isolated, rotating, low-mass core and ask whether a binary can form. This question is appropriate for the case of star formation by monolithic collapse (Chap. 5), where a particular molecular cloud core collapses to form a single stellar system without much interaction from the outside. We then consider the more complicated case of cluster formation.

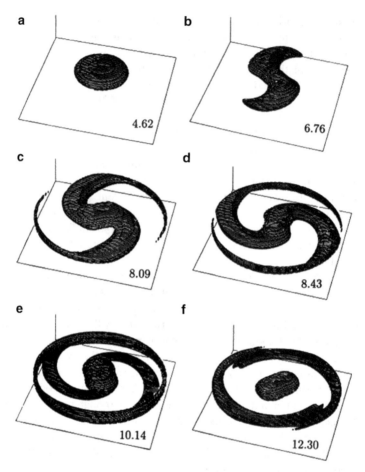

Fig. 6.7 Numerical simulation of the evolution of a rapidly rotating polytrope of index 0.8 with initial $\beta = 0.31$. Times are indicated on each frame in units of the initial central rotation period. Constant-density surfaces are shown at 10^{-3} of the central density. Reproduced, by permission of the AAS, from [550]. © 1988 The American Astronomical Society

Protostar collapse is divided into an early optically thin isothermal phase and a later optically thick adiabatic phase. Conditions for fragmentation are favorable during the isothermal phase. It has been known for some time that isothermal rings are unstable to fragmentation [393]. Collapse calculations starting from a molecular cloud core of a few solar masses are often done with the assumption of isothermality, although calculations have also been done under the assumption of an adiabatic collapse, or a collapse with cooling, or a collapse with radiation transport, or a collapse with turbulent initial conditions, or a collapse with a magnetic field. Three-dimensional hydrodynamical calculations start from initial conditions which involve a large number of parameters: (α_i, β_i), which are, respectively, the thermal and rotational energies divided by the absolute value of the gravitational potential

6.5 Fragmentation

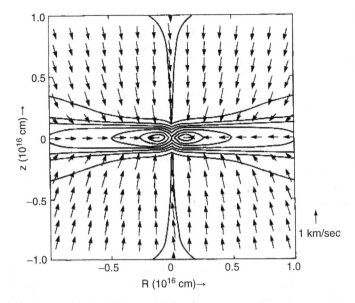

Fig. 6.8 Numerical 3D simulation of the evolution of a collapsing molecular cloud core. The meridional plane in (R, Z) cylindrical coordinates is shown after 1.21 free fall times from an initial cloud of uniform density and a size of 6.54×10^{16} cm. The Z axis is the rotation axis. *Curves* are contours of equal density, with a maximum density of 5×10^{-14} g cm^{-3}. *Arrows* are velocity vectors, with length proportional to speed. Initial $(\alpha_i, \beta_i) = (0.5, 0.1)$. The cloud is isothermal at 10 K. Reproduced, by permission of the AAS, from [69]. © 1980 The American Astronomical Society

energy, the density distribution (for example a power law), the angular momentum distribution (for example solid-body rotation), and the shape of the cloud, for example prolate or oblate with respect to the rotation axis. A small initial density perturbation, either ordered or random, is introduced. An example of an ordered perturbation, of mode $m = 2$ and 10% amplitude, is

$$\rho(r, \phi) = \rho_0(r)[1 + 0.1 \cos 2\phi], \qquad (6.3)$$

where r is the distance from the center and ϕ is the azimuthal angle about the rotation axis. Some of the parameters can be constrained by observations: a typical core has a mass of a few solar masses, a radius of 0.1 pc, $(\alpha_i, \beta_i) = (0.4, 0.01)$. The problem is computationally difficult because an increase in density by many orders of magnitude must occur before the fragmentation sets in. In the following examples, magnetic fields are not considered; however it is known that magnetic fields with strengths consistent with observations in molecular cloud cores can significantly suppress fragmentation [424].

Figure 6.8 illustrates the fact that before fragmentation can occur, the cloud must collapse and approach equilibrium in a disk-like configuration. This result follows from the fact that during a near-free-fall collapse the time scale for the growth of perturbations in density is nearly the same as the time scale for collapse of the overall cloud. Thus by the time pressure effects, which suppress fragmentation, become important, the perturbations will not have grown significantly with respect to the background. In the figure, the vertical axis is the rotation axis, and a shock forms on the upper and lower faces of the disk, decelerating the inflowing material there. The disk is supported against gravity in the vertical direction by gas pressure, and in the horizontal direction by rotation and gas pressure. Since β_i has to be relatively large (a few per cent) if rotational flattening (which is the necessary precursor to fragmentation) is to occur while the system is still isothermal, a long-period system is produced, typically with binary separation of 100–1,000 AU.

To include the transition to the adiabatic collapse phase in a simple way, one can modify the equation of state as follows:

$$P = \rho c_s^2 = \rho c_0^2 [1 + (\rho/\rho_{tr})]^{\gamma-1} \tag{6.4}$$

where c_s is the sound speed, c_0 is the sound speed in the isothermal collapse phase, ρ_{tr} is the approximate density where the transition to adiabatic collapse occurs, and the ratio of specific heats γ is 5/3 below about 80 K, making a transition to 7/5 above that temperature. Figure 6.9 shows an SPH calculation [106] in which the initial cloud did not have uniform rotation, but rather $\Omega \propto r^{-1/3}$. The initial density perturbation was small and random, and $(\alpha_i, \beta_i) = (0.60, 0.035)$. Nine different snapshots in time are shown, in the equatorial (x, y) plane. The equation of state switched from isothermal to adiabatic at $\rho_{tr} = 10^{-13}$ g cm^{-3}. About the time the configuration became adiabatic, an axially symmetric ring formed near the center. Then the ring fragmented into three equal mass objects in an unstable triangular configuration. The interaction of these fragments led to a fourth low-mass fragment, but as a result of mergers the final number of fragments was 2, in a binary system with separation of about 500 AU. Disks are evident around the individual binary components.

Figure 6.10 shows a simulation done with a 3-D grid code with adaptive mesh refinement. The initial condition is $(\alpha_i, \beta_i) = (0.26, 0.16)$ with uniform density and angular velocity and with a 10% $m = 2$ density perturbation. The adiabatic heating transition is used with $\rho_{tr} = 10^{-13}$ g cm^{-3}. Heating slows the collapse once the transition density has been reached. A binary forms with separation 10^{16} cm. Disks form around the individual components, having sizes of ≈ 100 AU, comparable to those observed around T Tauri stars. Also evident is a circumbinary spiral-arm structure. At the end of the simulation about half the mass is in the fragments plus the disks and spiral arms.

During the adiabatic phase fragmentation tends to be suppressed because pressure effects cause perturbations to decay unless α_i is very small. Nevertheless, calculations starting in the isothermal phase have demonstrated that fragmentation can still occur in the central regions after they have already entered the adiabatic phase and that orbital separations as small as ~ 1 AU can result [83]. However the main mechanisms for close binary formation are thought to be different from direct

6.5 Fragmentation

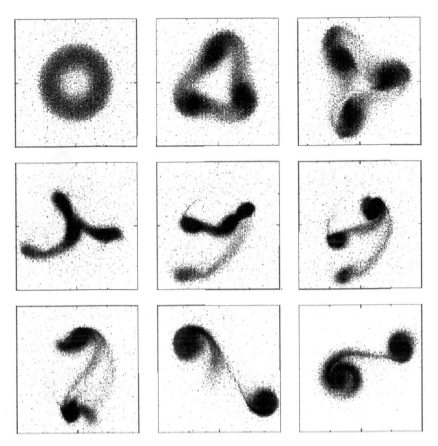

Fig. 6.9 Numerical 3D SPH simulation of the evolution of a collapsing molecular cloud core. Particle positions projected onto the (x, y) plane, where the z axis is the rotation axis, are shown at nine different times, starting (ending) at 1.15 (1.5) initial free-fall times. Initial $(\alpha_i, \beta_i) = (0.6, 0.035)$. The initial size of the cloud is 0.01 pc. The box size varies from 0.4 to 1.4×10^{16} cm. Credit: S.-H Cha, A. P. Whitworth: *MNRAS* **340**, 91 (2003). Reprinted with permission from John Wiley and Sons. © 2003 Royal Astronomical Society

fragmentation during collapse. The transition zone between isothermal and adiabatic collapse turns out to be important in determining the outcome of fragmentation, and to treat it properly requires 3D hydrodynamics including radiative transfer. Examples of work where radiative transfer has been included in collapse calculations for isolated cores may be found in [87] and [546]. In the latter study it is shown that including it results in temperature increases by up to an order of magnitude in some regions of the cloud, compared with results using (6.4).

The results may depend qualitatively on numerical resolution, particularly if the Jeans condition is not satisfied. This condition requires that the local Jeans length be resolved by several zones in a grid calculation [511, 512] or several smoothing lengths in an SPH calculation [42]. When the density has increased

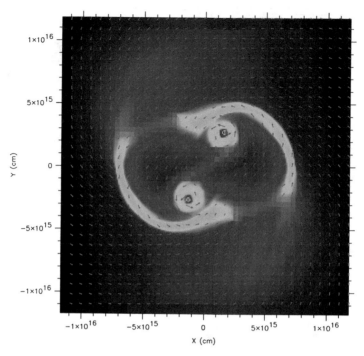

Fig. 6.10 Numerical 3D grid-based simulation, with adaptive mesh refinement, of the evolution of a collapsing molecular cloud core [65]. The density distribution in the equatorial plane is shown at a time of 1.5 initial free-fall times. The initial cloud had uniform density and an $m = 2$ density perturbation of 10% amplitude. Initial $(\alpha_i, \beta_i) = (0.26, 0.16)$. The density range is from log $\rho_{max} = -10.28$ (centers of fragments) to log $\rho_{min} = -16.48$ (violet). The maximum spatial resolution in the adaptive grid is 10^{14} cm. The *arrows* give velocity vectors with length proportional to speed, and a maximum velocity of 2.75 km s^{-1}. Credit: Color Plate 11 from *Protostars and Planets IV*, ed. by V. Mannings, A. P. Boss, S. S. Russell. © 2000 The Arizona Board of Regents. Reprinted by permission of the University of Arizona Press

to the point where fragmentation typically begins, the Jeans mass has decreased to about 0.01 M$_\odot$; thus very fine zoning is required. Even if the Jeans length is resolved, calculations must show numerical convergence, that is, a change in the number of zones or particles does not significantly change the outcome. A number of comparisons have been made between different numerical codes, and this is an important test for verifying the results of a particular simulation. For example, in [94] there is a comparison of the collapse of centrally condensed clouds as calculated with a grid code and with an SPH code. The transition to adiabatic collapse is included. The results are shown in Figs. 6.11 and 6.12. Four fragments form in the central regions. Apart from showing that the two codes give qualitatively the same results, these figures also demonstrate that the formation of an initial central binary in a close orbit can induce the formation of additional fragments farther out, and that the fragmentation process can lead to a small cluster. The fact

6.5 Fragmentation

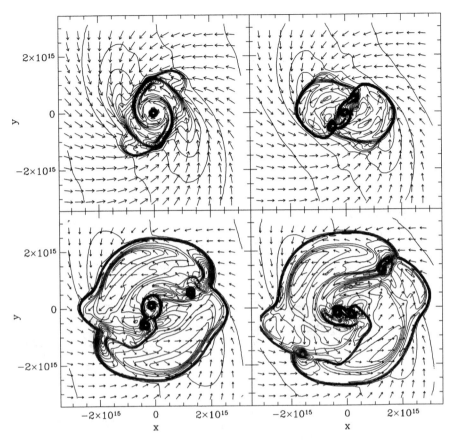

Fig. 6.11 Numerical 3D grid-based simulation, with a series of fixed nested grids increasing in resolution toward the center. The density distribution in the equatorial plane is shown at four different times. The initial cloud had a power-law density ($\rho \propto r^{-1}$) and an $m = 2$ density perturbation of 10% amplitude. Initial $(\alpha_i, \beta_i) = (0.25, 0.23)$. The maximum spatial resolution is $10^{-3} R_i$, where R_i is the initial cloud radius of 5×10^{16} cm. Velocity vectors are shown with length proportional to speed, with the maximum value ranging from $2 \, \text{km s}^{-1}$ (*upper left*) to $3.7 \, \text{km s}^{-1}$ (*lower right*). Maximum density in all frames is about 10^{-10} g cm^{-3}. The final time is 1.2×10^{12} s. Credit: A. Burkert, M. R. Bate, P. Bodenheimer: *MNRAS* **289**, 497 (1997). Reproduced by permission. © 1997 Royal Astronomical Society

that four fragments form is consistent with the fact that $\alpha_i = 0.25$; usually in such simulations the number of fragments is comparable to the number of Jeans masses in the initial configuration. We may conclude that different numerical codes with a similar degree of resolution produce about the same results; however the question of numerical convergence has not been investigated thoroughly enough, and there are noticeable differences in detail. The effect of the initial perturbation has also not been investigated extensively, but existing studies, for example [516] indicate that the occurrence of fragmentation depends only slightly on the form of the initial

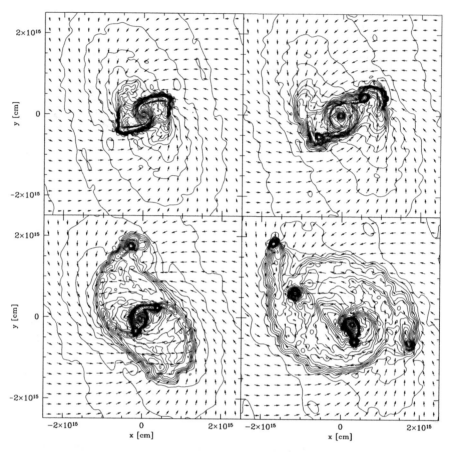

Fig. 6.12 Numerical 3D SPH simulation with 200,000 particles, of the same collapse that is shown in Fig. 6.11. The spatial resolutions in the centers of the two calculations are very nearly the same. The density distribution in the equatorial plane is shown at four different times. At the end of the simulation, a system of four fragments is produced, in agreement with the result shown in Fig. 6.11. If the simulations were continued, further interactions among the fragments would result. The density values, maximum speeds, and final time are about the same as in Fig. 6.11. Credit: A. Burkert, M. R. Bate, P. Bodenheimer: *MNRAS* **289**, 497 (1997). Reproduced by permission. © 1997 Royal Astronomical Society

perturbation, but the details of how the initial fragmentation phase proceeds is likely to be strongly affected.

One can reach several general conclusions from the results of the isolated-core fragmentation calculations. (1) Collapsing, rotating clouds are unstable to fragmentation, given a small initial perturbation, in the isothermal phase, subject to the criterion given below. If the calculation is started in the adiabatic phase, instability occurs if the initial thermal energy (α) is small. (2) Fragmentation begins only after one free-fall time, when rotational effects have become important. (3) The initial perturbation can contain less than one Jeans mass. Clouds with high α seem to break up into only 2 fragments. With smaller α the number of fragments seems to increase. Very roughly the number of fragments is comparable to the number of Jeans masses in the initial cloud. (4) Numerical spatial resolution is important, particularly for small initial perturbations or investigation of higher-order modes of fragmentation. (5) The properties of the fragments themselves are such that they are unstable to further collapse, generally have low α (so if they were produced in the isothermal phase they might still be able to subfragment in the adiabatic phase), and have a ratio of remaining spin to orbital angular momentum of roughly 0.2. The reduction in J/M of a fragment as compared with that of the initial rotating cloud is about a factor 15–20, so considerable progress can be made in solving the angular momentum problem. (6) Fragment interactions can play a role, as shown in Fig. 6.9. (7) Disk fragmentation, as well as spontaneous fragmentation, can occur during the collapse of an individual core. An example is the computation of the collapse of a core of $100\,M_\odot$ (Chap. 5) which initially resulted in the formation of a single fragment with a massive disk; the disk later fragmented and the end result was a massive binary [277]. An example of the initial phase of disk fragmentation is shown in Fig. 5.6, and the massive binary is shown in Fig. 5.7.

Various simulations of fragmentation have produced a number of different types of results, for example, (1) binary formation, (2) formation of a binary followed by induced fragmentation in a circumbinary disk, (3) formation of a small cluster, (4) fragmentation of filaments, or (5) formation of a binary plus low-mass fragments. The assumed initial conditions have some influence on the outcome.

However the typical fragmentation calculation has not been carried out long enough so that most of the material in the original cloud has collapsed to the equatorial plane. Thus the final outcome in these systems, as influenced by captures, mergers, further accretion, and escapes, is not known. For example, accretion of relatively high angular momentum material (compared with the orbital angular momentum) tends to equalize the mass ratio in a binary.

Is there a criterion for fragmentation? This question has been investigated only in particular cases, for example, the initial condition of a uniform density uniformly rotating sphere, with a small initial perturbation, either $m=0$ (ringlike) or $m=2$ (barlike). The results for various (α_i, β_i) are described in [516]. The general criterion is that fragmentation occurs if $\alpha < 0.4$, almost independent of β. The typical molecular cloud core with no turbulence is close to $\alpha = 0.4$. However other results show that if the core is centrally condensed, then the region in (α_i, β_i) where fragmentation takes place becomes smaller. For a power law $\rho \propto r^{-2}$, a linear analysis shows that the object is stable against fragmentation [513].

Simulations of the fragmentation of an individual core are very useful for examining the binary formation mechanism in detail, but they are less useful for explaining the overall statistical properties observed for a wide range of systems. The typical star, however, does not form from an isolated core but rather in a small cluster. Including the interactions between fragments in the process of cluster formation may be crucial for the explanation of these observations. Numerical simulations are beginning to be able to treat this problem. Small cluster formation involves an extension of the single-core case to a larger core which contains many thermal Jeans masses. The general picture is that this core is initially primarily supported against collapse by turbulent kinetic energy, that is turbulent energy is comparable to gravitational potential energy. However the turbulence, if not continuously driven by external effects, tends to decay. Once this occurs, the core begins to collapse and to develop into multiple fragments.

The process was simulated in [39], under the assumption that the initial condition was a cloud of $50 M_\odot$ with diameter 0.375 pc and temperature 10 K, containing ≈ 50 thermal Jeans masses. The cloud is initially supported against collapse by a turbulent velocity field, with an rms turbulent Mach number of 6.4, and with the velocity field consistent with the linewidth-size relation i.e. $\sigma \propto \lambda^{0.5}$, where σ is the velocity dispersion and λ is the size scale. A smoothed particle hydrodynamic code is used with 3.5×10^6 particles. The isothermal/adiabatic equation of state is included according to (6.4). The turbulence decays on a time scale of 1 initial free-fall time, dense self-gravitating cores form, and star formation begins. Figure 6.13 shows the configuration of gas and stars at the end of the simulation, 1.4 initial free-fall times. This particular simulation well illustrates some possible mechanisms for close binary formation. This problem will be discussed first, then the results of an even more detailed simulation [37] will be considered, with attention to the question of how well the simulation agrees with general observed properties of binary and multiple systems.

A number of binary and multiple systems are formed, as indicated in the figure. An important point is that relatively close binary systems are formed, down to separations of about 1 AU (below that limit the simulation is not resolved). These close systems are not formed by direct fragmentation of the collapsing cloud, because heating and the associated increase in pressure suppresses fragmentation on the small scales involved. Instead, wide binaries are forced to decrease their separation by three different processes: (1) material accretes from the surrounding gas that has lower angular momentum than that of the orbit, resulting in shrinkage ("hardening") of the orbit; (2) a circumbinary disk is affected by gravitational torques from the binary itself, so that angular momentum is transferred to it, again resulting in a decrease in the separation of the binary, and (3) close encounters with external third bodies, in the presence of the dissipative effects of the gas, result in transport of angular momentum from the binary to the third object and the resulting shrinking of the orbit. The particular simulation shown in Fig. 6.13 produced seven close binaries out of a total of about 50 stars and brown dwarfs. The binaries have mass ratios between 0.29 and 0.96 and separations in the range 1–10 AU. In general the simulations produce a larger proportion of high q-values than shown by the observations. The calculation also shows that more massive stars are more likely to have close companions, in rough agreement with observations. The calculation also suggests that formation of a hierarchical triple is a relatively common outcome.

Figure 6.14 shows a small section of the calculation starting at 1.32 initial free fall times, illustrating how multiple systems can form by the mechanism of disk fragmentation. A single protostar first forms and a very massive circumstellar disk forms around it (upper left). The disk develops spiral arms, which later fragment into

6.5 Fragmentation

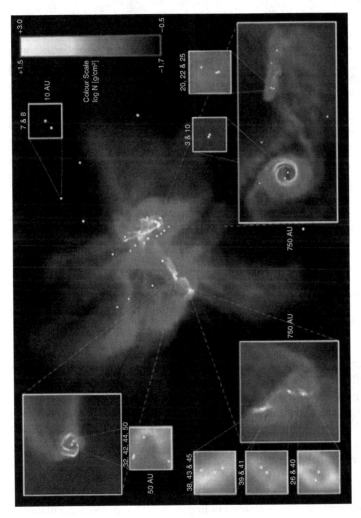

Fig. 6.13 Origin of close binaries in a numerical simulation of the formation of a small cluster. The color scale indicates the log of the column density through the cloud. The insets show the details of the seven close binary systems formed during the simulation. Most are members of multiple systems and are associated with disks. The main frame shows a region 80,000 AU across. Numbers above the insets are identifiers for particular objects. Note in particular the inset in the lower right, where, in the left half of the panel, a hierarchical triple has formed, surrounded by a disk. This system, labelled "3 & 10", consists of a binary with components of 0.73 and 0.41 $M_\odot$, separated by about 1 AU, and an outlying companion of 0.083 $M_\odot$, at a distance of 28 AU. Credit: M. R. Bate, I. Bonnell, V. Bromm: *MNRAS* **336**, 705 (2002). Reproduced with permission from John Wiley and Sons. © 2002 Royal Astronomical Society

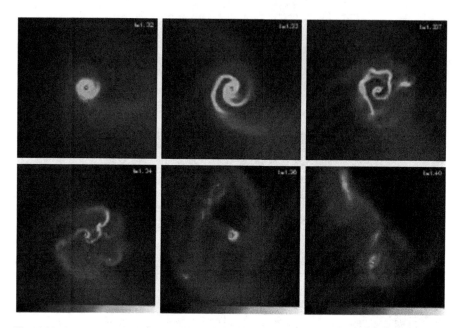

Fig. 6.14 Origin of a hierarchical multiple system in a numerical simulation of the formation of a small cluster [40]. The *grey scale* gives the log of the column density through the cloud, ranging from $\log N = 0$ (*dark*) to $\log N = 2.5$ (*light*). N has units of $\mathrm{g\,cm}^{-2}$. Times in the upper right of each frame are given in units of the initial free fall time of 1.9×10^5 yr; the values range from 1.32 to 1.4. At the end of the simulation a quadruple system has been produced, consisting of two binary pairs in orbit about each other (*center of last frame*). An unstable triple system (*upper left in last frame*) is orbiting the quadruple system. The box sizes are 600 by 600 AU. Credit: M. R. Bate, I. Bonnell, V. Bromm: *MNRAS* **339**, 577 (2003). Reproduced with permission from John Wiley and Sons. © 2003 Royal Astronomical Society

six additional protostars. These undergo gravitational interaction with themselves and are also influenced by the gas. At the end of the simulation the result is a triple system (upper left of lower right figure) which is unstable, in orbit around a quadruple system (center of figure), which is itself composed of two binaries. Each panel is 600 AU across. The disk fragmentation shown in the figure was studied earlier in a linear stability analysis [9] and through a numerical simulation [4]. The conditions required for fragmentation to occur are (1) the Toomre Q somewhere in the disk must be very close to 1; and (2) the radiative cooling time must be short, comparable to the orbital time [182]. For a massive disk in the isothermal phase these conditions are met, because cooling is in effect almost instantaneous, indicating that the process does contribute to binary or multiple formation. These authors conclude that most binaries form by fragmentation, either in the molecular cloud or in disks, and that close binaries form by hardening of wider binaries. Note that disk fragmentation can be an important process in the formation of massive stars [273].

6.5 Fragmentation

Cluster formation was further investigated in a similar calculation [37] in which the mass was increased to $500\,M_\odot$ and the radius to $0.4\,\mathrm{pc}$. The turbulent Mach number was increased to 13.6 to result in equality in absolute value of gravitational and turbulent energies. The numerical simulation included 3.5×10^7 SPH particles, allowing a mass resolution down to about $0.01\,M_\odot$. The main result of the calculation, which ran 2.85×10^5 yr, was the production of 1,254 stars and brown dwarfs, among them 90 binary systems, 23 triple systems, and 25 quadruples. The initial mass function (IMF) found was in agreement with that observed, above about $0.1\,M_\odot$ (Fig. 5.8). However too many brown dwarfs were produced as compared with the number of stars to be consistent with the observed ratio of about 1:4 [330,369]. It is very likely that this problem could be resolved with the incorporation of radiative transfer into the simulations. As shown by [38] for a lower-mass initial cloud, radiative transfer results in the heating of the gas in the vicinity of the stellar cores. Their disks are thus less likely to fragment, and in the simulation many of the brown dwarfs were produced by disk fragmentation, which typically results in low-mass companions.

The main calculation, without radiation transfer, produces encouraging results regarding the properties of binary and multiple systems. The increase in multiplicity fraction as a function of primary mass turned out to be in good agreement with observations, up to about $5\,M_\odot$, the highest mass produced in the simulation (Fig. 6.15). For example, for solar type stars this fraction was computed to be

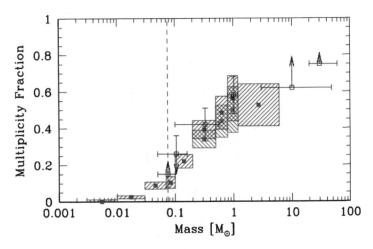

Fig. 6.15 The multiplicity fraction as a function of primary mass as calculated in a numerical simulation, compared with observations. The *open squares* give observed values, with the horizontal bars giving the mass range that applies, and the *vertical bar* giving the error and/or the upper or lower limit. The *blue filled squares* give the results from the numerical simulations, with the *shaded blue regions* giving the region of uncertainty. The *red filled squares* and shaded boxes give the numerical result excluding brown-dwarf companions, which are unlikely to be picked up in observational surveys. The *vertical dashed line* separates stars from brown dwarfs. Credit: M. R. Bate: *MNRAS* **392**, 590 (2009). Reproduced with permission from John Wiley and Sons. © 2009 Royal Astronomical Society

0.56 ± 0.12. Excluding brown-dwarf companions, which were overproduced in the simulation, the fraction was 0.5 ± 0.1, still reasonably consistent with the observed fraction of 0.58 ± 0.1 [147]. The agreement is good in the lower-mass range as well, where the fraction is 15–20%. The distribution of semi-major axes for the stellar-mass companions (Fig. 6.1) is also in good agreement with observations, with a peak at about 30 AU. The trend of decreasing mean separation with decreasing primary mass is also well represented. The distribution of mass ratios in binary systems (Fig. 6.3) was found to be fairly flat; it underproduced systems with low q. It has been argued that when a binary forms and continues to accrete gas, much of this gas will have higher angular momentum than that of the orbit, and therefore will preferentially accrete onto the secondary, which is the component farther removed from the center of mass. It is possible, however, that this disagreement with observations is a numerical effect [117]. The simulations are also in agreement with the observed mean value of the angle between the angular momentum vectors of the inner and outer orbits of triple systems. Although these conclusions may have to be changed as a result of future simulations that include radiation transfer, they do represent a major step forward in explaining the observed properties of binary and multiple systems.

We now consider the following question: Does fragmentation during a rotating collapse explain the observed properties of binary systems?

The positive answers to this question include the following points:

1. A wide range of orbital angular momenta is predicted, depending on the angular momentum of the initial cloud core or its turbulent structure, plus effects of captures after fragmentation into a small cluster as well as hardening of the orbits of some wide binaries
2. Calculations do indicate presence of circumstellar and circumbinary disks
3. This process predicts that components of young binaries are coeval, in agreement with most comparisons of young binaries with pre-main-sequence evolutionary tracks
4. Calculations predict eccentric orbits, in qualitative agreement with observations
5. Cloud core angular momenta are of the correct order of magnitude to explain wide binary orbits. Typical single-core collapse calculations produce binary systems with separations of a few hundred to 1,000 AU. The systems form either by direct fragmentation or by gravitational instability in a massive disk. Although Fig. 6.2 indicates that core angular momenta are somewhat higher than those of wide binaries, in fact simulations [41] show that not all of the angular momentum of the core goes into the orbit: a substantial fraction remains in a circumbinary disk. Thus when one considers the collapse of an individual rotating core, there is a built-in efficiency factor whereby some of the core's mass and some of its angular momentum do not become part of the resulting binary. This efficiency factor is in addition to that associated with mass loss and angular momentum loss in the bipolar flows from the individual components. Furthermore, very wide binaries are easily broken up during the evolution of a cluster

6.5 Fragmentation

6. Numerical simulations [39, 136, 269] indicate the early formation of hierarchical multiple systems in cluster formation, by gravitational interactions and captures, and the gradual breakup of such systems with time. This general trend is roughly supported by observations. Low-mass systems are more easily broken up by gravitational interactions than high-mass systems; this fact can explain the observed result that at the present day high-mass stars have a greater multiplicity fraction than low-mass stars.

However there are a number of unsolved problems:

1. For standard core parameters, fragmentation sets in just at the density where the center of the cloud is beginning to heat. Many calculations have not treated radiation transport but have assumed adiabatic heating starting from an assumed transition density. The precise treatment of radiation transfer has been shown to be important.
2. The role of magnetic fields in regulating the protostellar collapse and fragmentation phase has not been explored in sufficient detail. While the general effect of a magnetic field on cluster formation is to suppress fragmentation and star formation [423, 424], the numerical simulations that yield that conclusion are based on ideal MHD, while in fact at the densities where fragmentation becomes important one would generally expect the field to be largely decoupled from the gas because of a very low degree of ionization.
3. Since there are so many possible initial conditions, the statistics of mass ratio distributions, eccentricities, and multiplicity fraction will be difficult to derive from detailed three-dimensional calculations. It will be interesting to determine in the future how dependent the results of cluster-formation scenarios are on the details of the initial conditions. Although many observed properties of the binary population are explained by the limited number of existing cluster-formation simulations, the effects of improved spatial resolution and improved physics remain to be investigated. In particular, the mass ratio distribution in binaries remains to be explained, both theoretically and observationally.
4. It is difficult to explain close binaries, with separations of a few AU or less, by the fragmentation process. It appears that interactions in a cluster-forming environment may be the key to their formation. Even though the transition in observable properties between close binaries and wide binaries is smooth, there are likely to be a number of different physical processes that produce close binaries. The formation of very close binaries, with separation < 0.1 AU, has not been explained in theoretical simulations. However an intriguing possibility is the Kozai mechanism [165, 267], which can operate in a hierarchical triple system. If the inner and outer orbital planes are inclined by more than a critical angle i_c (about 39°) and less than $180° - i_c$, then an exchange between the inclination and the eccentricity of the inner binary can occur, with both of these quantities undergoing periodic oscillations. If the amplitude of the eccentricity oscillation is large enough, the periastron point of the inner binary can be close enough so that tidal effects reduce the semimajor axis of the inner orbit, gradually converting a relatively long-period orbit into a short-period orbit of a few days.

6.6 Summary

Detailed numerical simulations, combined with observational data, are beginning to provide some clues regarding the dominant physical processes that govern binary formation. However, the range of input parameters in such calculations is wide, and the results often depend sensitively on them. The theory is not yet at the point where it can fully explain the observed distributions of binary period, mass ratio, and eccentricity. However the simulation described above [37] does provide a significant step forward in this respect. The initial condition of this simulation, however, represents only one of a large number of possible cases. Furthermore there remain significant physical effects to be explored, including feedback on the cloud from outflows and radiation from the stars that have been produced [281]. Nevertheless, a few conclusions can already be reached.

First, fission has essentially been ruled out as a viable process for the formation of close binaries. Second, capture is unlikely to have produced many of the observed systems, although a few captures may have occurred in the cores of dense young clusters where circumstellar disks can act as the dissipation mechanism. Third, protostellar disks have been shown to be unstable, in the linear regime, to the growth of one-armed or two-armed spiral modes as long as the Toomre Q value is below a critical value and the disk can cool efficiently. Non-linear numerical calculations show that under the right conditions this instability can lead to formation of multiple stellar systems. Fourth, fragmentation during the isothermal phase of isolated-core protostar collapse has been demonstrated to produce binaries with long periods. Whether fragmentation occurs or not depends on initial conditions, in particular the density distribution and the value of α_i. However the majority of stars form in clusters, and the properties of binary and multiple systems depend on the complicated interactions betweeen stars and between stars and gas during the formation phase of the cluster. Thus massive numerical simulations are required to account for the details of binary properties deduced from observations.

The formation of close binaries is still a major issue, and it may require a combination of processes. At least three interesting suggestions should be further explored. First, at least some fragments that form during the isothermal collapse are suitable for subfragmentation as they evolve into the adiabatic phase. Thus, hierarchical fragmentation is likely to produce some close systems, and there is observational evidence to support this view. Second, an initial fragmentation stage during the isothermal collapse could produce a number of fragments in eccentric orbits. Gravitational interactions, disk dissipation, gravitational instabilities in disks, and captures in such a system could produce binary orbits with a wide range of periods and eccentricities. Third, the orbit of a long-period binary, formed during the isothermal collapse of a core or in a cloud–cloud collision, can decay as a result of angular momentum transport to a circumbinary disk or through dynamical interactions in which a close encounter can lead to hardening of the orbit. Thus the general mechanism of fragmentation could either directly or indirectly be responsible for the formation of a wide range of binary and multiple systems.

6.7 Problems

1. Two M stars (half a solar mass each) are in a circular orbit with period 15 days. A G2V star (1 $M_\odot$) is in a circular orbit around the pair with a period of 10 yr. Find the specific angular momenta of each of the orbits and compare with that in molecular cloud cores. Suggest one or more scenarios by which this system could have formed.

2. A uniform-density molecular cloud core has a mass of $2.2\,M$, a radius R, and a rotational frequency Ω. It collapses and splits up into an equal-mass binary, on a circular orbit, each component having mass M. The total angular momentum of the binary is half that of the original cloud. Each component has uniform density, uniform rotation in synchronism with the orbit, and radius 0.1 times the binary separation, which is $2d$.

    (a) Find the orbital frequency of the binary, Ω_b, in terms of the quantities given.
    (b) What is the ratio of the specific angular momentum of spin of one of the fragments to the specific angular momentum of spin of the original cloud?
    (c) Put in numbers to find d: $M = 2\,M_\odot$, $R = 3 \times 10^{17}$ cm, $\Omega = 2 \times 10^{-14}\,\mathrm{s}^{-1}$. Is this system considered to be a wide binary or a close binary?
    (d) What does this result imply with regard to the solution of the angular momentum problem?

3. A T Tauri star of mass M and radius R is spinning, with uniform Ω at its maximum rate, the so-called "breakup" velocity. Assume it fissions and forms a binary in a circular orbit with equal mass $(M/2)$ components. The components are spinning at the orbital frequency. The problem is to figure out whether this result is reasonable.

    (a) What is the binary separation in units of R?
    (b) What is the ratio of the spin specific angular momentum of one component to the specific orbital angular momentum of the system?
    (c) What assumed mass ratio for the system would be needed to get a reasonable result?

4. Assume that both the primary (m_1) and secondary (m_2) masses in a binary are chosen at random from an initial mass function, and that function is given by (1.7). Plot the resulting distribution of mass ratios $q = m_2/m_1$. Choose a mass range $10^{-2} - 10\,M_\odot$.

    (a) for all primary masses
    (b) for primary masses only in the range $0.5 - 1\,M_\odot$.

    How do the results of (b) compare with observations? The best comparison with observations would be to consider a range of q between 0.3 and 1.

5. During a certain phase of the formation of a star, the central star has a mass of $5\,M_\odot$, and it is surrounded by a disk of $2\,M_\odot$. The disk's outer radius is at $R = 1{,}000$ AU, its distribution of surface density is $\Sigma \propto R^{-1}$, and its temperature

distribution is $T \propto R^{-1/2}$ where $T = 1{,}000$ K at 1 AU. What is the minimum radius in the disk where it is likely to fragment? Estimate the mass of the fragment, assuming it forms at that radius. The most unstable radial wavelength in a gravitational instability is roughly [391]

$$\lambda_{\text{crit}} = \frac{2c_s^2}{G\Sigma}.$$

6. A cloud core of radius 2.5×10^{17} cm and mass $2\,M_\odot$, of uniform density and uniformly rotating, collapses and begins to fragment near the transition to adiabatic collapse, when the density in the central regions is 3×10^{-14} g cm^{-3} and the temperature is 10 K. The two equal-mass fragments in the binary each have one Jeans mass and are separated by 2 Jeans lengths. Assume that the total mass and angular momentum of an inner sphere at the given density has gone into the binary orbit.

  (a) What is the angular frequency Ω of the initial cloud? Mass and angular momentum of the inner sphere are conserved during collapse.
  (b) What is the specific angular momentum of the initial cloud and its ratio of rotational energy to gravitational energy? Compare the J/M with those of observed molecular cloud cores (Fig. 6.2).
  (c) Assume that half of the mass of the initial cloud eventually ends up in an equal-mass binary, conserving mass and angular momentum. What is the period and separation of the resulting orbit?

Chapter 7
The Formation of the First Stars

The first stars formed in the very early Universe at times in the range 10^8–10^9 years after the Big Bang. Conditions then were very different from those in the Galaxy at the present time. (1) Only insignificant amounts of elements heavier than H and He were present; (2) dust grains were not present, so opacities and optical depths were quite different from those in present-day protostars; (3) heating and cooling mechanisms in prestellar clouds and protostars were also quite different, leading to altered conditions of temperature and density; (4) dark matter, the exact nature of which is still unknown, played an important role in the formation process; (5) magnetic fields were probably not present, or at most were very weak compared to present-day interstellar values. The first stars are known as Population III and have not been directly observed. They may be divided into two sub-classes: Pop. III.1 and Pop III.2. The former refers to the true first generation of stars, where the composition is that produced in the Big Bang and there has been no prior star formation. The latter refers to the second generation, in which the composition is still primordial, but star formation is influenced by the kinetic energy and radiation deposited into the interstellar medium by the first generation. In contrast, the oldest observed stars have iron abundances between 10^{-3} solar and less than 10^{-5} solar [114, 174, 471]; they are known as Population II and clearly belong to a still later generation of stars but still very old, ≈ 13 Gyr. The carbon and oxygen abundances in these stars are less depleted relative to solar than is the iron. The details of the abundance patterns in these objects give important clues regarding the nature of the earliest generations of stars whose supernovae produced these elements. This chapter concentrates primarily on Population III.1.

The properties of the first stars have an important impact on the early evolution of the Universe, as their production of UV photons affects reionization of the Universe, and their supernovae produce the first enrichment of the heavy elements in the interstellar medium, which substantially affects the formation of the later generations of stars. If, as it is likely, the first stars were very massive, they also could have contributed to black-hole formation in the early Universe; however, the nature of the initial mass function, and the upper mass limit, are critical and as yet unanswered questions. This chapter summarizes some of the important physical processes relevant to the formation of the first stars and sketches the evolutionary phases through which they must progress.

7.1 Physics of the First Stars

Many of the basic processes previously discussed for present-day star formation still apply to the first stars.

- Hydrodynamic collapse of material under gravity, although the importance of magnetic fields is probably negligible
- Equation of state of the gas, including heating, cooling, ionization, dissociation
- Molecular chemistry and the determination of abundances of molecules, particularly H_2, whose cooling radiation is important
- Radiative transfer, both in the continous and line spectra
- Turbulence and convection in the prestellar gas as well as in the interiors of protostars; however turbulence is thought to play a less important role than in present-day star formation, because it is probably sonic or subsonic, rather than supersonic
- Shock waves, both those involving accretion of material onto forming stars and disks, and those generated by the radiation or explosion of the first stars, which can induce the formation of the second stars.

In some ways, the physics here is somewhat simpler than that in present-day star formation, because of the virtual absence of magnetic fields and dust grains, because the cooling mechanism is straightforward, because there are no previously-formed stars to influence the physical conditions at the formation site, and because the initial conditions can be fairly well defined through cosmological simulations.

Of all of these physical processes, the formation and excitation of H_2 is the most crucial, since it is the cooling that arises from this molecule that allows the Jeans mass in primordial gas clouds to become low enough to allow the formation of stellar-mass objects. However, at the time of formation of the first stars the hydrogen gas in the Universe has recombined and is primarily neutral, and molecular formation on dust grains, the principal process in present-day molecular clouds, is clearly not possible. Also, direct attachment of two H atoms to form a molecule, with the binding energy given off as a photon, is not possible because vibration-rotation transitions in the H_2 molecule are forbidden for electric dipole radiation.

Thus the following reactions must occur, given a few e^- and H^+ left over from recombination:

$$H + e^- \rightarrow H^- + h\nu \quad (7.1)$$

$$H^- + H \rightarrow H_2 + e^- \quad (7.2)$$

7.1 Physics of the First Stars

or

$$H + H^+ \rightarrow H_2^+ + h\nu \qquad (7.3)$$
$$H_2^+ + H \rightarrow H_2 + H^+. \qquad (7.4)$$

The first set of reactions generally dominates. The abundance of electrons and ions is very low, so it turns out that these reactions, below a density $n = 10^8 \, \text{cm}^{-3}$, produce only a small amount of H_2. Above that density, three-body reactions are possible

$$H + H + H \rightarrow H_2 + H \quad \text{and} \qquad (7.5)$$
$$H_2 + H + H \rightarrow H_2 + H_2 \qquad (7.6)$$

and the density of H_2 increases considerably.

Once the temperature reaches about 2,000 K and the density is relatively high, the hydrogen molecules undergo dissociation, mainly by collisional processes

$$H_2 + H \rightarrow 3H \qquad (7.7)$$
$$H_2 + H_2 \rightarrow H_2 + H + H. \qquad (7.8)$$

The main heating mechanisms are gravitational compression and heating of free electrons by the cosmic microwave background. When H_2 forms the heat of formation must be taken into account. Cooling occurs from atomic and molecular line cooling and Compton cooling, which involves transfer of energy from high-energy electrons to low-energy photons by scattering. At relatively low densities, the collapse proceeds nearly adiabatically. At higher n and T, collisions between H atoms and H_2 molecules excite the H_2 molecules, primarily to the lowest-lying rotational levels. Then the molecules de-excite with the emission of a photon, which, up to a density of about $10^{13} \, \text{cm}^{-3}$, is able to escape from the cloud, thus cooling it. However the excitation potential of the first level of H_2 corresponds to an excitation temperature of 512 K; thus the gas is able to cool via this process down to 200 K but not lower; it can cool to that value through collisions with particles in the high-energy tail of the Maxwell velocity distribution. When H_2 dissociates, the dissociation energy must be provided by the gas, so that is effectively a cooling mechanism.

Figure 7.1 shows a sketch of the path in a ($\log n$, $\log T$) diagram which a metal-free protostar is expected to take. At low densities, to the left of point A, the molecular hydrogen fraction is less than 10^{-4}, all cooling mechanisms are ineffective, and the gas heats adiabatically by compression. Between points A and C, H_2 forms through reactions (7.2) and (7.4), the H_2 fraction is in the range 10^{-4}–10^{-3}, and cooling acts to reduce the temperature to 200 K. Point B represents the range where H_2 cooling is most effective. At point C, the cooling rate saturates, as the level populations in H_2 approach local thermodynamic equilibrium. Between

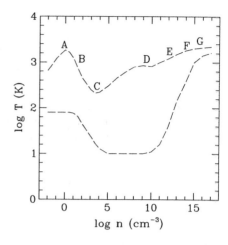

Fig. 7.1 Evolutionary path of protostars in the $(\log n, \log T)$ plane. *Upper curve:* a metal-free protostar in the early Universe; *lower curve:* a protostar with solar metal abundance forming at the present time. *Upper curve* is adapted from [568]

points C and D the gas again heats nearly adiabatically. Near point D, three-body reactions ((7.5) and (7.6)) become important. Between points D and E the molecular fraction rises to almost unity. However there is only slight cooling, because in that region the optical depth in the H_2 cooling transitions becomes larger than one. Because the probability of escape of the photon becomes smaller, cooling by that process becomes less and less effective. An additional cooling process comes in at densities above 10^{14} cm^{-3}, point F. The process of collision-induced emission involves a transition within a molecule from one state, characterized by a rotational quantum number J and a vibrational quantum number v, to another state with J', v', resulting in the emission of a photon:

$$H_2(v, J) + H_2 \to H_2(v', J') + H_2 + h\nu. \tag{7.9}$$

Such transitions result in a continuum-like spectrum [180]. However beyond point G, at $n \approx 10^{18}$ cm^{-3} the gas becomes opaque to the continuum emission as well, and further increase in density is nearly adiabatic, as no cooling processes are important. At point G, the H_2 begins to dissociate, resulting in a decline of the H_2 fraction.

Note the considerable differences between the tracks in the $(\log n, \log T)$ diagram between the metal-free protostar and the one with solar metallicity. In the latter case, CO molecules and dust grains provide much more efficient cooling than H_2, and even at low densities the gas can cool to 10 K. The gas stays at that temperature over several orders of magnitude in n, because the cooling time is shorter than the free-fall time. The gas becomes optically thick in the continuum at $n \approx 10^{11}$ cm^{-3}, much earlier than in the metal-free case. The compression beyond that density is adiabatic. Note that the conditions for onset of dissociation are very similar in the two cases.

7.2 Sequence of Events During Formation

Several phases can be identified in the formation of the first stars, with size scales ranging from very large (Mpc) cosmological scales down to a few $R_\odot$. The work describing this process includes semi-analytical and one-dimensional numerical simulations, but the most detail is provided by three-dimensional numerical hydrodynamic simulations. Nevertheless the problem is sufficiently difficult, owing partly to the large range of length scales involved and partly to the complications of radiation transport, so that such simulations have not been carried all the way to the point where the star has obtained its final mass. Observations of the first stars have so far been not possible, but with improved instrumentation in the future this goal may be achieved, particularly if they end up as very bright supernovae with gamma-ray bursts.

7.2.1 Cosmological Phase

The start of the process involves a mixture of dark matter and baryons, with the baryons providing only about 15% of the density. Their composition, a result of big-bang nucleosynthesis, is about 76% hydrogen and 24% helium by mass. The dark matter particles, whose nature is unknown, are considered to be collisionless, while the baryons are collisional and can dissipate energy. The Universe at the start of the simulations has a spectrum of very tiny density perturbations on a smooth background. Detailed numerical simulations, starting at a redshift $z \approx 100$, are usually based on the Cold Dark Matter picture, which, along with the observations of the cosmic microwave background radiation, determines the nature of these perturbations. Those which have above-average amplitude tend to grow the fastest, forming small overdense regions composed mainly of dark matter. These regions grow larger in mass through encounters with each other and mergers. The gravitational interactions and mergers result in a net spin angular momentum for each of these dark-matter-dominated haloes. The first region to grow up to a mass of 10^5–$10^6 \, M_\odot$ is the probable site of the formation of the first star. A dark-matter halo, in general, consists of an infall region and a virialized region, where, in the latter, the kinetic energy of random motions is one-half the absolute value of gravitational energy. At redshifts of 10–20, the virialized region contains 10^5–$10^6 \, M_\odot$ and the corresponding temperature is about 1,000 K. Within the dark-matter halo of say $10^6 \, M_\odot$, the baryonic matter, under the influence of the dark-matter gravitational potential, settles to the center, forming a core of about $1,000 \, M_\odot$, close to hydrostatic equilibrium. At about $z = 18$, corresponding to 200 Myr after the Big Bang, the baryonic core begins to collapse, marking the onset of the protostellar phase.

7.2.2 Protostellar Collapse

One requirement for the baryonic core of a dark-matter halo to collapse is that the cooling time must be less than the free-fall time. When the halo mass approaches $10^6 \, M_\odot$ [92], temperatures in the core are about 1,000 K, and atomic transitions in the neutral hydrogen are ineffective at cooling; they require about 10^4 K. The condition can be satisfied only when a fraction of about 10^{-4} of the hydrogen has been converted to H_2, corresponding approximately to point A in Fig. 7.1. The cooling rate increases with n.

When a sufficient amount of H_2 has formed, the baryonic material cools and compresses, reaching conditions (point C in Fig. 7.1) where $T \approx 200$ K and $n \approx 10^4 \, \text{cm}^{-3}$. This density corresponds to the critical density where the excited levels of H_2 are de-excited by collisions (which produce no photons) at about the same rate as radiative de-excitations. The cooling time, previously inversely proportional to the density, becomes independent of density. Thus further compression is held up. The condition now needed to get the core to collapse is that its mass must be greater than the Jeans mass, which [92] can be written

$$M_J \approx 700 M_\odot \left(\frac{T}{200 \, \text{K}}\right)^{3/2} \left(\frac{n}{10^4 \, \text{cm}^{-3}}\right)^{-1/2}. \tag{7.10}$$

Thus once about $1,000 \, M_\odot$ has accumulated with conditions appropriate to point C, collapse can begin. The total mass of the baryonic matter in the halo is at least $10^5 \, M_\odot$, and this matter gradually accretes into the center. However the final mass of the first star is probably not determined by the Jeans mass of the initial core, but rather by feedback effects that occur later on.

The protostellar core of the dark-matter halo has angular momentum, determined by tidal interactions with the neighboring halos. However at the initial state of the core, the rotational effects are small, with rotational kinetic energy less than 1% of the gravitational energy. Also, magnetic and turbulent effects are expected to be small, so the thermal Jeans mass, (7.10), is probably appropriate.

The question arises whether this core of $1,000 \, M_\odot$ will fragment into a cluster of stars, or at least into a binary system. This question is still unresolved. For solar-metallicity clouds, the likely fragmentation regime is the isothermal collapse stage, where the Jeans mass decreases with density and pressure effects are relatively unimportant. As discussed in Sect. 6.5, rotational effects are important in slowing down the collapse during this stage and promoting fragmentation. However in the metal-free case the isothermal stage does not occur (Fig. 7.1) and pressures at a given density are much higher than in the present-day case. The high pressures tend to suppress perturbations that could lead to fragmentation. Further, the angular momentum of the baryonic material that falls to the center of the dark-matter halo is relatively small. For example, a three-dimensional simulation, carried up to a central density of $10^{16} \, \text{cm}^{-3}$ [398], showed that indeed fragmentation did not occur, and an analytic stability analysis [568] verified this result. Three-dimensional calculations

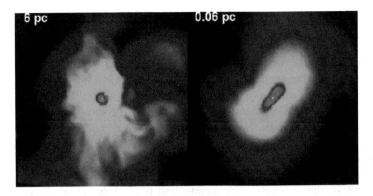

Fig. 7.2 Numerical simulation, in three space dimensions, of the first protostar [1]. On two different scales but at the same time, a two-dimensional density slice through the collapsing object is shown. The number density is shown on a logarithmic color scale, at the end of the simulation (redshift $z = 18.18$, corresponding to a time of about 1.5% of the current age of the Universe). The scales indicated on the plots correspond to the side length of the box. At that time the central density is 3×10^{13} cm^{-3}, the outer density on the 6 pc scale is about 100 cm^{-3}, and the outer density on the 0.06 parsec scale is about 10^7 cm^{-3}. The enclosed mass on the 0.06 parsec scale is about 100 $M_\odot$. The small *yellow dot* in this frame corresponds to a protostar of 1 $M_\odot$, which is still in the collapse phase, before the formation of the hydrostatic stellar core. From T. Abel, G. Bryan, M. Norman: *Science* **295**, 93 (2002). Reprinted with permission from AAAS. © American Association for the Advancement of Science

carried to even higher densities [567] also did not show fragmentation. However, the amount of mass included in the high-density central regions at the end of these simulations was small, as was the corresponding angular momentum. At later times, not yet fully explored by numerical simulations, the material farther out, with higher angular momentum, can collapse, and a relatively massive disk could form. Cooling in this disk could be sufficient to give the right characteristics for it to become gravitationally unstable and to fragment.

If the central region of a protostar, once it has reached the stage represented in Fig. 7.2, is converted into a sink particle, the evolution can be followed to later stages, and the question of fragmentation can be investigated. One such simulation [480] treats a sink particle, initially of approximately 1 $M_\odot$, as unresolved, although mass is allowed to accrete onto it. Thus the very short time steps required to follow its evolution are avoided. The result of this simulation was the formation of a disk with mass $\approx 30 M_\odot$, and radius 2,000 AU. The disk became gravitationally unstable and fragmented. The end result was two major fragments of 40 and 10 $M_\odot$, respectively. The final fate of these fragments is unknown – for example, it is possible they could merge and/or continue to accrete to high mass – but the possibility that the first star is a binary remains open.

Returning to the single-star case, once the Jeans mass is reached, the cloud goes into collapse, with gradually increasing temperatures. At first the H_2 fraction is still quite low (10^{-3}) and the cooling efficiency is low, resulting in a temperature rise back to around 1,000 K. Then, at a density of above 10^8 cm^{-3} (point D in Fig. 7.1) the three-body reactions for the formation of H_2 become significant, and the H_2

fraction rises quickly, approaching 100%. About 1 $M_\odot$ at the center becomes fully molecular. The more efficient cooling then results in a brief phase of temperature decline. However at only slightly higher density the optical depth in the lines corresponding to the cooling transitions becomes greater than unity, and the cooling photons become trapped in the cloud, having a smaller and smaller chance to escape with increasing density. By the time the density reaches 10^{14} cm^{-3} the escape probability is only 1% that at 10^{10} cm^{-3}. Figure 7.2 shows the structure of the protostar at about this time. Note that protostar at this stage has not yet developed an equilibrium stellar core. However, above that density, collision-induced transitions of molecular H provide further cooling through continuum radiation, slowing down but not stopping the upward trend in temperature. When $n \approx 10^{16}$ cm^{-3} and $T \approx 2{,}000$ K, the H_2 starts to dissociate, and beyond that point the gas becomes completely optically thick in the continuum as well as in the lines. The collapse then becomes nearly adiabatic, because the time for transfer of radiation out of the central region is longer than the collapse time.

Once hydrogen dissociation has begun at the center, that region goes into dynamical collapse once more. The dissociation is not complete until the temperature is above 20,000 K. The collapse is fast, nearly free-fall, with a time scale of only a year between points A and B (Fig. 7.3). Then a hydrostatic core forms, with a

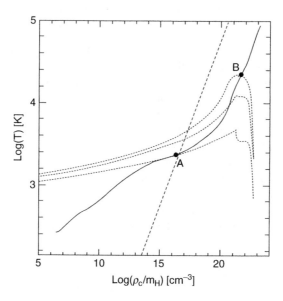

Fig. 7.3 One-dimensional hydrodynamic simulation of the collapse of the first protostar [436]. The evolution of the center (*solid line*) is shown in the ($\log n, \log T$) plane. The *dashed line* and point A indicate where the cloud becomes optically thick in the continuum. (In 3-D calculations, this point is shifted to higher densities, because non-spherical effects allow escape of photons in the polar direction). Point B indicates where the initial hydrostatic "stellar" core forms. The *dotted lines* indicate the molecular hydrogen dissociation zone; the *upper, middle*, and *lower lines* correspond to equilibrium H_2 fractions of 0.01, 0.1, and 0.9, respectively. From E. Ripamonti et al.: MNRAS **334**, 401 (2002). Reproduced with permission from Wiley. copyright Royal Astronomical Society

7.2 Sequence of Events During Formation

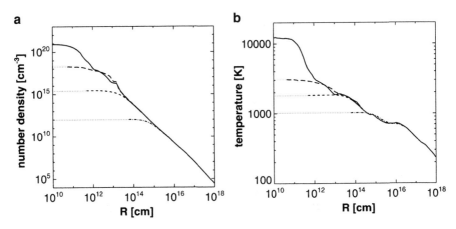

Fig. 7.4 Spherically averaged density (**a**) and temperature (**b**) distributions from three-dimensional simulations of the evolution of the first protostar at several different times [567]. This protostar was formed at the time corresponding to redshift $z \approx 14$ and was carried to higher central density than that shown in Fig. 7.2. The density structure approaches a power law $n \propto r^{-2}$. The four times shown (*bottom* to *top*) correspond to central densities of 10^{12}, 10^{15}, 10^{18}, and 10^{21} cm^{-3}, corresponding, respectively, to the points where the core becomes optically thick in molecular lines, where cooling by collision-induced emission sets in, where the core becomes optically thick in the continuum, and where H$_2$ dissociation is complete in the core. The time intervals between curves are roughly the free-fall time at the given density. From N. Yoshida, K. Omukai, L. Hernquist: *Science* **321**, 669 (2008). Reprinted with permission from AAAS. © American Association for the Advancement of Science

shock wave on its outer edge, in a process that is similar to that in a protostar with solar metallicity. The initial core mass is about 10^{-2} M$_\odot$ with a radius of about 5×10^{11} cm, and it is accreting mass rapidly, a few $\times 10^{-2}$ M$_\odot$ yr^{-1}. Rotational effects have become relatively important at this time, and a disk forms outside the stellar core, with a radius of about 10^{13} cm and a mass of 0.1 M$_\odot$. In a clustered environment, it is possible that close encounters with other protostars could eject the core from its parent cloud with the result of cutoff of accretion and a rather low final mass. However in the case of the first star there are no such disturbances, and the core is expected to grow to high final mass, as discussed in the next subsection.

The evolution of the density- and temperature structures and the characteristics of the protostar at time of stellar core formation, according to three-dimensional calculations [567] are shown in Figs. 7.4 and 7.5. The density distribution becomes extremely centrally condensed, and at the final time the central density is 21 orders of magnitude higher than it was when cooling from H$_2$ began (point A in Fig. 7.1). The temperatures maintain a generally upward trend despite various cooling processes. Exceptions are noted at 10^{10} cm^{-3}, where 3-body reactions convert all of the hydrogen to H$_2$, and at about 10^{13} cm^{-3}, where the gas is still optically thin to collision-induced transitions. The initial stellar core, at the final time, is that region with $R < 5 \times 10^{11}$ cm.

Figure 7.5 shows that in that core (10^{-2} M$_\odot$) only a few percent of the hydrogen is still molecular, most of it is neutral, and it is beginning to ionize. Only when the

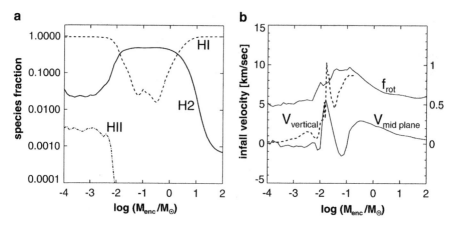

Fig. 7.5 Characteristics of the protostar [567] at the final time shown in Fig. 7.4. *Left:* the number fractions of molecular hydrogen (H2), neutral hydrogen (HI), and ionized hydrogen (HII) as a function of enclosed mass. *Right:* Infall velocities as a function of enclosed mass (left scale) in the direction parallel to the rotation axis (V_{vertical}) and azimuthally averaged in the equatorial plane (V_{midplane}). The right-hand scale shows the degree of rotational support $f_{\text{rot}} = (L/r)/v_K$, where L is the specific angular momentum at radius r and v_K is the Keplerian velocity $\sqrt{GM_{\text{enc}}/r}$. From N. Yoshida, K. Omukai, L. Hernquist: Science **321**, 669 (2008). Reprinted with permission from AAAS. © American Association for the Advancement of Science

core has built up to considerably higher masses will it become completely ionized. Outside the core, the region where H_2 dominates is only a few $M_\odot$, while the bulk of the mass, out to $100 M_\odot$, is primarily neutral H. But even the very small amount of H_2 that remains there is sufficient to cool the gas to about 200 K. The right-hand panel shows that high velocities of infall are generated just outside the core, induced by molecular dissociation, in both the polar and equatorial directions. At the edge of the core there is a sudden deceleration caused by a shock wave, and inside the core the velocities reduce to near zero, signalling the onset of hydrostatic equilibrium. The curve labelled f_{rot} shows that at the time of stellar core formation rotational effects have become important throughout the inner $100 M_\odot$. They are particularly important just outside the core, where a disk forms. The reduction in f_{rot} in the core itself is caused by angular momentum transport induced by spiral waves in the disk.

7.2.3 Accretion Phase

The next phase of the formation of the first star involves the accretion of material from the low-density infalling envelope onto the high-density stellar core (initially $n \approx 10^{21}$ cm^{-3}). Although this core of $10^{-2} M_\odot$ is similar for the first star and for solar-metallicity protostars, the accretion phase differs in two important respects in the two cases. First, the first-star low-density core contains at least $1,000 M_\odot$ of

7.2 Sequence of Events During Formation

baryonic material which has collected at the center of a dark-matter halo, while the typical molecular cloud core in the solar-metallicity case contains only a few $M_\odot$. Second, a rough estimate of the accretion rate is given by $\dot{M} \approx c_s^3/G$, where c_s is the sound speed in the low-density core. In the case of a typical molecular cloud core at 10 K, $\dot{M} \approx 1.6 \times 10^{-6}\, M_\odot\, \text{yr}^{-1}$, while in the first-star case at 200 K the rate is two orders of magnitude higher at about $1.5 \times 10^{-4}\, M_\odot\, \text{yr}^{-1}$, and, at 1,000 K, three orders of magnitude higher at $\sim 10^{-3}\, M_\odot\, \text{yr}^{-1}$.

The actual accretion rate can differ significantly from the rough (dimensional) estimate, and in fact it can vary, depending on the detailed properties of the dark-matter haloes in which the first stars form. As we have seen, at the time of stellar core formation the rate is above $10^{-2}\, M_\odot\, \text{yr}^{-1}$. However, as in the case of solar-metallicity protostars, the complete 3-dimensional evolution of the first star has not been calculated all the way up to the final mass. Thus, various estimates have been made for the accretion rate. In any case, if the accretion can be approximated as infall onto a metal-free main-sequence star of a given mass, the infall rate cannot exceed the Eddington limit for that mass, which is the luminosity above which the radiation pressure force exceeds gravity at the surface of the star. Assuming that the luminosity is dominated by accretion, then

$$L_{\text{acc}} \approx \frac{GM\dot{M}}{R}, \qquad (7.11)$$

where R is the main-sequence radius of the star with mass M. The Eddington limit is given by

$$L_{\text{ed}} = \frac{4\pi c GM}{\kappa}, \qquad (7.12)$$

where c is the velocity of light and κ is the Rosseland mean opacity at the stellar surface. Thus the limiting mass accretion rate is

$$\dot{M}_{\text{lim}} = \frac{4\pi c R}{\kappa}. \qquad (7.13)$$

This limit is plotted in Fig. 7.6, under the reasonable assumption that κ is dominated by electron scattering and that the main-sequence radii are given by standard stellar structure calculations [448] for zero metals.

A reasonable estimate of the actual accretion rate can be made, based on the structure of the protostar, as determined from 3-D calculations [398, 568], when the central density has reached about $10^{15}\, \text{cm}^{-3}$. The estimates of this rate are displayed in Fig. 7.6. The rate is calculated by obtaining

$$\dot{M} = 4\pi r^2 \rho v_r \qquad (7.14)$$

as a function of the enclosed mass, where ρ is the gas density in g cm^{-3} and v_r is the radial infall velocity. This formula essentially gives the rate at which gas in the low-density protostellar core is delivered to the point of onset of dissociation,

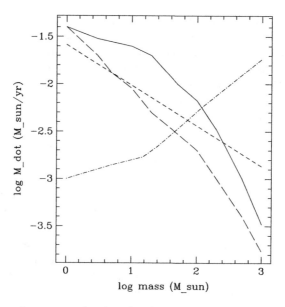

Fig. 7.6 Mass accretion rates as a function of enclosed mass for simulations of the formation of the first stars. (*Dot-dashed line:*) the Eddington limit for accretion onto main-sequence stars with zero metals (Z=0.). (*Solid line:*) accretion rate as estimated by [398]. (*Long-dashed line:*) accretion rate as estimated by [568]. (*Short-dashed line:*) Accretion rate as estimated by [498]. Below the mass range 50–100 $M_\odot$ the calculated rates are higher then the Eddington limit for main-sequence radii, therefore the actual radii during this phase must be larger than main-sequence values

from which it free-falls onto the surface of the stellar core. It is clear that this rate is very fast initially and then decreases with time, which can be roughly understood as reflecting the increase in dynamical time as one proceeds outwards in the core to lower-density material, as well as the decrease in c_s in the estimate $\dot{M} \approx c_s^3/G$.

An analytical estimate [498] which agrees well with the above numerical results is based on the assumption that the low-density core has a hydrostatic, polytropic structure with $P = K\rho^\gamma$. The value of K is obtained from the numerical results. Then the accretion rate is estimated to be

$$\dot{M} = \phi \frac{M}{t_{\text{ff}}}, \tag{7.15}$$

where M is the central mass, t_{ff} is the free-fall time measured in the initial core where the mass interior is M, and ϕ is a numerical parameter of order unity that depends on the structure of the low-density core. The accretion rate that results is

$$\dot{M} \approx 0.026 \left(\frac{M}{M_\odot}\right)^{-3/7} M_\odot \, \text{yr}^{-1} \tag{7.16}$$

7.2 Sequence of Events During Formation

for typical parameters of the initial core, assuming all the collapsing mass ends up on the star rather than in an outflow, and that the inflow is spherical, without a disk. The exponent of $-3/7$ arises from the assumption $\gamma = 1.1$, approximately correct for the evolution beyond point C in Fig. 7.1. If accretion through a disk is taken into account, the accretion rate is multiplied by a factor of about 2/3 [498]. This formula is also plotted in Fig. 7.6. The time to accrete up to a given mass is then

$$t_{\rm acc} \approx 27 \left(\frac{M}{M_\odot} \right)^{10/7} \; {\rm yr}, \qquad (7.17)$$

which gives 20,000 yr for $100\,M_\odot$ and 94,000 yr for $300\,M_\odot$. If disk accretion is included, the times increase by about a factor 1.5. These mass values are chosen because they bracket the probable range of final masses according to the picture we have been describing, in which feedback, to be described in the next subsection, determines the final mass. In the absence of feedback, the star could in principle accrete all the baryons in the initial dark-matter halo, more than $100{,}000\,M_\odot$.

In Fig. 7.6 it is clear that for the early part of the accretion phase, at low stellar masses, the estimated accretion rate exceeds the estimated Eddington limit. The actual accretion rate will be slowed down partly because of the effects of angular momentum, which would tend to result in the formation of a disk around the forming star. Also, the rapid accretion will feed back on the structure of the star. If the time scale for accretion $t_{\rm acc} \approx M/\dot{M}$ is shorter than the thermal adjustment time t_{KH}, which is true for low M, then the star cannot accept the material as fast as it is being supplied, and it must expand in radius. As shown in Fig. 7.7 this expansion is significant, resulting in radii greater than $200\,R_\odot$.

Although approximate three-dimensional calculations have been performed for the accretion phase [92], the essence of the character of the evolution of a proto-first star during the accretion phase can be obtained [568] in the spherically symmetric approximation through the use of standard stellar models in hydrostatic equilibrium. Accretion through a disk can also be taken into account [498]. The star is simply assumed to add mass as given by the appropriate curve in Fig. 7.6. While the mass is less than $10\,M_\odot$, as expected, the thermal adjustment time t_{KH} is relatively long and the accretion rate is fast, so the star is unable to accept the mass at its main-sequence radius: it must expand. The radius as a function of mass is shown in Fig. 7.7. As a result of the large radius, the accretion luminosity $(GM\dot{M}/R)$, also shown in the figure, drops well below the Eddington limit. Around $10\,M_\odot$ the radius makes a small upward spike as a result of the details of the transition of the interior of the star from a convective to a radiative structure. At masses between 10 and $100\,M_\odot$ the accretion rate drops and the thermal adjustment time decreases, so a decrease in radius results and the star is able to contract to the main sequence and start nuclear burning at about $100\,M_\odot$. At masses above this value, the star is able to accept the mass while remaining at its main-sequence radius. The evolution is then simply up along the main sequence with increasing mass and increasing radius. The increase in radius along with the decrease in $\dot{M}$ result in a fairly abrupt shift from

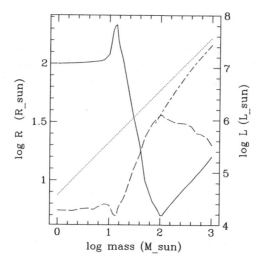

Fig. 7.7 Radius (*left scale*) and luminosity (*right scale*) as a function of total mass from an approximate model of a spherically accreting primordial star [568]. *Dotted line:* the Eddington limit for accretion (7.12; *right scale*). *Solid line:* radius of the accreting star. The increasing radius above $100\,M_\odot$ corresponds to the zero-metal main sequence. *Long-dashed line:* the accretion luminosity, based upon the mass accretion rate given as the long-dashed line in Fig. 7.6. *Short-dashed long-dashed line:* the nuclear luminosity from main-sequence stellar models with $Z = 0.$, starting at $100\,M_\odot$ [448]. Feedback effects were not included; they would tend to shut off accretion at about $140\,M_\odot$ (Fig. 7.8)

the luminosity being dominated by the accretion luminosity to being dominated by nuclear luminosity, as shown also in Fig. 7.7. The plot continues up to $1{,}000\,M_\odot$, but in practice feedback effects of the stellar photons upon the infalling material will limit the mass to values less than that, as discussed in the next section. As shown in the figure, the total luminosity never exceeds the Eddington limit. An evolution similar to that shown in Fig. 7.7 results if the presence of a disk is included [498]. The general conclusion is that, depending upon parameters, the first star arrives on the main sequence in the mass range 50–$100\,M_\odot$, at which point it is still accreting. A higher accretion rate corresponds to a slightly higher mass at the arrival point.

7.2.4 Feedback Phase

When the first star actually arrives on the main sequence, its surface temperature is approaching $100{,}000\,\mathrm{K}$. As a result there is a high luminosity of UV photons beyond the Lyman limit. The precise mass at which the effects of these photons cut off accretion from the dark-matter halo is important, because it affects the reionization of the Universe, and the initial enrichment of heavy elements when it becomes a supernova. The pattern of elements produced depends on the mass of the supernova

7.2 Sequence of Events During Formation

progenitor, and in some mass ranges no metals are produced at all because the entire star collapses and becomes a black hole.

In present-day star formation, the main feedback effect is radiation pressure on the infalling dust, which becomes important for masses as low as $10\,M_\odot$ (see Chap. 5.) But in the early Universe, this effect is absent. Stellar winds and outflows are likely to be important in limiting accretion in present-day star formation, but the magnetic fields that play a crucial role in producing bipolar flows in low-mass stars, and the effect of radiation pressure in metal absorption lines that is important in driving winds in massive stars, are also very likely to be of little importance in the case of the first stars. Instead, effects resulting from ionization and dissociation dominate the feedback processes.

The main effects that are likely to be important [355] are (1) dissociation of H_2, (2) Lyman α radiation pressure, (3) creation of an HII region, and (4) photoevaporation of a disk. Rotational effects and the presence of a disk are important in determining the final mass, and they have been considered in an approximate way. These four effects are discussed briefly in the following paragraphs.

Once the forming first star has reached a mass of 25–$30\,M_\odot$ it radiates substantial energy in the far-ultraviolet (FUV) with energies less than the $13.6\,\text{eV}$ necessary to ionize hydrogen. Although during protostellar collapse the main mechanism for dissociating H_2 is by collisions, in low-density regions the primary mechanism is photodissociation by photons above $11.2\,\text{eV}$ in energy ($H_2 + h\nu \rightarrow H + H + h\nu' + KE$). These photons are able to excite the molecule from the ground electronic state to the first excited electronic state (Lyman band) or to the second excited electronic state (Werner band). The net result is either decay back to the ground state, or, a fraction of the time, dissociation. In the latter case the energy in excess of the $4.48\,\text{eV}$ binding energy of the molecule goes into radiation and kinetic energy of the hydrogen atoms.

Once the flux in these bands becomes significant, the molecules at low density in the baryonic core of the dark-matter halo can become dissociated. Without the molecules, the primary cooling mechanism for the core disappears. One might expect that the low-density core would heat up and stop its contraction, but by the time the dissociation becomes significant, there is already a significant amount of collapsed mass interior to the core. In most cases a detailed analysis shows [355] that accretion onto the forming star will continue, at a rate that is at most reduced by 20% from that before dissociation. Thus this feedback effect has a negligible effect on shutting off accretion. However, the Lyman–Werner radiation from the first star could result in dissociation in nearby dark-matter halos with baryonic cores that have not yet reached the point of collapse. Here suppression of star formation could occur, but only if the affected core is relatively nearby, ~ 100 parsec or less [495], and then only if the core is still relatively early in its evolution, with a central density of 10^3–10^4 particles cm^{-3} or less.

The FUV radiation from the central star can also be absorbed in the Lyman α and other Lyman lines in the infalling material, resulting in a radiation pressure which could decelerate and perhaps reverse the collapse. The complicated problem of calculation of the transfer of momentum from the outgoing photons to the infalling gas [355] leads to an estimate of the conditions under which the accretion can be stopped. The result in this case depends on the rotation of the flow, because for moderate rotation the mass inflow rate is low enough in the polar direction so that radiation pressure can easily reverse the infall. At about 20 $M_\odot$ this effect becomes important, and mass is blown off primarily at the rotational poles. Once the mass in the infalling region in that direction is lost, a low-opacity cavity is created, and the Lyman photons tend to escape through the cavity, reducing the effect of their radiation pressure in other directions. Thus the end result is that the radiation pressure effect is not particularly important. The accretion rate can be somewhat reduced because of outflow in the polar direction, but inflow in most directions, involving most of the mass, cannot be stopped.

Once the central star becomes hot enough, it produces substantial radiation in the extreme UV (EUV), with photon energies above 13.6 eV. The radiation ionizes and heats the infalling gas. The balance of heating and cooling processes in an HII region with no metals gives a temperature of about 25,000 K. Thus the pressure is substantially increased with respect to that of neutral H, and in particular the pressure gradient between ionized and neutral material at the outer edge of the HII region can suppress infall. Again the rotation of the infalling gas, which modifies the density distribution in the gas near the protostar, results in a non-spherical HII region. The radius of the outer edge of the HII region thus depends on the angle with respect to the rotation axis, the infall $\dot{M}$, and the mass of the central star, which determines the production rate of ionizing photons.

The accretion onto the star will tend to be suppressed when the HII region has expanded to the point where the thermal energy of the ionized gas is comparable to its gravitational potential energy, and thus a particle is only marginally bound. This so-called *gravitational radius* is given by

$$r_g = \frac{G\phi_{ed} M}{c_i^2}, \tag{7.18}$$

which is on the order of 100 AU for a star of 100 $M_\odot$. Here M is the total mass of the central star plus disk, c_i is the sound speed in the ionized gas, and the quantity ϕ_{ed} corrects for the effects of radiation pressure

$$\phi_{ed} = 1 - \frac{L}{L_{ed}}. \tag{7.19}$$

It can then be determined [355] at what mass the HII region will cut off accretion. The suppression will occur first at the pole, because the reduced density of the rotating flow in that direction results in a larger HII radius than in other directions. For moderate rotation it is found that polar inflow will be suppressed at 50–100 $M_\odot$. However the presence of the relatively high-density disk near the equatorial plane shields some of the infalling material from the ionizing photons. The fraction of infalling material shielded depends on the geometry and detailed structure of the

7.2 Sequence of Events During Formation

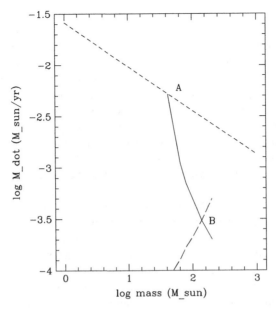

Fig. 7.8 Schematic illustration of the effects of feedback on limiting the mass of the first star. *Short-dashed line*: the mass accretion rate onto a forming star based on the properties of a polytropic low-density baryonic core [498], without feedback. This line is the same as the *short-dashed line* in Fig. 7.6. *Solid line:* The modified accretion rate onto the forming star, taking into account the effects of the HII region in limiting the rate. The point A gives the mass at which this effect begins to be important. *Long-dashed line*: the rate at which the disk surrounding the forming star is photoevaporated, as a function of mass. Once this rate exceeds the infall rate (point B) then the disk, which has been shielding some of the infalling matter from becoming ionized, disappears, the infall stops, and the final mass is determined. Reprinted in adapted form from [355] by permission of the AAS

disk, but could be about 0.3. Overall, therefore, the accretion rate is reduced because of the presence of the HII region but not shut off altogether.

The disk shielding effect is important at about $100\,M_\odot$, but eventually the disk itself will be photoevaporated; this process will determine the final mass of the star. The critical point will occur when the photoevaporation rate exceeds the mass accretion rate, which declines with time (Fig. 7.8). Once this happens, the disk is eroded, and its shielding effect becomes negligible. The entire infalling region becomes ionized, and the accretion flow stops.

The disk evaporation rate can be estimated according to the theory developed for present day star formation [232] and modified to account for the properties of the first star [355]. The disk evaporation $\dot{M}$ can be obtained as a function of mass and ionizing flux, which also is a function of mass. Then this rate can be equated to the infall rate as a function of mass (Fig. 7.6) and the limiting mass, at which accretion is shut off, determined. The results of a more detailed numerical study, which also takes into account the prior effects of the HII region, are shown in Fig. 7.8. At about 45 $M_\odot$ the effect of the HII region, especially in the polar directions, becomes important in reducing the inflow rate. At about 140 $M_\odot$ the accretion is shut off completely. This final mass naturally is subject to some uncertainty, for example in the properties of the low-density protostellar core, and it could range from 35 to 300 $M_\odot$. One of the uncertainties is the core rotation rate which is parameterized by $f_{Kep} = v_{rot}/v_{Kep}$, the ratio of orbital velocity of rotation compared with the Keplerian orbital velocity $[(GM/r)^{1/2}]$, evaluated at the point where the infall velocity in the low-density core exceeds the sound speed. The curves shown in Fig. 7.8 are calculated for $f_{Kep} = 0.5$, which is a value typically obtained in the three-dimensional simulations of core formation [1, 398]. Another uncertainty is the entropy parameter in the low-density core, that is, the value of K in the polytropic approximation $P = K\rho^\gamma$.

The surprising result of the theory of the formation of the first stars is that there are likely to be quite a few of them, one per dark-matter halo, and that they are all going to be in a restricted mass range around 140 $M_\odot$, unless fragmentation of the low-density cores is likely. The typical mass is much higher than that produced at the present day, and the IMF is completely different (although it has not been well determined and is not observable for the first stars). In addition, there is a potential problem: according to calculations of stellar evolution of metal-free stars, most of them in the range 140–260 $M_\odot$ become so-called pair-instability supernovae, in which rapid oxygen burning and silicon burning eject most or all of the star [212]. But the computed abundance pattern of the elements is not in accord with that observed in the oldest, low-metal stars in our Galaxy [517]. For stars in the mass range 25–140 $M_\odot$, the outcome is collapse of the entire star into a black hole, with little, if any, ejection of heavy elements. Also, observations of element abundances in high-redshift, low metallicity, damped Lyman Alpha systems, which are a useful probe of the early universe [412], find disagreement with abundances predicted from the pair-instability supernovae. Both observed patterns are more nearly consistent with that produced by metal-free supernovae in the 8–40 $M_\odot$ range, or possibly by stars more massive than 300 $M_\odot$. Is there a way of producing the first star in a different mass range? Clearly the first possibility is fragmentation into a binary or multiple system, with one component less than about 25 $M_\odot$. Theoretical calculation of binary formation requires three-dimensional hydrodynamic simulations, and those which have been carried through the accretion phase to the point where a massive disk surrounds the protostar [116, 480, 518] suggest that binary formation, or even multiple fragmentation, is a possible outcome. In the next section we examine an alternative possibility – that the masses of the first stars could have been much larger.

7.3 Dark Matter Annihilation in the First Stars

The first stars form in dark-matter halos, but the dark matter itself plays a relatively passive role, simply providing the gravitational potential in the center of which the baryonic material can accumulate. However, although the nature of dark matter is

7.3 Dark Matter Annihilation in the First Stars

unknown, certain types of dark matter candidates, known as Weakly Interacting Massive Particles, are their own antiparticles. Thus when two of them collide, their entire mass is converted into energy, which can heat a forming star under the right conditions. A typical proposed particle mass is about 100 proton masses or ≈ 100 GeV in energy. In the present-day Universe, the volume density of such particles is low enough so that they are not a significant energy source. However at the centers of the dark-matter halos in which the first stars form, their density is significantly higher, to the point where they could influence the formation of the first stars through their interactions.

The dark-matter candidates with the right properties to affect the first stars are also known as supersymmetric (SUSY) particles. The lowest-mass particles of the class are known as neutralinos. These particles have an annihilation cross section $\langle \sigma v \rangle = 3 \times 10^{-26}$ cm^3 s^{-1} and a mass of about 100 GeV, although other cross sections and masses are possible. The expected abundances of such particles in the early Universe, and this value of $\langle \sigma v \rangle$, lead to a calculation of the expected relic density of such particles at the present time, which in fact agrees with current observations regarding dark matter. Thus these particles are the favored candidate for dark matter. But several conditions must be satisfied if heating from dark matter annihilation is to be significant in a star.

- The dark-matter density must be high enough. This condition is actually not satisfied in the centers of 10^5–10^6 M$_\odot$ dark matter halos in the early Universe when the baryonic matter starts to collect there. What is required is, that as the baryonic matter contracts as a result of molecular hydrogen cooling and eventually collapses to high density, the dark matter, as a result of gravitational interaction with the baryons, must also contract to high density. Simulations of this process show that the required high densities of dark matter can be attained.
- The products of dark-matter annihilation must be trapped inside the star. For typical densities (10^4 cm^{-3}) in low-density protostellar cores, this condition is not met. The main products are neutrinos, photons, electrons, and positrons. Once the gas density reaches about 10^{13} cm^{-3} (for a 100 GeV dark-matter particle) the electrons, positrons, and photons can deposit a significant amount of their energy into the gas and heat it; however the neutrinos escape.
- The heating from dark-matter annihilation must dominate over all other heating and cooling mechanisms. The main cooling process is from molecular H. Simulations of the various heating and cooling processes [478] show that for a typical protostar contraction track, this condition is also met at a density of about 10^{13} cm^{-3} for a 100 GeV particle. On the evolutionary path shown in Fig. 7.1 this condition is satisfied near point E, when the mass of the equilibrium region is about 0.6 M$_\odot$. However, as more mass accretes, the luminosity of the protostar exceeds that provided by the dark matter, and it contracts, reaching densities (10^{15} cm^{-3}) where H$_2$ dissociates, acting as an energy sink. A true hydrostatic and thermal equilibrium, where all of the radiative luminosity is provided by annihilation, is not reached until the density is much higher, and all of the H

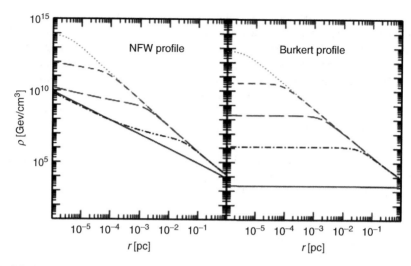

Fig. 7.9 Density profiles, according to adiabatic contraction, with increasing central density as a function of time, of dark matter in a halo, as influenced by the gravitational effects of the contracting baryonic matter. *Left panel*: the starting point (*lowest curve*) is a Navarro-Frenk-White profile [390]. *Right panel*: the starting point is a Burkert profile [93]. The enhancement of the dark-matter density, as shown here, is crucial for the effects of dark-matter annihilation to be important in the first stars. (1 GeV cm^{-3} = 1.065 H atoms cm^{-3} = 1.778 × 10^{-24} g cm^{-3}.) Reprinted with permission, from D. Spolyar, K. Freese, P. Gondolo: *Phys. Rev. Lett.* **100**, 051101 (2008). © 2008 The American Physical Society

has been dissociated and ionized. Once the dark matter heating dominates, the configuration is known as a *dark star*.

Calculations showing the increase in the dark-matter density at the center of a halo, as a result of the contraction and collapse of the baryonic component, are shown in Fig. 7.9. This process is known as *adiabatic contraction*[60]. As the baryons continue to contract they pull the dark matter in with it; however the dark matter density doesn't increase quite as fast as the density of ordinary matter. An approximate result [478] of the adiabatic contraction of the dark matter, verified by more detailed calculations, gives the dark-matter density

$$\rho_{dm} \approx 5 n_H^{0.81} \text{ GeV cm}^{-3}, \qquad (7.20)$$

where n_H is the local hydrogen volume density in cm^{-3}. For example, if the hydrogen density is 10^{13} cm^{-3} at a given radius, then $\rho_{dm} \approx 10^{11}$ GeV cm$^{-3} \approx 1.8 \times 10^{-13}$ g cm^{-3}. If it were not for adiabatic contraction, dark matter heating would not be a significant source of heating in the first stars. Note that an important restriction on the applicability of adiabatic contraction is that the orbital time of the dark-matter particles that are being drawn in must be short compared to the collapse time of the baryonic matter.

7.3 Dark Matter Annihilation in the First Stars

The quantity ρ_{dm} is important because the rate of heating per unit volume goes as its square:

$$Q_{dm} = \frac{\langle \sigma v \rangle \rho_{dm}^2}{m_{dm}} \approx 10^{-29} \frac{\text{erg}}{\text{cm}^3 \text{ s}} \left(\frac{\langle \sigma v \rangle n_H^{1.6}}{3 \times 10^{-26} \text{cm}^3/\text{s}} \right) \left(\frac{100 \text{GeV}}{m_{dm}} \right), \quad (7.21)$$

where again n_H is in cm^{-3} and m_{dm} is the mass of the dark matter particle. About 1/3 of the energy is lost in the form of neutrinos, so the total energy production in the star, in erg s^{-1} is

$$L_{dm} \approx \frac{2}{3} \int Q_{dm} dV, \quad (7.22)$$

where dV is the volume element.

The formation of a star in which dark-matter annihilation is taken into account follows the curve shown in Fig. 7.1. As previously mentioned, once the protostar reaches point E, dark-matter heating becomes important. Beyond that point, the dissociation collapse proceeds and the equilibrium stellar core is formed. The accretion phase follows, and its evolution has been calculated starting at the point where most of the hydrogen in the core has been ionized, which corresponds to a core mass of about 3 M$_\odot$. Beyond that point the core is in thermal and hydrostatic equilibrium, and the accretion rate onto it can be estimated from the curves in Fig. 7.6.

The buildup of a dark star in thermal and hydrostatic equilibrium can be calculated in a simple way. Thermal equilibrium implies that the energy radiated at the surface is entirely supplied by dark matter annihilation integrated over the interior:

$$L_{dm} = L_* = 4\pi R_S^2 \sigma_B T_{eff}^4, \quad (7.23)$$

where R_S is the outer radius and T_{eff} is the surface temperature. The hydrostatic structure can be approximated under the assumption that the star is a polytrope

$$\frac{dP}{dr} = -\rho \frac{Gm}{r^2}; \quad \frac{dm}{dr} = 4\pi r^2 \rho(r) \quad (7.24)$$

with

$$P = K\rho^{1+1/n}, \quad (7.25)$$

where P is the pressure, ρ is the density, m is the mass enclosed within radius r, and the constant K is determined once the total mass and radius are specified [110]. The dark stars are adequately described by polytropes in the range $n = 1.5$ (fully convective) to $n = 3$ (fully radiative). Given P and ρ at a point, the temperature T is defined by the equation of state of a mixture of gas and radiation

$$P(r) = \frac{R_g \rho(r) T(r)}{\mu} + \frac{1}{3} a T(r)^4 = P_g + P_{rad}, \quad (7.26)$$

where $R_g = k_B/m_u$ is the gas constant, m_u is the atomic mass unit, k_B is the Boltzmann constant, a is the radiation density constant, and the mean atomic weight $\mu = (2X + 3/4Y)^{-1} = 0.588$. The primordial composition consists of a hydrogen mass fraction $X = 0.76$ and a helium mass fraction $Y = 0.24$. In the resulting models $T \gg 10{,}000$ K except near the very surface, so the approximation for μ assumes that H and He are fully ionized. In the final models with masses near $800\,M_\odot$ the radiation pressure is of considerable importance.

The procedure for obtaining a model of a given mass M is to make a first guess for R_S and to assume that the polytrope has a given index n. These quantities determine the central density ρ_c. A second central boundary condition is obtained from (7.24) from which one can show that $d\rho/dr = 0$ there because $m \propto r^3$. One then integrates the ordinary differential equations outward from the center. Given $\rho(r)$ and $T(r)$ at a given point one can determine the Rosseland mean opacity $\kappa_R(r)$ from a table calculated for zero-metal gas [238]. The integration stops at the surface which is defined by the photospheric condition

$$\kappa_p P_p = \frac{2}{3} g_p, \quad (7.27)$$

where the subscript p denotes the photosphere and g is the acceleration of gravity. At the surface the radius is set to R_S and the temperature determined from (7.26) is set to T_{eff}. The radiated luminosity is then determined from (7.23).

To see whether the thermal equilibrium condition is now satisfied, one must determine the annihilation luminosity L_{dm} (7.22). Given the polytropic density distribution one can apply the adiabatic contraction model to determine the dark-matter density distribution ρ_{dm}; the result is approximately given by (7.20). Then (7.22) is integrated to determine L_{dm}, assuming a particle mass m_{dm}. This quantity isn't known, so it enters as a parameter into the calculations. The condition $L_* = L_{\text{dm}}$ must now be met. To do so one revises the first guess radius R_S, recalculates the distributions ρ and ρ_{dm}, and iterates on the radius until the luminosities agree.

To progress from the first converged model, which will have $M \approx 3\,M_\odot$, through an evolutionary sequence in which the mass increases through accretion, one defines a mass increment ΔM and then assumes an accretion rate $\dot{M}(M)$, as determined from one of the curves in Fig. 7.6, which are based on the properties of the low-density core of baryonic matter within the dark-matter halo. The associated time increment

$$\Delta t = \frac{\Delta M}{\dot{M}}. \quad (7.28)$$

Note that $\dot{M}$ is quite high, $> 10^{-2}\,M_\odot\,\text{yr}^{-1}$ for $M < 10\,M_\odot$, but it decreases considerably for higher masses. The dark matter annihilated during Δt is removed. It then must be determined which polytropic index n is appropriate. The lowest-mass models turn out to be fully convective, so that $n = 1.5$. At higher mass, the interior heats up, the opacity drops, and there is a transition to an interior radiative structure, for which $n = 3$ approximately holds. To determine how to make the transition one uses the Schwarzschild criterion for convection

7.3 Dark Matter Annihilation in the First Stars

$$\frac{d\ln T}{d\ln P} = \frac{3\kappa_R L_r P}{16\pi G a c T^4 m} > 0.4 \qquad (7.29)$$

for an ideal gas. One calculates this quantity at all points in a given model and thus determines when the shift to a radiative structure is necessary. Given the new M, n, one takes a new guess for R_S and goes through the same iteration procedure as described for the first model to obtain the thermal equilibrium for the new mass. The added mass and resulting increase in the central density of the gas results in the pulling in of more DM by the adiabatic contraction process. Thus the dark matter is resupplied at a rate that somewhat exceeds the rate at which it is destroyed, at least for the first few 100 $M_\odot$.

One might expect that the mass could build up until all of the baryonic matter in the dark-matter halo, more than 10^5 $M_\odot$, has accreted. A typical evolutionary track is shown in Fig. 7.10 (solid line) for $m_{dm} = 100$ GeV. Up to a mass of 600 $M_\odot$ the surface temperature increases gradually from 4,000 to 9,000 K, the radius stays relatively constant at 8×10^{13} cm (a few AU), and the luminosity increases from 10^5 to 5×10^6 $L_\odot$. Thus the "dark star" is a large, luminous, but relatively cool object. At these surface temperatures, feedback effects (described in Sect. 7.2.4), which could potentially limit the mass, are negligible. However at about 600 $M_\odot$ two separate physical effects come into play, the second a consequence of the first.

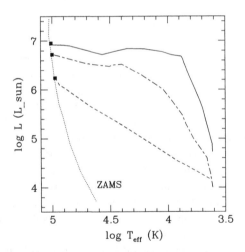

Fig. 7.10 Schematic evolution in the Hertzsprung–Russell diagram of first stars during the accretion phase. *Dotted line:* Zero-age nuclear-burning main sequence of zero-metallicity stars without mass accretion [448], in the mass range 10–1,000 $M_\odot$. *Solid line*: evolutionary track of an accreting star including the effects of dark-matter annihilation as an energy source [477], assuming a dark-matter particle mass of 100 GeV. *Short-dashed long-dashed line:* the same with a dark-matter particle mass of 10 TeV. *Short-dashed line:* Evolutionary track of a star without dark-matter annihilation in which the luminosity is derived primarily from spherical accretion (data from [568]). The final masses, where accretion stops, are indicated by filled *squares*. From *top* to *bottom*, these masses are 780, 550, and 140 $M_\odot$, respectively

The first is connected with the dark matter supply. Because of the high luminosity, which is generated by dark-matter burning, and because $\dot{M}$ is decreasing with time, eventually the point is reached where the dark matter mass reaches a maximum, and from then on it is used up faster than it can be resupplied. Note that this mass is less than 1% of the baryonic mass; nevertheless it is able to account for the entire radiated luminosity because annihilation is about 67% efficient at converting mass into heat energy of the star (in stellar nuclear burning less than 1% of the mass is converted into energy).

Once the dark-matter mass starts to decrease, the star must contract to increase the rate of energy production by annihilation. The central temperature is still only 2×10^6 K, not nearly enough to supply the required energy by nuclear burning (deuterium from the Big Bang is present, but its burning supplies only a small fraction of the luminosity for a brief period). Once the contraction starts, some of the dark matter that previously was inside the star is now outside it, in optically thin layers, and it cannot contribute to the heat generation. In a short time the star contracts by 2 orders of magnitude and the dark matter runs out completely. For a time, the energy is provided by gravitational contraction, the luminosity remains nearly constant, the surface temperature increases, and the second mechanism, feedback, begins to operate. In the range $T_{eff} = 50{,}000{-}100{,}000$ K the ionizing radiation and the heating of the infalling gas reverses the inflow at a final mass of $\approx 780\,M_\odot$. At this point the star has reached the metal-free zero-age main sequence with $T_{eff} = 100{,}000$ K and with all of the energy supplied by nuclear burning. The overall time for evolution through the dark-star phase is about 400,000 yr.

One might ask how sensitive this result is to the main parameter m_{dm}. One might expect quite different results because Q_{dm} is inversely proportional to m_{dm} and the plausible range in m_{dm} is more than 4 orders of magnitude! However, surprisingly, the final stellar mass is only weakly sensitive to m_{dm}. The numerical simulations show that a range in m_{dm} from 1 GeV to 10 TeV translates into a range of final masses of $820{-}550\,M_\odot$. For example, in the 10 TeV case the reduced heating results in a smaller radius and smaller luminosity but similar T_{eff}, and the evolutionary time to the main sequence is also somewhat shorter (250,000 yr). Thus in spite of the greatly reduced energy supply the final mass is still $550\,M_\odot$. The end result is that the effect of dark-matter annihilation is to produce masses considerably higher than those ($\approx 140\,M_\odot$) produced without dark-matter heating. Of course the final masses are undoubtedly dependent on other parameters, such as the details of the adiabatic contraction model and the precise history of $\dot{M}$.

What are the consequences of this difference in mass? The main point is the eventual fate of the star once it runs through its nuclear evolution [212], which will not be discussed in detail here. Acccording to spherically symmetric evolution calculations, metal-free stars in the mass range $500{-}1{,}000\,M_\odot$ do not become supernovae but simply collapse to become black holes, ejecting no material enriched in metals for incorporation in future generations of stars. But they could provide the seeds for the $10^9\,M_\odot$ black holes that are thought to have existed in the early Universe [168]. On the other hand the stars around $140\,M_\odot$ reach the pair instability supernova, which blows up the entire star leaving no black hole at all. Furthermore, the chemical abundances of the ejected material do not agree with those observed in the oldest stars in the Galaxy. Thus actually both scenarios have problems with the metal enrichment. Perhaps the problem can be resolved by including non-spherical effects in the higher-mass models [396]. First results of such calculations show that about half of the mass collapses to a black hole, and the remained is ejected, with chemical abundances that better match those observed.

Even with dark-matter annihilation, the final mass of a first star is limited to less than $1,000\,M_\odot$, because the supply of dark matter near the center of the dark-matter halo is used up, and it is very difficult to resupply it. However there are two further speculative avenues by which the first stars could have even higher mass. If this mass could approach $10^5\,M_\odot$, the objects would be so luminous that they could potentially be observationally detectable. They also could provide even more massive seeds for primordial black holes, which could be built up to $10^9\,M_\odot$ by gas accretion triggered by galaxy–galaxy mergers. The first possibility is that the amount of dark matter available in the star might be higher than previously calculated. The initial estimates, on which the results shown for $780\,M_\odot$ in Fig. 7.10 are based, assumed a spherical dark-matter halo. In fact numerical simulations of the formation of the haloes show that they are not spherical, and if they are not, the supply of dark matter to the very center could be considerably enhanced. The actual amount depends on complicated calculations of the trajectories of dark-matter particles in the halo. The second possibility is that even if the supply of dark matter delivered by adiabatic contraction runs out, once the star contracts to relatively high densities, the dark matter particles from the surroundings can scatter elastically off the nuclei in the star and be captured, eventually sinking to the center of the star and annihilating. Various laboratory experiments have aimed at the measurement of this scattering cross section, but so far only upper limits are available. In either case, calculations show [177] that dark stars can be built up to at least $10^5\,M_\odot$, with luminosities of $2 \times 10^9\,L_\odot$. The surface temperatures are cool enough so that the feedback effects discussed above do not limit the mass; the only limit is the available supply of baryons in the dark-matter halo.

7.4 Summary

Clearly many problems remain to be solved concerning the first stars. Since none of them have been observed, understanding progresses through a complex web of theory and indirect observation. For example, the oldest known observed objects already have been enriched somewhat in heavy elements such as carbon and iron. The first stars form from the growth of tiny density fluctuations in the baryonic and dark matter of the very early Universe. The usual cosmological model applied is ΛCDM, where typically 70% of the initial mass energy is "dark energy", 26% is cold dark matter, and 4% is baryonic (ordinary) matter. The nature of the dark matter and dark energy is unknown. The density fluctuations grow, filamentary structure develops, gravitational instability results in gravitationally bound dark matter halos, and mergers of the halos bring them up to a mass of about $10^6 \, M_\odot$. At this mass the temperature is about 1,000 K, and the baryons are settling toward the center. A small amount of molecular hydrogen forms, and once the cooling from this molecule gives a cooling time less than the dynamical time, the baryons cool to about 200 K and collapse to higher densities.

At a density of about $10^4 \, \text{cm}^{-3}$ the cooling saturates and the gas gradually heats up. Once about $1,000 \, M_\odot$ has accumulated, the Jeans mass has been reached and the condition of gravitational collapse for the baryons is met. Once the density reaches about $10^9 \, \text{cm}^{-3}$ molecular hydrogen can form by three-body reactions, and the hydrogen becomes essentially 100% molecular. Cooling is limited however, because at a density of about $10^{11} \, \text{cm}^{-3}$ the H_2 cooling lines become optically thick. At somewhat higher densities the gas becomes optically thick in the continuum as well. A small core reaches a density of $10^{16} \, \text{cm}^{-3}$ and a temperature of 2,000 K, where dissociation of the H_2 sets in, triggering a collapse. The collapse doesn't stop until a very low-mass region, with less than 1% of a solar mass, heats to 20,000 K and a density of $10^{21} \, \text{cm}^{-3}$. This material forms the initial stellar core, and the rest of the formation process involves the accretion of the remaining baryonic material in the core of the dark-matter halo onto this stellar core.

Accretion rates onto the stellar core are estimated from the properties of the contracting matter in the low-density core. They are high at first, more than $10^{-2} \, M_\odot \, \text{yr}^{-1}$ and decline with time. The first stars are thought to be massive, at least $100 \, M_\odot$, but they can form in 10^5–10^6 yr. They have sufficient angular momentum so that much of the accreted stellar material passes first through a disk. Initially the accretion rate is high enough so that the time scale for accretion is less than the thermal adjustment time of the core that is receiving the matter. Thus the radius remains large, well above the main-sequence radius, for some time. Once the accretion rate is reduced and the thermal time becomes relatively short at higher masses, the star is able to contract to the main sequence, but this does not occur until the star has grown to $\approx 100 \, M_\odot$. The surface temperature is now high, $T_{\text{eff}} \approx 100,000$ K, and the ionizing radiation creates an HII region in the infalling matter. The thermal pressure is able to halt and reverse accretion in the polar direction. Later on the disk is photoionized, and accretion stops in the equatorial plane as well. The final mass is in the range 50–300 $M_\odot$, and the first stars probably did have a range in mass, primarily dictated by varying properties of the initial low-density core in the dark-matter halo.

An important issue is whether all the mass from the low-density core forms one star, or whether fragmentation into a binary or small cluster can occur. The relatively high temperature of the protostar, and the resulting high pressure that tends to suppress growing density perturbations, suggests that fragmentation does not occur. However the question is not settled, and some numerical simulations do show fragmentation. If fragmentation did occur, then the supernovae in the 10–40 $M_\odot$ range could provide the metal enrichment needed for the later generations of stars, with abundance ratios appropriate as observed for the oldest stars in the Galactic halo. If not, a 140–260 $M_\odot$ star later blows up completely as a pair-instability

7.4 Summary

supernova, producing metal enrichment but not with the correct abundance ratios. Or if the final mass is 40–100 $M_\odot$ or above 260 $M_\odot$, the star becomes a black hole and doesn't produce any enrichment at all. In the mass range 100–140 $M_\odot$ a "pulsational" pair-instability supernova occurs, in which part of the mass, but essentially only hydrogen and helium, is ejected and the rest becomes a black hole.

An alternate path to the formation of the first stars includes the effects of annihilation of dark matter particles and the resulting heating of the star. This heating is a significant energy source, and it can result in a stellar structure in hydrostatic and thermal equilibrium at densities and temperatures well below those on the main sequence. Models of these "dark stars" show that they are large, with radii of a few AU, they are cool, with $T_{\text{eff}} = 5{,}000$–$10{,}000$ K, and they have high luminosities, more than 10^6 $L_\odot$. They build up in mass by accretion of the baryons in the dark matter halo. As the baryons contract, they pull in more dark matter through their gravitational effects, thus maintaining the fuel supply. Because the objects are cool, the feedback effects which ordinarily would shut off the mass accretion at around 150 $M_\odot$, are delayed until the mass reaches about 800 $M_\odot$, determined by the point where the dark-matter fuel supply has effectively run out. This final mass, when the star has reached the main sequence, is weakly dependent on the assumed dark-matter particle mass. But even at a fixed particle mass there will be a spread in masses of the first stars, because cosmological simulations show that there are variations in the properties of the dark-matter halos in which are produced the first stars, for example in their density distributions. The eventual fate of the $\sim 800 M_\odot$ stars is not well known. They may collapse to black holes and provide the seeds for supermassive black holes known to exist at early times. They may also explode as supernovae, leaving part of their mass in a black hole but expelling the rest, with chemical abundances that do match the observations of the oldest stars.

Whether dark-matter annihilation is included or not, the initial mass function of the first stars is completely different from that of the present stars, or even of the Population II stars in the Galaxy. Thus there must have been a transition, relatively early in the history of the Universe, when the mass scale for star formation was reduced considerably from the 100–1,000 $M_\odot$ range. The key physical process involved is the cooling in the low-density baryonic core of the dark matter halo. Beyond cooling by H_2 there are two other known mechanisms. First, if the molecule HD is produced ($D^+ + H_2 \rightarrow HD + H^+$), it can result in cooling to temperatures well below the 200 K allowed by H_2. But to produce HD an ionized medium is required, above 10,000 K. To get the virial temperature in the dark-matter halo into this range, a more massive halo, say $10^8 M_\odot$, is required, which could be produced at a later stage of cosmological evolution than the $10^6 M_\odot$ needed for the first stars. In that case, stars in the mass range 10 $M_\odot$ could be produced [196], and these supernovae would be responsible for producing most of the early metal enrichment. The second process is metal enrichment, which through various atomic and molecular transitions can result in cooling down to the 10–20 K range and allow star formation in the solar mass range. There is considerable debate on exactly how much metal enrichment is required, but a typical figure is $Z_{\text{crit}} \sim 10^{-4} Z_\odot$.

7.5 Problems

1. A dark star accretes at $10^{-3} M_\odot$ yr^{-1} up to $10^5 M_\odot$. During most of the accretion, beyond $500 M_\odot$, the star is radiation-pressure dominated, so luminosity $L \propto M$, at half the Eddington limit for electron scattering. What is the total energy radiated by the star up to the time when it reaches its final mass? What is the total time? How many solar masses of dark matter have to be annihilated to produce this total energy?
2. A protostar (without dark matter) accretes at the rate given by (7.16). Assume that the accretion luminosity is equal to the Eddington luminosity at all times, and that the opacity is given by electron scattering. Assume that the radius cannot be less than the main-sequence radius for a given mass at zero metallicity: $R_{ms} = 4.3[M/(100 M_\odot)]^{0.55} R_\odot$. Plot the evolution in the Hertzsprung–Russell diagram starting at $1 M_\odot$ and ending at $150 M_\odot$.

Chapter 8
Pre-Main-Sequence Evolution

The time for the completion of protostellar evolution is effectively 2–3 times the free-fall time of the outer layers, which lies in the range 10^5–10^6 yr, depending on the initial density. As the accretion rate tapers off in the later phases and the infalling envelope becomes less and less opaque, the observable surface of the protostar declines in luminosity and increases in temperature. Once the infalling envelope becomes transparent, the observer can see through to the stellar core at optical and near IR wavelengths, which by now can be identified in the H–R diagram as a star with a photospheric spectrum. The locus in the diagram connecting the points where stars of various masses first make their appearance is known as the *birth line*; it is shown as the upper envelope of the tracks in Fig. 1.21.

Once internal temperatures are high enough (above 10^5 K) that hydrogen is substantially ionized, the star is able to reach an equilibrium state with the pressure of an ideal gas supporting it against gravity. Rotation, according to observations, has a relatively small effect on the structure of the central star during pre-main-sequence evolution. The angular momentum problem has largely been solved by the beginning of this phase. Much of the angular momentum of the collapsing cloud has been taken up by the circumstellar disk, which lasts 1–10 Myr before being photodissipated. It is thought that magnetic coupling between star and disk has substantially reduced the angular momentum of the material that went into the star. Also there has been significant angular momentum removal through bipolar outflows, which are strongest during the protostellar phase. Before the disk disappears, giant planets may have a chance to form, and the left-over solid material, including dust grains and planetesimals, could evolve into terrestrial planets all the way through the pre-main-sequence phase.

The star itself radiates from the surface, and this energy is supplied by gravitational contraction. This contraction does imply a very slight deviation from strict hydrostatic equilibrium, but the acceleration term is completely negligible and the normal hydrostatic equation can be used. The evolution can be regarded as a passage through a series of equilibrium states, and the process is often referred to as *quasi-hydrostatic contraction*.

The Virial Theorem thus simplifies to

$$2E_{th} + E_{grav} = 0. \qquad (8.1)$$

The total energy (of a spherical star) is

$$E_{tot} = E_{grav} + E_{th} = -E_{th} = E_{grav}/2. \qquad (8.2)$$

In a small contraction $\Delta E_{th} = -\Delta E_{grav}/2$, so the internal kinetic energy increases by half the change in gravitational energy. The total energy drops by $\Delta E_{grav}/2$ so the remainder is radiated away. For the simple case of an ideal, nondegenerate gas, $E_{th} = (3/2) R_g T M/\mu$, where μ is the mean atomic weight per free particle, so an increase in internal kinetic energy corresponds to heating.

This chapter describes the equations that are solved to calculate the structure and evolution of stars in this phase, summarizes the results of such calculations, and shows how these results may be compared with observations. Although the stellar evolution calculations are straightforward, they are based on simplifying assumptions, and the observations regarding the phase bring out a number of puzzles. For example, what causes the irregular variability and activity of many T Tauri stars? How do the angular momentum and magnetic field evolve during pre-main-sequence evolution? How are FU Orionis outbursts explained? What is the role of the stellar wind? Are the masses for young objects derived from evolutionary tracks a good measure of their actual masses? Does the Initial Mass Function derived from very young stars agree with that derived from older, main-sequence stars?

8.1 Physical Relations

8.1.1 Basic Equations of Structure and Evolution

The calculation of the evolution can now be accomplished by solution of the standard structure equations, which assume spherical symmetry, no rotation or magnetic fields, and no mass loss or accretion. Although young stars do show evidence of accretion, the amount accreted during this phase is quite small. The equations of hydrostatic equilibrium (3.1) and mass distribution (3.2) are rewritten with m, the mass enclosed within radius r, as the independent variable:

$$\frac{dP}{dm} = -\frac{Gm}{4\pi r^4} \qquad (8.3)$$

and

$$\frac{dr}{dm} = \frac{1}{4\pi r^2 \rho}, \qquad (8.4)$$

8.1 Physical Relations

where the dependent variables ρ and r are, in general, functions of m and time t, but the ordinary derivative is used to emphasize the fact that the system is Lagrangian.

The third condition is that of conservation of energy. In the pre-main-sequence phase the star is not in thermal equilibrium: its total energy decreases with time, as only half of the released gravitational energy goes into internal heat. The equation becomes

$$\frac{dL_r}{dm} = \epsilon_{\text{nuc}} - \frac{dE}{dt} - P\frac{dV'}{dt}, \tag{8.5}$$

where $V' = 1/\rho$, ϵ_{nuc} is defined as the nuclear energy generation rate per unit mass, and E is now the internal energy per unit mass including non-kinetic forms of energy such as ionization energy. The luminosity $L_r = 4\pi r^2 F_r$, where the net outward energy flux F_r is the amount of energy per unit time per unit area crossing a spherical surface at radius r. If $\epsilon_{\text{nuc}} = 0$, the star contracts and obtains its energy from the third term on the right hand side. The nuclear energy term can become important during pre-main-sequence evolution because deuterium, a small amount of which is present in the material from which the star forms, burns at temperatures of 10^6 K, well before the main sequence is reached.

A fourth differential equation describes the energy transport. The transport of energy outward from the interior of a star to its surface depends in general on the existence of a temperature gradient. Heat will be carried by various processes from hotter regions to cooler regions; the processes that need to be considered include (1) radiative transport and (2) convective transport. Conductive transport is not important during this evolutionary phase. In each case a relation must be found between the energy flux F_r and the temperature gradient dT/dr. The diffusion approximation for radiative transport (to be discussed in Sect. 8.1.3) can be rewritten, with m as the independent variable, in a form which can be used for all three types of transport [251]:

$$\frac{dT}{dm} = -\frac{GmT}{4\pi r^4 P}\nabla, \tag{8.6}$$

where, if the energy transport is by radiation,

$$\nabla = \nabla_{\text{rad}} = \frac{3}{16\pi Gac}\frac{\kappa_R L_r P}{mT^4} = \left(\frac{d\ln T}{d\ln P}\right)_{\text{rad}}, \tag{8.7}$$

where the derivative refers to the actual temperature–pressure variation in the structure of the star, and κ_R is the Rosseland mean opacity (8.40).

The Schwarzschild criterion for the onset of convection in material with uniform chemical composition is

$$\nabla_{\text{rad}} > \nabla_{\text{ad}} = (d\ln T/d\ln P)_{\text{ad}} \approx \frac{\gamma - 1}{\gamma}, \tag{8.8}$$

where the factor involving γ holds for an ideal gas. An equivalent relation is given as (7.29). Here γ is the ratio of specific heats c_P/c_V and ∇_{ad} is the so-called *adiabatic gradient*. Each point in the star must be tested to see if this condition is satisfied and if so, ∇ in (8.6) is replaced by the "convective gradient" ∇_{conv}. In most of the interior of a star it can be shown that convection is very efficient, and that only a very small excess of the actual temperature gradient over the adiabatic gradient is required for the transport of the entire flux by convection. In this case, $\nabla_{conv} = \nabla_{ad}$ to a high degree of accuracy.

However, if convection occurs in the surface layers of a star, for example just below the photosphere as in the case of the Sun, then ∇_{conv} must be calculated in more detail. A simple one-dimensional hydrodynamic formulation known as the "mixing-length theory" [71, 251] is generally used, but it contains an arbitrary parameter, the ratio (l/H) of the mixing length, which is effectively the mean free path of the largest convective elements, to the local pressure scale height $H = P/|dP/dr| = P/(g\rho)$ where g is the acceleration of gravity. The ratio l/H, which is of order unity, can be determined empirically by fitting a stellar model to the known properties of the Sun, or numerically through two-dimensional or three-dimensional simulations of the turbulent, convective motions in the outer layers of a star [328, 489]. This effect is important for pre-main-sequence stars in their convective phase, and the treatment of the mixing-length approximation affects the location of the evolutionary tracks in the H–R diagram.

The four differential equations for stellar structure require four boundary conditions, which are split between surface and center. At the center of the star the boundary condition is simple: at $m = 0$, the radius r and luminosity L_r also must vanish. At the surface, the boundary conditions are more complicated, and they play an important role for pre-main-sequence stars in the phase where the energy transport is mainly by convection.

A simple way to apply them is to take first the definition of the effective (or photospheric) temperature:

$$L = 4\pi R^2 \sigma_B T_{eff}^4, \tag{8.9}$$

where L is the surface luminosity, σ_B is the Stefan-Boltzmann constant, and R is the surface radius. Second, one can obtain the equation of hydrostatic equilibrium for an atmosphere with a fixed value of the gravitational acceleration g by combining (8.3) and (8.4):

$$\frac{dP}{dr} = -\frac{Gm\rho}{r^2} = -g\rho \tag{8.10}$$

if m (the interior mass) $= M$ (the total mass) in the thin atmospheric layer. Then from the definition of optical depth $d\tau = -\kappa_R \rho dr$, where κ_R is the Rosseland mean opacity, we obtain

$$\frac{dP}{d\tau} = \frac{g}{\kappa_R}. \tag{8.11}$$

Integrating this expression approximately from a small value of τ inwards to $\tau = 2/3$, which defines the photosphere,

$$\kappa_p P_p = \frac{2}{3}g, \tag{8.12}$$

8.1 Physical Relations

where the subscript p refers to the photosphere. An adequate but still approximate representation of the two surface boundary conditions is given by (8.9) and (8.12).

In order to make detailed comparisons of models with observations, it is desirable to have a more accurate atmospheric calculation. In some cases frequency-dependent radiative transport has been employed in the atmosphere so that colors and fluxes could be calculated to compare with observed objects [30]. Another reason to make a detailed atmosphere model is that the dominant elements H and He tend to be only partially ionized in the outer layers, while in most of the deep interior they are fully ionized. Furthermore, the convective temperature gradient in the outer convection zone of cool stars is usually superadiabatic rather than adiabatic in a fairly thin surface layer; therefore the mixing-length approach must be employed. It is convenient to confine the complicating frequency-dependent effects as well as the effects of partial ionization and non-adiabatic convection to the thin surface layer where they are important, and thereby to allow the simplification of the physics in the deep interior. The strategy, therefore, is to apply atmospheric physics in the outermost mass zone, between the surface $m = M$ and a deeper layer $m = m_{atm}$, and to apply the outer boundary condition for the interior at m_{atm}.

8.1.2 Equation of State

The equation of state in the interior can for the most part be taken to be that of an ideal gas (3.23) plus radiation pressure

$$P_{rad} = \frac{1}{3}aT^4, \qquad (8.13)$$

where a is the radiation density constant ($\sigma_B = ac/4$). For a fully ionized gas the mean molecular weight μ can be approximated by

$$\mu^{-1} = 2X + \frac{3}{4}Y + \frac{1}{2}Z, \qquad (8.14)$$

where X, Y, Z are, respectively, the mass fractions of H, He, and heavier elements. For the very lowest-mass stars, as they approach the main sequence, non-ideal gas effects as well as partial degeneracy of the free electrons begin to become significant. Partial degeneracy occurs when quantum effects begin to affect the free-electron pressure (see below).

The transition from an ideal to a partially degenerate equation of state is important in determining whether the end point of star formation is an actual star or a brown dwarf, in which nuclear reactions never become fully established as an energy source. A simple argument [251] is based on the assumption that the contraction is homologous, that is, the same mass fraction resides within the same radius fraction at all stages during contraction. Then, approximating the derivative of the pressure in (3.1) by P_c/R, where P_c is the central pressure and R is the total radius, and the average and central densities, respectively, by $\bar{\rho} \approx C_1 M/R^3$ and $\rho_c \approx C_2 M/R^3$, one obtains

$$P_c \approx C_3 \frac{GM^2}{R^4} = \frac{C_4}{R^4} \text{ and } \rho_c \approx \frac{C_5}{R^3} \quad (8.15)$$

for constant total mass M, where the various C's are constants. Differentiating (8.15)

$$\frac{dP_c}{P_c} = -\frac{4}{R} dR; \quad \frac{d\rho_c}{\rho_c} = -\frac{3}{R} dR. \quad (8.16)$$

Now consider a general equation of state

$$\rho(P, T) = K_1 P^\alpha T^{-\delta} \quad (8.17)$$

and differentiate it

$$\frac{d\rho_c}{\rho_c} = \alpha \frac{dP_c}{P_c} - \delta \frac{dT_c}{T_c}. \quad (8.18)$$

Then eliminate dP_c/P_c to obtain

$$\frac{dT_c}{T_c} = \frac{4\alpha - 3}{3\delta} \frac{d\rho_c}{\rho_c}. \quad (8.19)$$

For an ideal gas, $\alpha = 1$ and $\delta = 1$ so $dT_c/T_c = (1/3)d\rho_c/\rho_c$, giving a slope in the $(\log \rho, \log T)$ plane of $1/3$ for a homologously contracting object, that is, one in which the density distribution, normalized to the central value, does not change.

However if the density increases for a fixed temperature for an ideal ionized gas, eventually the point is reached where the available momentum quantum states for free electrons, according to the Pauli exclusion principle, begin to be filled, forcing electrons into higher and higher momentum states, thus increasing the pressure beyond that of an ideal gas. When this increase is significant the gas is said to be approaching the condition of electron degeneracy. The equation of state for a completely degenerate gas – one in which all electron quantum states are filled up to some limiting momentum – but in which the electrons are non-relativistic – is given by [110]

$$P_e = 1.004 \times 10^{13} (\rho/\mu_e)^{5/3} \text{ dyne cm}^{-2}, \quad (8.20)$$

where the total pressure is well approximated by the electron pressure P_e. Here μ_e is the mean atomic weight per free electron: $\mu_e = 2/(1 + X)$ for a fully ionized gas with hydrogen mass fraction X. Thus $\alpha = 3/5$ and $\delta = 0$. As a result, in a degenerate gas,

$$\frac{dT_c}{T_c} \to -\frac{1}{5\delta} \frac{d\rho_c}{\rho_c} \quad (8.21)$$

and the slope becomes large and negative as $\delta \to 0$, so the gas cools upon contraction.

It is easy to show that a contraction track in the (log ρ, log T) plane must eventually reach the degenerate region. Simply equate the ideal gas electron pressure to the fully degenerate electron pressure to define a line in that plane where degeneracy effects start to become important:

$$\rho > 2.4 \times 10^{-8} \mu_e T^{3/2} \text{ g cm}^{-3}. \quad (8.22)$$

The slope of this expression (see Fig. 8.4) is 2/3 in the (log ρ, log T) plane; thus contraction tracks with slope 1/3 must eventually intersect it.

8.1.3 Opacity

An important quantity in pre-main-sequence evolution is the opacity. When the star is mostly convective, there is still a thin radiative zone at the surface that controls the loss of energy to space. When the star is largely radiative, the opacity throughout the interior controls the evolution. This subsection provides a simple derivation of the radiative transfer equation in the interior, defines the Rosseland mean opacity, and discusses the main opacity sources.

Numerous atomic processes contribute to this quantity, and in general the structure of a star can be calculated only with the aid of detailed tables of the opacity, calculated as a function of ρ, T, and the chemical composition. Starting at the highest temperatures characteristic of the stellar interior and proceeding to lower temperatures, the main processes are

1. Electron scattering, also known as Thomson scattering, in which a photon undergoes a change in direction but no change in frequency during an encounter with a free electron.
2. Free-free absorption, in which a photon is absorbed by a free electron in the vicinity of a nucleus, with the result that the photon is lost and the electron increases its kinetic energy.
3. Bound-free absorption on metals, also known as photoionization, in which the photon is absorbed by an atom of a heavy element (e.g. iron) and one of the bound electrons is removed.
4. Bound-bound absorption of a heavy element, in which the photon induces an upward transition of an electron from a lower quantum state to a higher quantum state in the atom.
5. Bound-free absorption on H and He, which generally occurs near stellar surfaces where these elements are being ionized.
6. Bound-free and free-free absorption by the negative hydrogen ion H^-, which forms in stellar atmospheres in layers where H is just beginning to be ionized

(example: the surface of the sun). The bound-free process is $H^- + h\nu \to H + e^-$ where the photon energy must exceed 0.75 eV.
7. Bound-bound absorptions by molecules, which can occur only in the atmospheres of the cooler stars ($T_{\text{eff}} < 4{,}000$ K, although even the Sun shows a few molecular features in its spectrum).
8. Absorption by dust grains, which can occur in the early stages of protostellar evolution and possibly in the atmospheres of brown dwarfs, at temperatures below the evaporation temperature of the more refractory grains (1,400–1,800 K).

A simple derivation of how the Rosseland mean opacity is calculated follows (for more details see [119, 364]). Energy transport by radiation depends on the emission of photons in hot regions of the star and absorption of them in slightly cooler regions. The radiation field may be characterized by the *specific intensity* I_ν, which is defined so that

$$dE_\nu = I_\nu \cos\theta \, d\nu \, d\Omega \, dt \, dA, \qquad (8.23)$$

where dE_ν is the energy carried by a beam of photons across an element of area dA in time dt in frequency interval ν to $\nu + d\nu$ into an element of solid angle $d\Omega$, in a direction inclined by an angle θ to the normal to dA (Fig. 8.1).

Another fundamental quantity is the radiation flux density F_ν. In a spherical coordinate system (r, θ, ϕ) the flux in the radial direction is the net energy crossing a given surface, per unit area per unit time, per unit frequency interval, integrated over all directions, thus

$$F_\nu = \int_0^{2\pi} \int_0^{\pi} I_\nu \cos\theta \sin\theta \, d\theta \, d\phi. \qquad (8.24)$$

A third fundamental radiation quantity is the *energy density*, which is the amount of energy per unit volume per unit frequency interval in the radiation field

$$u_\nu = \frac{1}{c} \int_{4\pi} I_\nu \, d\Omega. \qquad (8.25)$$

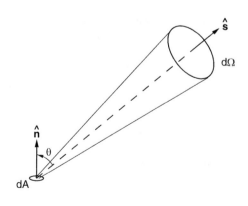

Fig. 8.1 Illustration of the definition of the specific intensity. The radiation is flowing through an element of area dA into an infinitesimal cone with solid angle $d\Omega$, in the direction $\hat{s}$, which is inclined by an angle θ with respect to the normal, $\hat{n}$, to dA

8.1 Physical Relations

If we integrate u_ν over all frequencies we get, in the case of black-body radiation, where $I_\nu = B_\nu$, the Planck function

$$u = \frac{4\pi}{c} B(T) = \frac{4\pi}{c} \frac{\sigma_B}{\pi} T^4 = aT^4, \tag{8.26}$$

where the radiation density constant $a = 7.65 \times 10^{-15}\,\mathrm{erg\,cm^{-3}\,{}^\circ K^{-4}}$. Also, for black-body radiation,

$$u_\nu = \frac{4\pi}{c} B_\nu = \frac{8\pi h}{c^3} \frac{\nu^3}{\exp\left(\frac{h\nu}{kT}\right) - 1}. \tag{8.27}$$

The derivation to follow relies heavily on the concept of *thermodynamic equilibrium*. Strict thermodynamic equilibrium (TE) [364] refers to the situation where material in an enclosure is adiabatic, homogeneous, and isothermal at a constant temperature T. The relevant properties of a gas in TE are: (1) the radiation field I_ν is isotropic, (2) $I_\nu = B_\nu(T)$, and (3) across any given surface in the region, there is no net flux of energy. However, a star in which the energy transport is by radiation cannot be in strict TE, because there must be a net energy flow through it. However the star's interior is very close to TE, and it can be approximated by the condition of *local thermodynamic equilibrium*, or LTE. In this case, all the thermodynamic properties of the gas that apply in TE still apply, for example the degree of ionization and the opacity. The main difference is that in LTE, the radiation field is not precisely isotropic, the intensity I_ν departs slightly from $B_\nu(T)$, and the radially outward flux across a given surface is slightly greater than the inward flux. The main condition that is required for LTE to be an adequate approximation is that the temperature at which a photon is emitted must be very close to the temperature at which it is later absorbed, or, in other words, the temperature change must be negligible over the mean free path of a photon.

The equation of transfer shows how the intensity of a beam is changed as it interacts with matter. The *mass emission coefficient* j_ν is defined so that $j_\nu\,\rho\,dV\,d\nu\,d\Omega\,dt$ is the energy emitted by the volume element dV into $d\Omega$ in time dt in the frequency range $d\nu$. The corresponding *mass absorption coefficient* κ_ν is defined so that the energy absorbed in the same intervals is $\kappa_\nu\,\rho\,I_\nu\,dV\,d\nu\,d\Omega\,dt$. If we consider both the absorption and the emission of the radiation passing through a cylinder with length ds and cross section dA, in the direction perpendicular to dA, the equation of transfer becomes

$$\frac{dI_\nu}{ds} = -\kappa_\nu \rho I_\nu + j_\nu \rho. \tag{8.28}$$

Suppose that the direction ds is inclined by an angle θ to the radial direction in the star so that the projected distance element $dr = ds\cos\theta$. Also define the optical depth (in the radial direction) τ_ν by

$$d\tau_\nu = -\kappa_\nu \rho dr \tag{8.29}$$

and the equation of transfer becomes

$$\cos\theta \frac{dI_\nu}{d\tau_\nu} = I_\nu - \frac{j_\nu}{\kappa_\nu}. \tag{8.30}$$

Note that this equation only considers the radial variation of I_ν. In stellar layers with significant curvature, in principle a term in $dI_\nu/d\theta$ should also be included. However this effect in the stellar interior can be shown to be negligible [451].

In the stellar interior the mean free path of a photon $(\kappa_\nu \rho)^{-1}$ before it is absorbed is only 1 cm or less. Thus a typical photon is absorbed at practically the same temperature as it is emitted. These conditions are so close to strict thermodynamic equilibrium that the approximation of LTE applies, and the ratio j_ν/κ_ν can be shown to be the same as it is in strict thermodynamic equilibrium, namely $j_\nu/\kappa_\nu = B_\nu(T)$. This approximation holds very well in the interior of a star, but it breaks down near the surface where the mean free path can be long.

The equation of transfer in stellar interiors can thus be expressed as

$$\cos\theta \frac{dI_\nu}{d\tau_\nu} = I_\nu - B_\nu(T). \tag{8.31}$$

This equation can be integrated over all frequencies and solved for the flux in terms of the temperature gradient for conditions appropriate to the stellar interior, that is, if LTE holds. Multiply (8.31) by $\cos\theta$ and integrate over all directions:

$$\frac{d}{d\tau_\nu} \int_{4\pi} I_\nu \cos^2\theta \, d\Omega = F_\nu \tag{8.32}$$

using the definition of the flux F_ν, and noting that $d\Omega = \sin\theta d\theta d\phi$ and that B_ν is isotropic (I_ν is not assumed to be isotropic). But the radiation pressure is defined by

$$P_{rad,\nu} = \frac{1}{c} \int_{4\pi} I_\nu \cos^2\theta \, d\Omega \tag{8.33}$$

an expression which, when integrated over all directions and all frequencies, given the LTE assumption $I_\nu \approx B_\nu$, yields (8.13). Using the definition of optical depth $d\tau_\nu = -\kappa_\nu \rho dr$ we obtain

$$\frac{dP_{rad,\nu}}{dr} = -\frac{\kappa_\nu \rho}{c} F_\nu. \tag{8.34}$$

If we now integrate over all frequencies, we obtain

$$\frac{dP_{rad}}{dr} = -\frac{\bar{\kappa}\rho}{c} F. \tag{8.35}$$

8.1 Physical Relations

Since we require that $\int F_\nu d\nu = F$, where F is the total flux, the mean opacity $\bar{\kappa}$ must be defined so that

$$\frac{1}{\bar{\kappa}} = \frac{\int_0^\infty \frac{1}{\kappa_\nu}\frac{dP_{rad,\nu}}{dr}d\nu}{\int_0^\infty \frac{dP_{rad,\nu}}{dr}d\nu}. \tag{8.36}$$

We now use LTE and a slightly non-isotropic intensity, so we can assume that $P_{rad,\nu} = \frac{1}{3}u_\nu$, where u_ν is the energy density, and also $u_\nu = \frac{4\pi}{c}B_\nu$. Note that this approximation is acceptable because the integral for $P_{rad,\nu}$ over direction involves $\cos^2\theta$, so the integrand is always positive and the actual radiation field deviates only slightly from a black body. On the other hand, this approximation cannot be made for calculation of the flux, because the $\cos\theta$ factor is positive of $0 \le \theta \le \pi$ and negative otherwise. An assumption of isotropy for that calculation would lead to zero net flux.

Then the expression for $\bar{\kappa}$ can be written

$$\frac{1}{\bar{\kappa}} = \frac{\int_0^\infty \frac{1}{\kappa_\nu}\frac{dB_\nu}{dT}\frac{dT}{dr}d\nu}{\int_0^\infty \frac{dB_\nu}{dT}\frac{dT}{dr}d\nu} = \frac{\int_0^\infty \frac{1}{\kappa_\nu}\frac{dB_\nu}{dT}d\nu}{\int_0^\infty \frac{dB_\nu}{dT}d\nu}. \tag{8.37}$$

Then from (8.35)

$$\frac{d}{dr}\left(\frac{1}{3}aT^4\right) = -\frac{\bar{\kappa}\rho}{c}F. \tag{8.38}$$

or

$$F = -\frac{4ac}{3\bar{\kappa}\rho}T^3\frac{dT}{dr} \tag{8.39}$$

which gives the relation between the flux and the temperature gradient. Note $ac = 4\sigma_B$ where σ_B is the Stefan-Boltzmann constant, c is the velocity of light and a is the radiation density constant. Here $\bar{\kappa}$ is the absorption coefficient averaged over frequency according to the so-called *Rosseland mean*; its unit is cm^2 g^{-1}. A more detailed derivation, taking into account pure absorption processes, as done above, as well as scattering processes, gives the general form of the Rosseland mean

$$\frac{1}{\bar{\kappa}} = \frac{1}{\kappa_R} = \frac{\int_0^\infty \frac{dB_\nu(T)/dT}{\kappa_{\nu,a}[1-\exp(-h\nu/kT)]+\kappa_{\nu,s}}d\nu}{\int_0^\infty dB_\nu(T)/dT\, d\nu}. \tag{8.40}$$

Here $\kappa_{\nu,a}$ refers to processes of true absorption which have to be corrected for induced emission, and $\kappa_{\nu,s}$ refers to scattering processes. This form of the mean opacity is useful because it can be calculated as a function of density, temperature, and composition and stored in a table for general use; it does not depend on the radiation field at a particular layer in the star.

Equation (8.39) is known as the diffusion approximation for radiative transfer, an appropriate nomenclature because a photon is absorbed almost immediately after it is emitted, so that an enormous number of absorptions, re-emissions, and scatterings must occur before the energy of a photon is transmitted to the surface.

The quality of the radiation changes during this process. As the photons diffuse to lower temperatures, their energy distribution corresponds closely to the Planck distribution at the local T, because matter and radiation are well coupled. The appropriate time scale for significant changes in the thermal profile of a radiative star is known as the *thermal adjustment time*. A reasonable approximation for this time scale is the Kelvin-Helmholtz time scale, as shown by [251], their Sect. 5.3.

8.2 Pre-Main-Sequence Evolutionary Tracks

A theoretical evolutionary track is specified by a mass, chemical composition, l/H, and an initial model. The starting point is generally close to the birth line, where for stars of 1.5 $M_\odot$ or less the object is nearly fully convective and can be represented by a polytrope of index 1.5. For higher masses, the prior protostellar evolution must be taken into account to determine the initial distributions of ρ, T, P, L_r, r with mass. The birth line itself is determined by the physics of the accretion phase of protostellar evolution, as described in more detail in [481]. There it is shown that the radius of the stellar core during that phase is only a few $R_\odot$ and that it is relatively insensitive to the infall $\dot{M}$.

The solutions are shown in the H–R diagram in Figs. 8.2 and 8.3. These tracks start at the birth line for each mass. For earlier times, a hydrostatic stellar-like core may exist, but it is hidden from view by the opaque infalling protostellar envelope. Note that the sun arrives on the standard H–R diagram with about 5 times its present luminosity, The higher-mass stars actually do not appear on the normal H–R diagram until they have already reached the main sequence. For accreting protostars that reach these higher masses, the mass at which the star arrives at the main sequence depends on the accretion rate. If the accretion rate is constant with time at the appropriate value for a typical low-mass protostar (10^{-5}–10^{-6} $M_\odot$ yr^{-1}) then the corresponding main-sequence arrival mass is 5–10 $M_\odot$. However at higher accretion rates, the arrival mass is higher (Sect. 5.2).

The results of the calculations show in general that a star of a given mass first passes through a convective phase, known as the Hayashi track [209], during which the evolution in the H–R diagram is nearly vertical and downwards. Later, for all masses greater than about 0.5 $M_\odot$, the evolution goes into a radiative phase, known as the Henyey track [216], which is relatively horizontal in the diagram. The relative importance of the two phases depends on the stellar mass. During the convective phase, energy transport in the interior is quite efficient, and the rate of energy loss is controlled by the thin radiative layer right at the stellar surface. The opacity is a very strongly increasing function of T in that layer, and that fact combined with the photospheric boundary conditions ((8.9) and (8.12)) can be shown to result in a nearly constant T_{eff} during contraction. As the surface area decreases, L drops and T_{eff} stays between 2,000 and 4,000 K, with lower masses having lower T_{eff}.

As the star contracts, the interior temperatures increase and in most of the star the opacity decreases as a function of T. The star gradually becomes stable against

8.2 Pre-Main-Sequence Evolutionary Tracks

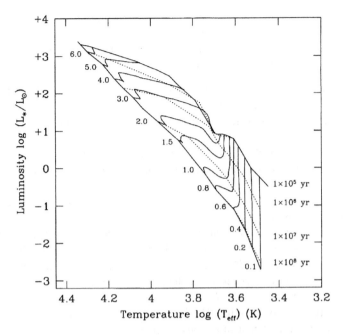

Fig. 8.2 Overview of pre-main-sequence evolutionary tracks in the Hertzsprung–Russell diagram. The end point of each track, on the main sequence, is labeled with the corresponding mass, in $M_\odot$. A given track starts at the birth line (*upper solid line*) and ends on the zero-age main sequence (*lower solid line*). Loci of constant age are indicated by *dotted lines*. Reproduced by permission of the AAS, from [407]. © The American Astronomical Society

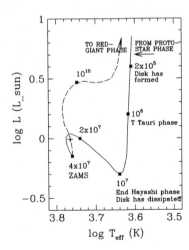

Fig. 8.3 Sketch of the pre-main-sequence (*solid line*) and the beginning of the post-main-sequence (*dashed line*) phases for 1 $M_\odot$. Evolutionary times in years are given for the *filled squares*. The present Sun is indicated by a *plus sign*. The ZAMS (Zero-Age Main Sequence) is the point where nuclear reactions completely take over the energy production. Adapted from [63]

convection, starting at the center, because the radiative gradient (8.7) drops along with the opacity and soon falls below ∇_{ad}. When the radiative region includes about 75% of the mass, the rate of energy release is no longer controlled by the surface layer, but rather by the opacity of the entire radiative region. At this time the tracks in the 1 $M_\odot$ range make a sharp bend to the left, and the luminosity increases gradually as the average interior opacity decreases.

The following points can be made regarding the Hayashi tracks, named after the Japanese astrophysicist who discovered them:

- The Hayashi line, strictly, represents the locus in the H–R diagram for fully convective stars of a given mass at various radii. This location depends on mass and composition; the effective temperature generally decreases as mass decreases.
- The region to the right of the Hayashi track is "forbidden" for a star of a given mass in hydrostatic equilibrium, at any phase of evolution, including the red giant phase.
- An object not in hydrostatic equilibrium, such as a variable star or a collapsing protostar, can exist in the forbidden region.
- The exact location of the Hayashi line in T_{eff} depends on the opacity in the radiative surface layer. The opacity in the relatively cool (3,000–4,000 K) surface layers is mainly due to H^- and molecules. An increase in the metal abundance produces more free electrons from elements, such as Fe, with low first ionization potentials and thus more H^-. The result is higher opacity, and thus lower luminosity for a given radius. The Hayashi line shifts to the right.
- The location of the Hayashi line also depends on the efficiency of convection in the superadiabatic zone below the surface. An assumed increase in the mixing length results in more efficient convective energy transport and thus higher luminosity for a given radius. The Hayashi line shifts to the left.
- A decrease in the radius of a star of given mass along the Hayashi line corresponds to decreasing (more negative) total energy.
- The entropy of a monatomic ideal gas is given by $S = \text{const} + \frac{R_g}{\mu} \ln \frac{T^{3/2}}{\rho}$. Since the temperature scales approximately as M/R, and the density as M/R^3, the entropy also decreases as the star contracts.

Tracks for four different masses in the ($\log \rho_c$, $\log T_c$) plane are shown in Fig. 8.4. As one would expect from the Virial Theorem, at a given density a star of higher mass has a higher temperature. At the end points of the tracks for 7, 1, and 0.1 $M_\odot$ nuclear burning has become dominant, but before that time the slope of the tracks is in fact close to 1/3. The tracks for 7 and 1 $M_\odot$ reach nuclear burning, which stops the contraction, well before they enter the degenerate region. The track for 0.1 $M_\odot$ does enter the degenerate region and the slope begins to decrease, but the object is able to reach nuclear burning temperatures. However the evolution of 0.04 $M_\odot$ reaches the degenerate region at a relatively low temperature, significant nuclear burning does not develop, and the interior begins to cool; this is the characteristic track of a brown dwarf.

Contraction times to the main sequence, starting at the birth line, for various masses are given in Table 8.1. These numbers can vary considerably for a given mass among the various published tracks, as a result of different physical assumptions and parameters and, for the highest masses, the location of the birth line. They also

8.2 Pre-Main-Sequence Evolutionary Tracks

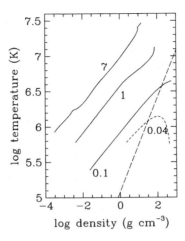

Fig. 8.4 Overview of pre-main-sequence evolutionary tracks in the ($\log \rho_c$, $\log T_c$) diagram, where the subscripts refer to the center of the star. Stars (*solid lines*) of 7.0, 1.0, and 0.1 $M_\odot$ and a brown dwarf (*short-dashed line*) of 0.04 $M_\odot$ are shown. The *upper ends* of the stellar tracks correspond to the zero-age main sequence. The *long-dashed line* shows the approximate boundary in the equation of state between the ideal gas region (*left*) and the electron-degenerate region (*right*). Data from [464] (*stars*) and [96] (*brown dwarf*)

Table 8.1 Evolutionary times (years)

Mass ($M_\odot$)	Pre-main-sequence time	Mass ($M_\odot$)	Pre-main-sequence time
0.1	1.2×10^9	1.2	3.4×10^7
0.2	5.1×10^8	1.4	1.6×10^7
0.3	3.8×10^8	1.6	1.1×10^7
0.4	2.3×10^8	1.8	9.0×10^6
0.5	1.5×10^8	2.0	7.0×10^6
0.6	1.0×10^8	2.5	4.0×10^6
0.7	7.5×10^7	3.0	2.0×10^6
0.8	6.5×10^7	4.0	5.0×10^5
0.9	5.5×10^7	5.0	2.0×10^5
1.0	4.0×10^7	6.0	1.0×10^5

depend on how the zero-age main sequence is precisely defined, because the phase of transition to full nuclear burning takes a substantial fraction of the time to reach it. The times given in the table correspond to the point where at least 99% of the energy radiated by the star is supplied by nuclear burning. Stars above 6 $M_\odot$ are not included because they do not have a well-defined quasi-static contraction phase. They remain in the protostellar accretion phase until, in some cases, they are well onto the main sequence. The time to reach the final mass depends on the accretion rate. That time is best estimated from the accretion time, $M/\dot{M}$, which, for example, is 5×10^5 yr for a final mass of 5 $M_\odot$ accreting at 10^{-5} $M_\odot$ yr^{-1}.

The stars in the range 1–5 $M_\odot$ have relatively high internal temperature and therefore relatively low internal opacities and are able to radiate rapidly. The contraction times are short, and because the luminosity is relatively constant during the radiative phase, during which they spend most of their time, the contraction time is well approximated by the Kelvin-Helmholtz time

$$t_{\text{KH}} \approx \frac{GM^2}{L}\left(\frac{1}{R} - \frac{1}{R_b}\right), \tag{8.41}$$

where R is the main-sequence radius and R_b is the radius at the birth line. A star of 1 $M_\odot$ spends 10^7 yr on the Hayashi track. For the next 2×10^7 yr, the star is primarily radiative, but it maintains a thin outer convective envelope all the way to the main sequence. The final 10^7 yr of the contraction phase represents the transition to the main sequence, during which the nuclear reactions begin to be important at the center, the contraction slows down, and, as the energy source becomes more concentrated toward the center, the luminosity declines slightly.

As the mass decreases below 1 $M_\odot$ the convective phase begins to dominate. For a star of 0.5 $M_\odot$ a radiative core forms in the center and increases in mass to about 60% of the total mass. The track remains approximately vertical all the way to the main sequence. Stars of 0.3 $M_\odot$ or less remain fully convective all the way to the main sequence. Because the luminosity varies continuously during the contraction, t_{KH} is determined by integration along the track. It takes a star of 0.1 $M_\odot$ about 10^9 yr to reach the main sequence. At 0.075 $M_\odot$, depending somewhat on chemical composition, occurs the borderline between stars and brown dwarfs. As shown in Fig. 8.4, below this mass, because of the onset of electron degeneracy, the central temperature never becomes high enough so that nuclear reactions can provide the energy needed to power the star. After reaching a maximum, the central temperature begins to decline, even though the slow quasi-static contraction continues. The released gravitational energy no longer is able to heat up the star, as it would for an ideal gas, but instead it is required to lift the degenerate electrons into the higher energy states, as determined by the Pauli exclusion principle.

Another view of the low-mass star/brown dwarf transition is shown in Fig. 8.5. The stars are able to reach hydrogen burning through the proton-plus-proton reaction, and their luminosity levels off with time as they reach the main sequence. The brown dwarfs (masses below 0.075 $M_\odot$) are able to burn deuterium ($^1\text{H} + {}^2\text{D} \rightarrow {}^3\text{He} + \gamma$), which accounts for the leveling off of the luminosity tracks (known as *cooling curves*) at the earlier phases of the evolution. This energy supply is soon exhausted, and the proton-plus-proton reaction is never able to supply the entire radiated energy. Note also that there is a mass limit below which Li does not burn ($^1\text{H} + {}^7\text{Li} \rightarrow {}^4\text{He} + {}^4\text{He}$) because internal temperatures never reach 2.5×10^6 K, the characteristic burning temperature. Thus brown dwarfs below about 0.065 $M_\odot$ should show Li in their surface layers, a very useful test to show that an object actually is a brown dwarf. Figure 8.5 can be used to estimate the mass of an observed low-mass object, if the luminosity and age can be determined.

8.3 Comparison with Observations

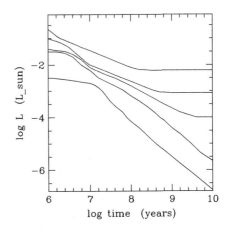

Fig. 8.5 Luminosity vs. time for low-mass stars and brown dwarfs. From *top* to *bottom*, the curves are for masses 0.2, 0.1, 0.075, 0.05, and 0.02 $M_\odot$. The *dividing line* between stars and brown dwarfs in these calculations is at 0.075 $M_\odot$. Data from [31, 32, 96]

8.3 Comparison with Observations

The comparison of observed properties of young stars with the theoretical contraction tracks reveals a number of properties of these objects. The observations of the young-star population known as the T Tauri stars are reviewed, for example by [33, 51]. The class was defined by Joy [245] on the basis of irregular variability, presence of emission lines, and association with nebulosity. They are now classified on the basis of the strength of the Hα line in emission, with equivalent width greater than 10 Å defining a classical T Tauri star (CTTS), and a width less than that amount defining a weak-line T Tauri star (WTTS) [221]. The CTTS are generally also known as Class II objects and are associated with infrared excess, indicating the presence of a disk. The WTTS would generally fall into Class III, with little IR excess. Sample spectra and spectral energy distributions [359] are shown in Fig. 8.6. The Hα in emission originates mainly in the magnetospheric cavity, generally within about 0.1 AU of the star but outside the actual photosphere. The details of the line strength and line profile depend in a complicated way on both the accretion characteristics and the outflow characteristics of the star [320]. Young stars in the mass range 0.1–2 $M_\odot$ fall within the T Tauri classification. The higher-mass analogues are known as Herbig Ae and Be stars.

The T Tauri stars have high lithium abundances, typically 10^{-9} that of hydrogen, comparable to the abundance in meteorites in the solar system and much higher than in the solar photosphere itself. These abundances are considered to be the primordial abundances of the material out of which the stars formed. The stars are associated with dark clouds where star formation is taking place and display irregular variability in light which can amount to up to 3 magnitudes in the visible spectrum. The CTTS in particular display mass loss that is much more rapid than that of most main-sequence stars, have excess infrared emission over that expected from a normal photosphere, indicating the presence of a surrounding disk, and excess ultraviolet radiation, indicating accretion onto the star.

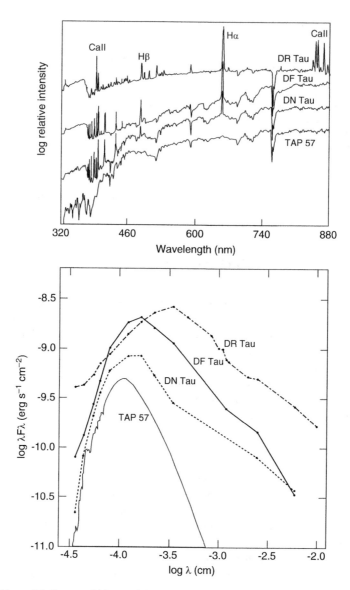

Fig. 8.6 *Upper*: Medium-resolution optical spectra of 4 T Tauri stars, ranging from an extreme CTTS (DR Tau) to moderate CTTS (DN Tauri and DF Tauri) to a WTTS (TAP57). Note that the spectrum of DR Tau consists mainly of emission lines; there is no absorption-line photospheric spectrum. Note also the ultraviolet excess around 350 nm. The Hα line decreases in strength from the upper to the lower spectra. *Lower*: The spectral energy distributions of the same objects, showing the increase in the IR excess from WTTS to extreme CTTS. These data are not corrected for extinction. Reproduced with kind permission of Springer Science and Business Media, from *The Origin of Stars and Planetary Systems*, ed by C. J. Lada, N. D. Kylafis, article by F. Ménard, C. Bertout: The nature of young solar-type stars, p. 341, Fig. 1. © 1999 Kluwer Academic Publishers

8.3 Comparison with Observations

In a few cases, surface magnetic fields have been measured through the Zeeman effect; the values are of order a kilogauss. In general T Tauri stars are thought to have magnetic fields considerably stronger than that of the Sun. The main piece of observational evidence that supports this claim is the almost ubiquitous detection of X rays in young stars [169, 185], both in CTTS and WTTS. The X-ray luminosity of a given star is highly variable, but on the average, in the 0.2–2 keV energy range it is 10^{-3}–10^{-4} times the total luminosity, whereas in the Sun the ratio is closer to 10^{-5}–10^{-6}. The X-ray properties differ little between CTTS and WTTS, except that the CTTS tend to be somewhat fainter. In fact a large number of WTTS have been discovered on the basis of their X-ray brightness, since their Hα is weak. The observations strongly suggest that in most cases stellar activity produces the X-rays, and they are unrelated to the presence of a disk. The source of the X rays is probably similar to that in the chromosphere and corona of the present Sun, but at an enhanced level, with complicated magnetic field structures evolving to produce rapid magnetic reconnection events, which lead to local heating to temperatures sufficient to produce X rays. In any case, although the inferred fields are strong, there must have been substantial magnetic flux loss during the protostellar phase.

From spectroscopic and photometric observations one is able to determine the rotational velocity at the surface of T Tauri stars; however there is no information on the distribution of angular momentum as a function of radius in the interior. It is often assumed that in the fully convective phase the star is uniformly rotating (constant angular velocity Ω); however it is known from helioseismology that this assumption is incorrect in the case of the solar convection zone. Nevertheless it is a useful first approximation.

From spectroscopic observations of line broadening arising from the rotational Doppler shift, one obtains $v \sin i$, where v is the rotational speed at the equator and i is the angle of inclination, that is, the angle between the rotation axis and the line of sight, with $\sin i = 1$ for a star viewed equator-on. Already in the 1950s the method was used [218] to determine that $v \sin i$ for 3 T Tauri stars fell in the range 20–65 km s^{-1}. Many subsequent observations give similar results, with most T Tauri stars in the range 1–150 km s^{-1}. From photometric observations of the stars one can often detect periodic variations superimposed on the general irregular variability; these variations are presumably due to active regions ("starspots") that are carried by rotation around the star. The direct observational result is the rotational period P (or $\Omega = 2\pi/P$), which, when combined with the stellar radius (obtained from T_{eff} and the luminosity), gives the actual rotational velocity, not dependent on $\sin i$. In general the rotational velocities are fast compared with that of the Sun. A T Tauri star, of 1 M$_\odot$, uniformly rotating with $v = 20$ km s^{-1} and a radius of 3 R$_\odot$ has an angular momentum about 100 times larger than that of the Sun. However the rotational velocities are generally at last a factor of 10 slower than the "breakup" rate, where the centripetal acceleration balances gravity at the equator. The typical angular momentum is 4 orders of magnitude smaller than that of a molecular cloud core of the same mass.

The evolution of angular momentum during pre-main-sequence evolution is complicated and is only beginning to be understood [223]. There are three main physical effects operating: (1) A star of constant mass and constant angular momentum will spin up during contraction, and the rotational period will decrease. (2) A star with a disk will interact magnetically with the disk, resulting in angular momentum transport to the disk at a rate depending on the strength of the magnetic field. This braking mechanism can be efficient, resulting in approximately constant Ω with time as the star contracts. It could also be only partially effective, resulting in some angular momentum loss during contraction but with Ω increasing slowly with time. (3) Even if the star does not have a disk, the ordinary stellar wind, expected to be a scaled-up version of the solar wind, will result in angular momentum loss. However this mechanism is inefficient, with the time scale for angular momentum loss longer than the typical Kelvin-Helmholtz contraction time.

Figure 8.7 shows evidence for these effects through comparison of two clusters of different age [297]. The rotational periods shown are based on photometric variation in the light from the stars as a result of starspots carried with the rotation of the star, so there is no uncertain factor of $\sin i$. On the average, both the higher-mass stars and the lower-mass stars show definite evidence of spin-up with age. Furthermore, the higher-mass stars show definite evidence of two peaks in period. In the Orion cluster these peaks fall around 8 days and about 2 days. A similar double peak, at 5 days and 1 day, is seen in the slightly older cluster NGC 2264. The shorter-period

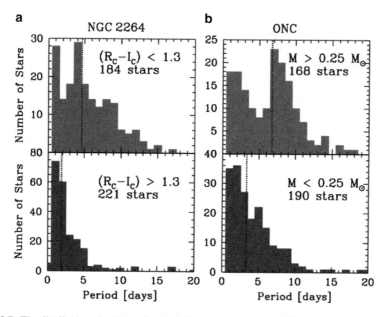

Fig. 8.7 The distribution of rotational periods for the young cluster NGC 2264 and for the Orion Nebula cluster is shown for two groups of stars in each case: mass greater than 0.25 $M_\odot$ (*upper*), and mass less than that value (*lower*). The quantity $(R_c - I_c)$ is a red color index, an approximate indicator of mass. Age estimates are uncertain, but the general understanding is that the Orion cluster, with a mean age of ≈ 1 Myr, is at least a factor 2 younger than NGC 2264. *Vertical lines* indicate median periods. Credit: Lamm et al. *Astron. Astrophys.* **430**, 1005 (2005). Reproduced with permission. © European Southern Observatory

peak is thought to represent stars which are contracting at fairly constant angular momentum and spinning up. The other peak is thought to represent stars which are magnetically locked to their disks, and thus have relatively slow rotation. At both ages shown there is a tail of relatively slowly rotating stars, indicating that in at least some cases there is significant angular momentum loss. Then, if disk locking is a significant effect one would expect a correlation between slow rotation and the presence of a disk accreting onto the star. The main signatures for accretion include a high equivalent width in Hα emission and mid-infrared excess. There is some evidence for such a correlation, but the observational results are ambiguous [90], the statistics are incomplete, and this point is subject to contention.

All the characteristics of T Tauri stars are consistent with youth; their locations in the H–R diagram fall along Hayashi tracks for stars in the mass range $0.1–2.0\,M_\odot$. The arguments that the properties of T Tauri stars clearly point to their identification as pre-main-sequence stars, as originally suggested by Ambartsumian [16], were summarized by Herbig [219]. He provided a further argument [220] by showing that the T Tauri stars in the Taurus-Aurigae region have the same radial velocities as the associated molecular clouds. The disk characteristics vanish after ages of a few million to 10^7 yr, putting a constraint on the time available for the formation of gaseous giant planets. Here we consider in somewhat more detail some definite connections between theory and observation.

1. As predicted by Hayashi, no stars in equilibrium are observed to exist in the forbidden region of the H–R diagram. In young clusters, the location of observed objects, corrected for extinction, in the H–R diagram is consistent with the coolest objects being on their Hayashi tracks and consistent with the decreasing value of T_{eff} on the tracks as the mass decreases. For old evolved clusters, the red giants are not fully convective, but their convective envelopes are deep enough so they fall close to, but slightly to the left of, their Hayashi lines. The observed cluster diagrams, for non-variable, extinction-corrected stars, are in agreement with this conclusion.

2. The location of the theoretical "birth line" in the H–R diagram is consistent with the upper envelope of the positions of stars in a young cluster. In Fig. 8.8 the color-magnitude diagram for one of the youngest clusters known, NGC 2264, is shown. The upper dashed-dot line, corresponding to an age of 10^5 years, corresponds approximately with the birth line for the range of masses shown. The birth line is based on protostar theory and, for a given mass, corresponds to the point in the H–R diagram where for the first time the infalling circumstellar dust has been cleared away and the photospheric spectrum of the underlying stellar core is visible. The line itself connects the points for the various masses. The birth line may be considered to correspond to the end of significant mass accretion, but in fact a slow accretion may still occur beyond it. Clearly none of the observed points lie significantly above the theoretical birth line. Another example is shown in Fig. 1.21, where Palla/Stahler evolutionary tracks are compared with the observed [226] positions of young stars in the Orion Nebula cluster. Again, practically no observed objects fall above the birth line.

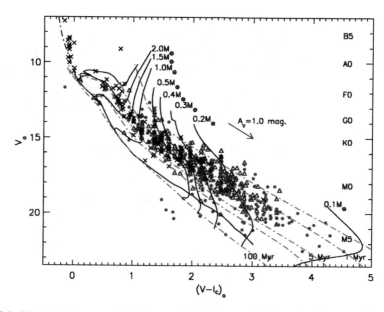

Fig. 8.8 The extinction-corrected color-magnitude diagram for the young cluster NGC 2264 is shown [129]. Visual magnitude is plotted against $(V-I)$ color. The *solid lines* are evolutionary tracks [132], converted to the observed quantities [226]. The *dashed-dot lines* are isochrones, with examples of ages marked. The *filled circles* are classical T Tauri stars (CTTS), the *open triangles* are weak-lined T Tauri stars (WTTS), and the *crosses* are X-ray sources. The *lower solid line* gives the observed main sequence of the Pleiades cluster; the *dot-dashed line* represents its upward extension. Reproduced by permission of the AAS, from [129]. © 2005 The American Astronomical Society

3. The distribution of stars in the H–R diagram in a young cluster is used to determine the age spread of the stars. Figure 8.8 shows that the stars in a young cluster are not co-eval; there is a spread in ages of about 10^7 years in NGC 2264. The median age of this cluster, based on a compromise among the discrepant results of the various pre-main-sequence evolutionary tracks [128] is about 3×10^6 yr, and most stars fall between 0.1 and 5 Myr. There is apparently no significant difference in the age spread among the CTTS and the WTTS. There is a similar spread in the Orion cluster. However the determination of ages in this manner is uncertain. Depending on which set of evolutionary tracks is used, the distribution can change because of differences in physics and parameters in the various calculations. Also somewhat uncertain is the conversion of the theoretical ($\log L$, $\log T_{\text{eff}}$) diagram into observed quantities, particularly for the cooler objects; also the presence of circumstellar disks must be accounted for. And of course some of the stars plotted in Fig. 8.8 may not actually be members of the cluster.

An accurate observed position of a young star in an H–R diagram can be used to determine not only its age but also its mass, in conjunction with evolutionary tracks. The distribution of masses in a young cluster is clearly an excellent probe of

8.3 Comparison with Observations

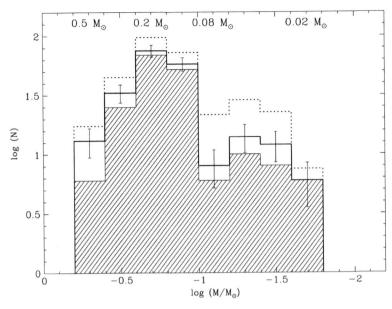

Fig. 8.9 A histogram showing the number of stars of a given mass as a function of log mass for low-mass stars in the inner region of the young Orion Nebula cluster. The evolutionary tracks used to determine the masses are from [132]. The *solid histogram* with *error bars* gives the results for all stars observed. The *dotted histogram* gives the results for all stars, corrected for incompleteness. The *hatched histogram* gives the results for stars with ages less than 5 Myr. Reproduced by permission of the AAS, from D. Spolyar, K. Freese, P. Gondolo: *Phys. Rev. Lett.* 100, 051101 (2008). © 2004 The American Astronomical Society

the IMF, a critical quantity for the understanding of star formation. Figure 8.9 is an example, giving the distribution of low-mass stars in the Orion Nebula cluster. The observed positions in the H–R diagram were obtained through use of both infrared spectroscopy and photometry. There is a peak at $0.2 M_\odot$ and a decrease in the function to both higher and lower mass. Below the peak there is a sharp drop near the stellar/brown dwarf boundary, then the function levels off in the brown dwarf region. The peak mass is in agreement with the general Galactic IMF given by (1.7); however the slopes away from the peak are not in precise agreement. Determining the IMF in this way is tricky, because there is no well-defined mass-luminosity relation as there is for main-sequence stars (see Chap. 5); the track masses must be used.

The question is often posed whether the IMF is universal. The cluster IC 348 has a similar age to the Orion cluster, and its low-mass IMF is similar [329]. Spectroscopic methods were used to derive the IMF, as in the case of Orion. However, in the Taurus-Aurigae star forming region, which is not regarded as a cluster, the IMF is quite different [329]. It was determined by similar methods to those used for IC 348, and the age (1–2 Myr) is also about the same. However the IMF peaks at about $0.8 M_\odot$ and then slowly declines toward lower masses and into the brown dwarf region. The clear difference between the functions for Taurus and

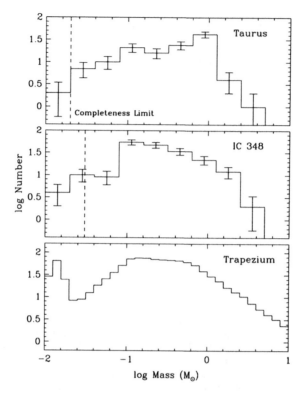

Fig. 8.10 A histogram showing the number of stars of a given mass as a function of log mass for low-mass stars in the Taurus star-forming region (*upper*) and in the young clusters IC 348 (*center*) and the Orion Nebula cluster (*lower*). The evolutionary tracks used to determine the masses in the upper and center frames are from [109] for mass $M < 0.1\,M_\odot$, from [30] for $0.1 < M < 1\,M_\odot$, and from [407] for $M > 1.0\,M_\odot$. The lower frame is derived from [374]. Reproduced by permission of the AAS, from [329]. © 2004 The American Astronomical Society

those in IC 348 and Orion, illustrated in Fig. 8.10, strongly suggests that the nature of the IMF depends on the star-forming environment. The reason for the discrepancy remains unresolved, but it has been suggested that either the Jeans mass [91, 331] or the turbulent velocity field [404] could have been different in the molecular cloud cores in Taurus, as compared to those in IC 348 or Orion. It is also possible that additional data in the future, as well as improved pre-main-sequence evolutionary tracks, will reduce the apparent discrepancy.

4. The masses derived in the previous paragraph are not fundamental mass measurements; they are based on theoretical evolutionary tracks, and they are subject to a number of uncertainties. It is a critical test of pre-main-sequence stellar evolution to compare the masses derived from the positions of stars in the H–R diagram with actual masses derived from binary orbits. Several methods are available to determine masses; one is from an eclipsing binary system, both of whose components have

8.3 Comparison with Observations

observable spectra from which the orbital radial velocity as a function of time can be determined. In such systems the individual masses can be measured, independent of the distance, which may not be known. There are not many eclipsing systems known in which both components are in the pre-main-sequence phase. One of the first to be discovered was that shown in Fig. 1.14 (RXJ 0529.3 + 0041), whose components are nearly coeval at $\approx 10^7$ yr. The masses derived from the positions in the H–R diagram and evolutionary tracks [30] are about 1.4 and 0.82 $M_\odot$. The binary observations give a period of 3.04 days and radial velocity amplitudes of 81 and 110 km s^{-1}. The binary is in Orion, and both components have lithium as well as Hα emission. The dynamical masses are $1.25 \pm .050$ and $0.91 \pm 0.05\,M_\odot$. Thus the track masses can be trusted to about 10% in this mass range. Other sets of pre-main-sequence tracks give about the same mass for the primary but disagree on the mass for the secondary.

Another method for determining the mass of a young star is to use the properties of its circumstellar disk. If the disk has low mass relative to the star, the Keplerian velocity measured in the disk is determined by the mass of the star; such velocities are usually measured by the Doppler shifts in CO lines in the mm part of the spectrum. Correction for the inclination of the disk can usually be obtained, if the disk is spatially resolved, from the observed morphology. The distance to the star must be known, since the radius at which the velocity is measured is needed to obtain the central mass.

An good example is GG Tau, which is a quadruple system, consisting of two binary pairs in orbit about each other. All components have Li in absorption, Hα in emission, and lie on approximately the same isochrone in the H–R diagram [545]. The more massive pair has an orbit with separation of about 30 AU. The components are surrounded by a circumbinary disk (Fig. 6.6) whose orbital velocities have been measured as a function of distance to the star, giving the sum of the masses of the two stars as $1.28 \pm 0.08\,M_\odot$ (and subject to an additional uncertainty of 15% because of the distance). The two stars have been also been placed in the H–R diagram, and the track masses agree or disagree, depending on which set of tracks is used and depending on the effective temperature scale used for T Tauri stars, which is uncertain. The best agreement is obtained from the Baraffe tracks [30] in which the ratio of convective mixing length to pressure scale height (l/H) is set to 1.9. The sum of the track masses in this case comes out to be $1.46 \pm 0.1\,M_\odot$, while if l/H is set to 1, the result is $2.04\,M_\odot$. However $l/H = 1.9$ is reasonably close to the solar value of 1.6, and the Baraffe models use non-grey model atmospheres as surface boundary conditions, which represents improved physics over the other sets of tracks.

A third method of obtaining masses is the combination of data for a spectroscopic binary with astrometric data for the system. Ordinarily spectroscopic binaries are close enough so that they cannot be spatially resolved to allow measurement of their relative positions as a function of time. However improvement in observational techniques, including the use of the Fine Guidance Sensors on the Hubble Space Telescope, adaptive optics on large telescopes, and ground-based interferometers, has resulted in both spectroscopic and astrometric data for a few pre-main-sequence

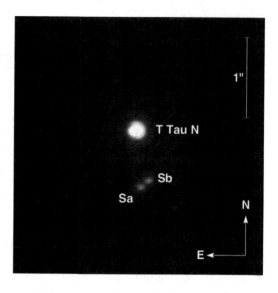

Fig. 8.11 A near-infrared image (K_s band) of the triple system T Tauri, taken with NACO, the adaptive-optics near-infrared camera on the ESO Very Large Telescope, on 1 Feb 2008. Reproduced with permission, from R. Köhler: Journ. of Phys.: Conference Series **131**, 012028 (2008). © 2008 IOP Publishing Ltd.

systems. If both spectra are visible, even though the system is not eclipsing, both masses can be determined without the necessity of an independent distance determination. The method has been used [61] to find masses of 0.70 and 0.58 $M_\odot$, with an uncertainty of about 8%, for the binary HD 98800Ba and Bb. The agreement with evolutionary tracks is not particularly good, unless the assumed metal abundance for the tracks is reduced below solar. Another example is NTT 045251 + 3016 [488] where the measured masses are 1.45 ± 0.19 and $0.81 \pm 0.09\,M_\odot$. The best agreement is obtained with Baraffe tracks [30] but with $l/H = 1.0$ rather than 1.9.

A particularly interesting system is T Tauri itself (Fig. 8.11). First observed in 1852 from London by Hind [230], it was long thought to be a single star. A surprising discovery [150] showed that in fact it had a companion, observable only in the infrared, which came to be designated as T Tau S, while the optically visible component was denoted T Tau N. Then it turned out that the infrared companion itself was a binary [266], whose components are known as T Tau Sa and Sb. The apparent separation of Sa and Sb is about 0.087 arcsec, with a probable semimajor axis of 13 AU. The apparent separation of N and S is about 0.65 arcsec (≈ 100 AU), but its orbit, and therefore the mass of N, is not well determined. The orbit of the Sa–Sb system is in the process of being determined by pure astrometry using radio and infrared observations [142, 261]. Although only a fraction of the 28 year orbit has been measured, the masses of the two components are reasonably well estimated at 2.28 and 0.41 $M_\odot$, respectively, for Sa and Sb, with an error of $\pm 0.2\,M_\odot$ in both cases. Both components of S radiate primarily in the infrared; the secondary because it is an M star, and the primary probably because it has an edge-on circumstellar disk. Because Sb is heavily extincted, the comparison of the dynamical mass with a track mass is very uncertain. On the other hand, the mass of T Tau N is not available from dynamical measurements, but its track mass is estimated to be 2 $M_\odot$.

8.3 Comparison with Observations

The general agreement between dynamical measurements of the masses of pre-main-sequence stars and the masses obtained from evolutionary tracks is not entirely satisfactory. In several cases, dynamical masses can be measured to an accuracy of a few percent or better. However the average deviation of track mass from dynamical mass is 20%, and in a few cases it is as large as 50% [228, 343]. Almost all of the measured dynamical masses are for stars above $0.5\,M_\odot$, and the agreement is generally better for stars above $1.0\,M_\odot$ rather than for the lower-mass stars. The tendency is for the theoretical tracks to under-predict the dynamically determined masses. The comparison is clearly most difficult for the cool, low-mass stars and the brown dwarfs, and very few dynamical measurements are available in that range. Thus derived quantities, such as low mass IMF's, should be viewed with caution.

Nevertheless, observational progress is expected in the future and already there has been a remarkable discovery of a double-lined eclipsing binary of a young brown dwarf in the Orion Nebula cluster [486, 487]. The respective masses are 0.054 and $0.034\,M_\odot$, with an accuracy of 8%, surface temperatures are 2,715 and 2,820 K, and radii are 0.67 and $0.51\,R_\odot$. These properties are generally consistent with theoretical Hayashi tracks at the ≈ 1 Myr age of the cluster [30, 132], but the track masses differ from the dynamical ones by about a factor 2. Also surprising is the fact that the more massive dwarf is the cooler, which is not consistent with the evolutionary tracks. This discrepancy is not explained, although it could possibly be a result of non-coevality of the components, or atmospheric magnetic effects which could affect the derived temperatures.

Given that dynamical masses, when available, are quite accurate, the main improvements needed for the comparison are first, better measurements of stellar properties, such as luminosity and T_{eff} for low-mass stars and brown dwarfs, and second, improvements in stellar evolutionary tracks and atmospheric models for comparing observed and synthetic spectra to determine T_{eff}. More dynamical measurements for stars below $0.5\,M_\odot$ are also highly desirable. These improvements are clearly important for the understanding of star formation.

5. Observations of the rare light element lithium provide a unique probe of the interior structure of pre-main-sequence stars and can serve to calibrate uncertain theoretical parameters, such as the convective mixing length. The main test is a comparison of observed lithium abundances of stars that have just reached the main sequence with calculations of the depletion of lithium in the surface layers during the contraction. The lithium abundance that present-day newly forming stars inherit from the interstellar medium is about 10^{-9} that of hydrogen. Yet it is observable in the red part of the optical spectrum through the presence of an absorption line at 670.7 nm. Lithium is easily destroyed by reactions with protons at temperatures above about $T = T_{\text{burn}} = 2.5 \times 10^6$ K ($^7\text{Li} + {}^1\text{H} \rightarrow {}^4\text{He} + {}^4\text{He}$). The youngest stars have internal temperatures less than this and they are fully mixed by convection, so they should show their initial Li abundances at the surface: most T Tauri stars in fact do.

During the contraction, the lower-mass stars, say $0.6\,M_\odot$ or less, remain fully convective at least until the time when $T = T_{\text{burn}}$ at the center. The Li will burn there, and the entire star will be mixed by convection, so even the surface value will be depleted. Thus low-mass stars will arrive at the main sequence with very little surface Li. However the higher-mass stars, say $1.2\,M_\odot$ or above, develop radiative cores before $T = T_{burn}$ at the center, and the inner edge of the convection zone retreats toward the surface. As a result, Li is burned in the central regions, but the inner edge of the convection zone always remains at $T < T_{burn}$, so there is no Li depletion at the surface. For masses in between these limits, the amount of depletion depends on the history of T at the base of the convection zone. For example, a $1\,M_\odot$ star reaches a maximum temperature at the base of the convection zone of about 4×10^6 K, and maintains this temperature for about 10^7 years, sufficient to burn only about half the Li at the surface.

Figure 8.12 shows the Li abundances on the main sequence of the Pleiades cluster, which is sufficiently young so that any main-sequence depletion mechanism should not have been important. The abundances clearly decrease as one proceeds from the F stars to the K stars, by 3 orders of magnitude. The relatively wide scatter in the observational points is only in part a result of observational uncertainties; stellar surface activity in these young stars could account for some of it. The theoretical curve is based on standard stellar models with solar abundances, which closely matches those in the Pleiades. A reduction in the metal abundance used for the theoretical curve would shift it upward. However, accounting for convective overshoot at the base of the convection zone would tend to move the curve downward, that is to higher Li depletion for a given mass. However the theory of overshoot is very uncertain. Although the theoretical result is sensitive to assumed physical parameters in the models, in general the agreement is good and gives reasonable certainty as to the presence of convection zones during pre-main-sequence evolution.

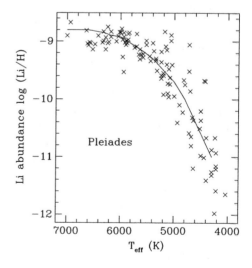

Fig. 8.12 The lithium abundance along the main sequence of the Pleiades cluster, by number relative to hydrogen, as a function of T_{eff}. At the cluster age of about 10^8 yr the depletion pattern seen at the lower temperatures is almost entirely the result of convective mixing during the pre-main-sequence evolution. The *crosses* are observations from [473] and the *solid line* is a model calculation from [416]

8.4 Summary

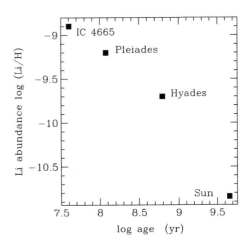

Fig. 8.13 The lithium abundance in a star of 1 M$_\odot$ as a function of age, as observed in various clusters and the Sun. At this mass the pre-main-sequence depletion, which ends at log age = 7.5, is roughly a factor 2, and the plot shows that nevertheless the lithium continues to be depleted on the main sequence

Figure 8.13 shows what happens to the abundance of lithium in a main-sequence star of a solar mass as a function of age, based on observations in three different clusters and the Sun itself. At the age of the Hyades (6×10^8 yr) the abundance has already dropped by a factor of 3 with respect to the Pleiades, and the general trend shows a gradual depletion on a time scale of 10^9 yr. But main-sequence models do not have surface convection zones deep enough to account for depletion due to mixing down to T_{burn}. The Sun's convection zone extends down to about 2×10^6 K during this phase. Overshoot, rotational mixing, and hydrodynamical instabilities have been considered to account for this effect, but there is no satisfactory solution. Clearly the time scale for this process must be very long, of order 1 Gyr.

8.4 Summary

For stars below about 6 M$_\odot$, once the protostellar infall has been completed, the internal temperatures are not high enough to supply the star's radiation through nuclear burning. Gravitational contraction is the main energy source, except that burning of primordial deuterium may hold up the contraction for ≈ 1 Myr. For stars above about 1 M$_\odot$ the interior of the star is mainly radiative during this phase, and the evolutionary track proceeds more-or-less horizontally and to the left in the H–R diagram towards the main sequence. For stars of 1 M$_\odot$ or less a substantial portion of the evolution is characterized by a fully convective phase, during which the evolution in the H–R diagram is mainly vertically downwards. The fully convective track, known as the Hayashi line, forms a limit to the right of which is a forbidden region for stellar models in hydrostatic equilibrium. The surface temperature along this line is about 4,000 K for 1 M$_\odot$, decreasing to about 3,000 K for 0.1 M$_\odot$. For stars below 0.3 M$_\odot$ the star remains fully convective on a nearly vertical track all the way to the main sequence. For stars above 0.3 M$_\odot$

a radiative core eventually develops. Contraction times from the birth line to the main sequence range from 2×10^5 years at $5\,M_\odot$ to 10^9 years at $0.1\,M_\odot$. Below about $0.075\,M_\odot$ the effects of increasing electron degeneracy prevent the onset of sufficient nuclear burning to supply the stellar energy, and the object becomes a brown dwarf, which eventually cools in the interior.

The observed properties of young stars can be combined with the theoretical evolutionary tracks to deduce some properties of the end-products of star formation. Precise observed locations in the color-magnitude diagram can be used to deduce masses and ages. In a typical young cluster, star formation apparently proceeds over a time of up to 10^7 years. In addition the distribution of masses can be estimated, which generally shows a peak at about $0.2\,M_\odot$, declining to lower masses. However these estimates are uncertain because of the generally anomalous observed properties: emission lines, excess infrared and ultraviolet radiation, irregular variability, strong X-ray flux. Some of the objects are accreting material from a disk as well as losing mass in a wind. It is difficult, especially for the cooler, lower-mass objects, to convert observed colors or spectra into T_{eff}.

There are two important tests on the validity of the theoretical results in pre-main-sequence evolution. The first is made possible by direct dynamical mass determination in a few cases, mainly from eclipsing binaries and stars with spatially resolved disks. In principle the dynamically determined masses should agree with the "track" masses determined from the positions of objects in the H–R diagram. For objects around $1\,M_\odot$ the agreement is reasonably good; for lower-mass objects in some cases the disagreement is at the 50% level and it varies according to which set of evolutionary tracks is used. The second test is based on observations of the rare light element lithium. Above $1\,M_\odot$ the stars are predominantly radiative in the interior, with surface convection zones that are not deep enough to reach the Li burning temperature. Thus these stars should reach the main sequence with close to the primordial Li abundance in their photospheres, a conclusion in agreement with observation. The lower-mass stars have deeper convection zones, and theoretical calculations show that the observable lithium should decrease as the mass decreases along the main sequence. This result is also in general agreement with observations, as long as the youngest possible main-sequence stars are used, to avoid the complication of the further long-term depletion of Li during main-sequence evolution.

8.5 Problems

1. Assume that the intensity in a slightly non-isotropic radiation field can be represented by $I_\nu = I_{\nu,0} + I_{\nu,1} \cos \theta$ where $I_{\nu,1} \ll I_{\nu,0}$ and θ is the angle with respect to the normal to a given surface. Calculate the radiation pressure from (8.33) and use the result to justify (8.13).

8.5 Problems

2. A star of 0.1 $M_\odot$ stays fully convective during its entire pre-main-sequence contraction and maintains a constant $T_{\text{eff}} = 3{,}000$ K. What is the contraction time to the main-sequence radius of 0.1 $R_\odot$, neglecting nuclear burning?

3. This problem provides an approximate solution for the Hayashi line in the H–R diagram, under the assumption that the star is a fully convective ideal gas, corresponding to a polytrope $P = K\rho^{5/3}$.

    (a) Find the relation between pressure and temperature.
    (b) The mass-radius relation for an $n = 1.5$ polytrope is [110] $K = 0.424 GM^{1/3} R$. K is not a constant along the Hayashi track for a given mass; as the star contracts the entropy decreases. Thus K needs to be eliminated from the expression for the pressure.
    (c) Assume the opacity in the radiative surface layers has the form $\kappa_p = \kappa_0 P^\alpha T^\beta$. Use the surface boundary conditions (8.9) and (8.12) to match the polytropic pressure and temperature to the surface conditions. Thus obtain an equation relating M, L, and T_{eff} along a Hayashi track.
    (d) Calculate the slope of the Hayashi line for a given mass, and show that it becomes steeper as β increases.
    (e) Plot the Hayashi line for two different masses, assuming $\alpha = 1$ and $\beta = 3$.

4. This problem provides a rough estimate of the transition mass between a star and a brown dwarf.

    (a) The central density ρ_c of a fully convective star is 6 times the mean density. The central temperature T_c of a star is roughly twice the mean temperature determined from the Virial Theorem (for an ideal gas). The gravitational energy of a fully convective star is $(6/7)GM^2/R$. Use these approximations to obtain a relation between ρ_c and T_c and mass.
    (b) Obtain a relation between ρ_c and T_c along which the gas is on the borderline between being an ideal gas and a fully degenerate electron gas (non-relativistic) by equating the ideal gas electron pressure ($R_g \rho T/\mu_e$) to the fully degenerate electron pressure. The quantity μ_e is the mean atomic weight per free electron, $2/(1+X)$, where X is the hydrogen mass fraction.
    (c) Assuming that $T_c = 4 \times 10^6$ K is the minimum temperature needed to support nuclear burning, find the corresponding minimum mass.

5. Show that a homologously contracting object has an effective energy generation per unit mass per unit time $\epsilon_{\text{grav}} = KT$ where T is the temperature and K is an unknown constant. Find the value of K. In practice it can be specified for an initial stellar model by a guess for the total luminosity. Use (8.5) with $\epsilon_{\text{nuc}} = 0$., and (8.16).

6. Combine (8.3) and (8.6) to obtain an equation for dT/dP. Assume that the opacity $\kappa_R = 0.2(1+X)$, where X is the hydrogen mass fraction; this expression holds for electron scattering as the main opacity source and is appropriate for hot stars. Now evaluate the expression in the surface layers where $L_r = L$, the total luminosity, and $m = M$, the total mass. You should have an equation relating T

and P, all other quantities being constants. Integrate this equation inwards from the surface, starting from $P = T = 0$, to get $P(T)$. Assume the equation of state is an ideal gas. Test the $P(T)$ relation, which assumes radiative energy transport, to see if it might actually be convective.

Chapter 9
Summary: Issues in Galactic Star Formation

Many new ideas, ground-based and space-based observations, theories, and numerical computations have considerably advanced our knowledge of star formation over the past few decades. This summary of the main physical ingredients in the process emphasizes the many questions that still remain unanswered or controversial. With more powerful telescopes and computers that will be available over the next decade, some of these questions will be further clarified, and further insights will contribute to this evolving subject. The topics covered in this text start with the assumed existence of molecular clouds over a range of masses, and end with stars in hydrostatic equilibrium, having a photospheric spectrum, and proceeding through the contraction phase toward the main sequence.

9.1 Molecular Clouds

Turbulence in molecular clouds is an important element in the physical picture. Turbulent elements are observed over a wide range of length scales L, and their velocities, as measured by molecular line widths, scale approximately as the square root of the size. The deduced supersonic velocities require that shocks be generated, and the result is dissipation of kinetic energy. Numerical simulations show that as a result the turbulence in a cloud should decay on approximately the turbulent crossing time L/v_{turb}, generally shorter than the lifetime of the molecular cloud itself. Magnetic fields are present in molecular clouds, with magnetic energy less than or comparable to gravitational energy, and with the deduced Alfvén speed, at least in some regions, similar to the turbulent speed. However the presence of the field does not appear to significantly lengthen the decay time. In many regions of molecular clouds the magnetic field appears to be weak, and the importance of magnetic fields, as compared with turbulence, in regulating star formation is still an open question.

The lifetime of a molecular cloud is not well determined and is subject to debate; most estimates fall in the range of a few million to 10^7 years or more. Thus it seems to be necessary to invoke some mechanism for maintaining the turbulence, assuming it is generated during the process of formation of the cloud. The maintenance of the turbulence requires that a significant amount of energy be injected on the larger scales. The role of the various possible mechanisms, for example continued infall of gas onto the cloud from the outside, or expansion of HII regions, or supernova explosions, remains to be clarified. There is a significant input of energy on small scales, from well-observed bipolar, collimated outflows from forming stars, but this mechanism by itself is probably not sufficient to maintain the overall turbulence.

The question of fragmentation at the molecular cloud level is still unresolved. Given a molecular clump of order $1,000\,M_\odot$ which is thought to be near virial equilibrium and not collapsing, how does it fragment into individual molecular cloud cores, and what are the properties of the fragments? This question is closely related to another fundamental question: how is star formation initiated, that is, how is molecular cloud material brought to the point where individual objects in the solar-mass range are gravitationally bound and are collapsing? There are two different avenues for star formation: stars are formed in clusters of typically a few hundred (but sometimes more) objects, as for example in the Orion Nebula cluster. Or they can form as more-or-less independent objects, such as in the Taurus-Aurigae region. In the case of individual objects, at least three different mechanisms have been actively considered for the initiation of star formation. First, in regions which are magnetically controlled, collapse is inhibited, but an object can contract quasi-statically until its density is high enough so that ambipolar diffusion increases the mass-to-flux ratio to the point where collapse can set in. Second, in supersonically turbulent regions, shocks of various strengths can compress randomly located regions to high enough density for a long enough time so they become Jeans unstable and begin to collapse. Third, collapse can be induced by outside events such as the expanding shells of HII regions or supernova remnants, as well as by cloud-cloud collisions. It still remains to be determined exactly what the role of these three different processes is. A key observational clue is simply the ratio of the number of starless cores (without protostars) to cores with infrared sources already embedded within them. The magnetic diffusion theory suggests that the ratio should be relatively large, close to ten, although turbulent effects could reduce it. The turbulence theory suggests that cores are relatively short-lived, and the ratio should be small, around unity. Observational estimates of this ratio have varied considerably in the past, but current determinations cluster about an intermediate value of about 3.

The situation for cluster formation is not as well defined. It is quite possible that cluster formation can be induced in an expanding HII region, where the increasing density in the shell of swept-up material can lead to gravitational collapse. There are observations which support this picture. Another mechanism for cluster formation is overall gravitational collapse of a molecular cloud clump of order $1,000\,M_\odot$, with fragmention into individual protostars occurring after collapse has started. Clumps are generally supported against collapse by turbulent and magnetic effects, but decay of the turbulence could allow collapse to proceed, and the initial turbulent fluctuations could induce fragmentation. However it is not clear why a few clumps undergo decay of turbulence while most others maintain it. A third alternative for cluster formation is for the individual cores to form first, perhaps initiated by the turbulent shocks. Then the Jeans-unstable cores collapse, and the clustering results from the overall properties of the turbulence, which has a hierarchical structure so that cores tend to form in clusters.

9.2 The Initial Mass Function

The theoretical understanding of the observed initial mass function is one of the important unsolved problems of star formation. Numerous theories and numerical simulations have produced mass functions in fairly good agreement with the observations, but it is not clear which mechanism dominates. The function itself is determined from observations of main-sequence stars in the field in the solar neighborhood and from stars in clusters with a well-defined main sequence, by use of the main-sequence mass-luminosity relation. It is also derived in very young clusters by means of comparison of the positions of stars in the H-R diagram with theoretical pre-main-sequence evolutionary tracks. A surprisingly uniform picture emerges. The IMF, in terms of number of stars per unit log mass, is fairly flat in the range 0.1–1 $M_\odot$, with a peak at about 0.2 $M_\odot$. It drops off at lower and higher masses, and above 1 $M_\odot$ the dropoff has the well defined (Salpeter) slope of $dN/(d \log M) \propto M^{-1.35}$. There are statistical fluctuations from cluster to cluster, and a notable exception, the Taurus-Aurigae region, shows a peak in the IMF at 0.8 $M_\odot$. But even the oldest stars, in globular clusters, show an IMF similar to that observed in the nearby field stars and open clusters.

A significant observational discovery regarding the IMF is that in various star-forming regions, and especially in the Pipe Nebula where the cores have not yet formed stars, the mass function of the dense molecular cloud cores ($n \approx 10^4 \,\mathrm{cm}^{-3}$) has a very similar shape to that of the IMF, although displaced to higher masses by a factor of about 3. A crucial question regarding the origin of the IMF is, then, how this core-mass function was produced. Theory suggests that turbulent shock compression and the resulting fragmentation could produce this mass function, or it could simply represent a statistical spread around the critical Bonnor-Ebert mass at the given temperature and pressure of the molecular cloud region. Then it still remains to be understood why there is a fairly universal efficiency factor of about 1/3, independent of mass, involved in the conversion of the cores into stars. The efficiency can qualitatively be explained by the facts that some of the core mass is later ejected in the form of bipolar flows, while other portions of it, with significant angular momentum, may be left behind in the form of a circumstellar or circumbinary disk, to be dissipated later. The similarity of the core mass function to the IMF suggests that cores form first, survive for a few free-fall times, and then collapse. This picture may well apply to stars that form in relative isolation. However, if stars form in relatively dense clusters, the picture becomes more complicated, because the individual protostars interact with each other, through encounters, mergers, outflows, ionized regions, competitive accretion, ejections of components of multiple systems, and mass segregation. If most stars are in fact formed under such conditions, it remains a puzzle why the core mass function is so well preserved as it evolves into the IMF.

9.3 Protostar Collapse

Once a molecular cloud core becomes gravitationally bound, with the gravitational force exceeding the effects of gas pressure, turbulence, rotation, and magnetic fields, a hydrodynamic collapse ensues, during which the object becomes extremely centrally condensed and the mean density increases by 16 orders of magnitude. One of the main issues associated with the collapse phase is the angular momentum problem. The angular momentum of a core is measured by observations of the change in Doppler shift of a molecular line as a function of position across the core. The origin of the angular momentum is most likely the random turbulent velocities in the molecular cloud. In the end products of the collapse, T Tauri stars, the rotation is best determined by measuring the photometric variations caused by starspots moving across the field of view as the star rotates. The specific angular momenta of the T Tauri stars are 3–4 orders of magnitude less than those of the cores.

Several mechanisms have been considered to explain this transition, and each one of them may contribute to some extent to the solution of the problem. In the low-density outer regions, where the magnetic field is still coupled to the gas, angular momentum can be transferred away by magnetic braking, as long as the travel time of an Alfvén wave across the core is shorter than the collapse time. However, at higher densities, the field effectively decouples from the gas as a result of the very low level of ionization. Once rotational effects become important as the core spins up during collapse, fragmentation into a binary or multiple system can occur; however the typical binary orbital angular momentum is still considerably less than those of cores. Mechanisms suggested for extracting angular momentum from binary orbits include angular momentum transfer to circumbinary disks, gravitational encounters with external third objects, which can carry away angular momentum and harden the binary, and accretion of low angular momentum material onto the binary components. Another process to be considered is the formation of circumstellar disks. The disk itself stores some angular momentum, and in addition, in the warm, partly ionized central regions, a magnetic field links the disk to the central star, resulting in angular momentum transfer out of the star. Some of the angular momentum thus delivered to the disk is probably lost from the system in the bipolar outflow from the forming star.

The magnetic field clearly plays a role in solving the angular momentum problem during both early and late stages of collapse, but the reduction of magnetic flux ($\approx \pi B R^2$) by 5 or 6 orders of magnitude between the initial state of a molecular cloud core and a T Tauri star has not been satisfactorily explained. At relatively high densities both ambipolar diffusion and Ohmic dissipation affect the magnetic

9.3 Protostar Collapse

field, but detailed numerical simulations that include these processes have not yet shown that the magnetic flux can be reduced to the values observed in young stars. The magnetic field is effectively decoupled from the gas once the density in the infalling cloud exceeds about 10^{11} cm^{-3}; some of the magnetic flux could simply be left in the circumstellar regions and the disk, as gas flows through it toward the protostellar core. The difficulty is that once the temperature in the central regions of the collapsing cloud reaches just above 1,000 K, the field recouples to the gas as a result of thermal ionization of heavy elements, and the magnetic flux in the recoupled region is expected to be conserved as that material accretes onto the star.

The evolutionary tracks for protostars in the 1 $M_\odot$ range or below are not nearly as well defined as those for stars. It is not even clear what the relative time scales are in the different phases; different studies give different results, and studies in different galactic regions give different results. The time scale for the Class II evolution, namely that for the T Tauri stars, is fairly well defined at an average of about 2–3 Myr. What needs to be determined is the ratio of the Class I time scale to the Class II time scale, and the ratio of the Class 0 time scale to the Class I time scale. It has generally been assumed that the Class I time scale is about 1/10 that of the Class II time scale, but recent Spitzer results put that ratio as closer to 1/5. The ratio of the Class 0 to Class I time scales is much less certain. If the Class 0 is defined as the phase during which the mass of the collapsing envelope is greater than the mass of the equilibrium stellar core, and if the accretion rate is constant through both phases, then the two time scales should be the same. However most existing observations suggest that the accretion rate is much higher in Class 0 than in Class I, so the Class 0 time scale is much shorter, perhaps by a factor 5. Further observations, especially of Class 0 objects, are needed to confirm this conclusion.

The accretion rate is determined observationally from the luminosity of the protostar, and its behavior as a function of time is difficult to determine. Is it nearly constant, at say 1.67×10^{-6} $M_\odot$ yr^{-1}, as needed to build an 0.5 $M_\odot$ star in 3×10^5 yr? Is it strongly decreasing as a function of time, as assumed in the relatively flat (in luminosity) tracks shown in Fig. 3.8? Is it episodic, with low accretion rates for much of the protostellar lifetime, punctuated by short bursts of very rapid accretion? The latter scenario may be necessary to explain the wide range of luminosities observed for protostars. It also could account for the sudden brightening associated with the FU Orionis event. Note that the accretion rate onto the disk can be fairly steady, but the accretion from disk to star can be irregular, perhaps as a result of gravitational instabilities; the latter rate is what primarily determines the luminosity. Alternatively, the turbulent nature of the inflow onto protostars could also result in fluctuations in the accretion rate onto the disk.

Theoretical calculations of collapse in a spherical or pseudo-spherical approximation, which provide accretion rates as a function of time, tend to show an increase in luminosity for a given protostar as a function of time. The luminosity L is approximately $GM_{\text{core}}\dot{M}/R_{\text{core}}$, and if $\dot{M}$ is approximately constant, as it would be in the $\rho \propto r^{-2}$ distribution set up during the isothermal phase of protostar collapse, then M_{core} increases with time, R_{core} remains the same, and L increases. This conclusion is not in agreement with observations, which show a large spread but a relatively constant average L in Class I protostars as a function of time. Also the actual average values of the luminosities observed in the Class I phase are considerably less than those predicted by theoretical tracks; this inconsistency is known as the 'luminosity problem'. However, full two-dimensional simulations of the protostar collapse, taking into account gravitational instability in the disk, do show a decreasing accretion rate from disk to star as a function of time. Furthermore, radiative transfer calculations in connection with such models show that the observable luminosity of a protostar varies considerably, by an order of magnitude or more, depending on the viewing angle with respect to the rotation axis and the wavelength of observation. This effect could explain at least part of the luminosity spread. Furthermore, protostellar outflows could modify the picture somewhat if they interact with and remove some of the infalling matter during the later stages of collapse. They also intensify the beaming effect of radiation from the protostar in the polar direction.

9.4 Binary and Multiple Systems

Once a core does form in a molecular cloud, under what conditions will it produce a single star, and under what conditions will it produce a multiple system? How does the outcome depend on the mass of the core? Theoretical and computational investigation of this question is needed to explain why the binary/multiple star fraction is much higher for high-mass stars than for stars below solar mass. Preliminary numerical simulations of cluster formation do demonstrate the latter property, in part because low-mass binaries are likely to be disrupted by gravitational encounters. It is possible that all stars formed in binaries or multiples, but the dissolution rate depends on mass. Stars formed in isolation should not show this effect; however most stars are believed to have formed in clusters.

The main binary formation mechanisms are fragmentation, capture, and fission. Fragmentation can occur either in a collapsing cloud or in a disk that becomes gravitationally unstable. A number of arguments indicate that fission is unlikely, but if it did work it could explain the closest observed binaries. Capture, which requires some mechanism of energy dissipation, is unlikely in the Galactic disk under present conditions, but it could play a role as a consequence of gravitational interactions in a collapsing, fragmenting cloud that is forming a cluster. Fragmentation, in both forms mentioned above, is likely to be the dominant mechanism. Fragmentation in a collapsing cluster cloud, or in the collapse of an individual protostar, is likely to produce a wide binary with a separation of 100–1,000 AU. Disk fragmentation will occur in the outer region of a disk, where cooling is effective; thus the binary separation is also likely to be wide.

In connection with fragmentation, it remains to be explained in detail what the primary mechanism is for brown dwarf formation. Although brown dwarfs are thought to have formed in the same way as stars, it is not clear how likely it is to form them by direct collapse of a molecular cloud core, which would have to be much denser and lower in mass than the typical observed case. Is fragmentation the primary mechanism, either in the protostar collapse or via disk fragmentation? If the latter, do the brown dwarfs retain their low mass by ejection from the system soon after formation?

The formation of close binaries, especially the ones with periods of only a few days and separations of only a small fraction of an AU, is particularly difficult to explain. The most likely process involves transferring angular momentum out of the orbit of a wide binary. The angular momentum could be transferred to a nearby third object in a gravitationally interacting system or to a circumbinary disk. Alternatively, the initial binary system could simply accrete material of low angular momentum from the surroundings. Nevertheless, periods of only a few days are still difficult to explain. An attractive possibility is the so-called Kozai mechanism. In a typical triple-star system, a pair of stars in a relatively close orbit has a companion in a wide orbit. Such systems are fairly common, with Polaris and α Centauri being familiar examples. If the inclination of the outer orbit to the inner orbit is greater than 39°, a combination of gravitational and tidal effects transfers angular momentum out of the inner orbit and allows that system to become a very close binary. The time scale for the process increases with the separation of the wide companion. Nevertheless, the quantitative explanation of the full observed range of binary periods, as well as the full range of mass ratios, has not been accomplished.

9.5 Star Formation in the Early Universe

As telescopes become more powerful and able to discern events occurring in the very early history of the universe, it becomes important to understand how the first stars form and to determine whether their IMF is significantly different from that of stars in our galaxy. On the one hand, if the first stars are very massive, they could be very luminous and even detectable by the most sensitive telescopes. Furthermore their final collapse could lead to the formation of black holes in the range $10^3 - 10^5$ $M_\odot$. Later mergers of these objects or rapid gas accretion onto them could explain the black holes of $\sim 10^9$ $M_\odot$ which are believed to have formed in the early Universe. On the other hand, if the IMF of the first stars is similar to that observed today, then the supernovae in the $10 M_\odot$ range would produce element abundances in agreement with those observed in the oldest and most metal-poor stars observed in the Galaxy. If the first stars become supernovae in the mass range 140–260 $M_\odot$, they do not produce element abundances in the observed range. In other mass ranges above $100 M_\odot$ it is not clear what the outcome is, but in many cases they just collapse to black holes without ejecting any synthesized elements at all. A major unsolved problem in first-star formation is to explain the observed abundance pattern in the oldest stars, and at the same time account for the massive black holes deduced to have formed in the early Universe. Most current theories for the formation of the first stars indicate that the typical star has over $100 M_\odot$, and that the IMF is quite distinct from that observed today.

The conditions for the formation of the first stars are different from those for stars forming at the present time. Cosmological evolution of very small density fluctuations in the very early universe leads to the formation of relatively dense halos, composed of about 85% dark matter and 15% ordinary matter (baryons) by mass. Once the mass of a halo reaches about $10^6 \, M_\odot$, conditions are suitable for forming stars. The baryons settle to the center of the halo, cool by radiation from H_2 molecules, and reach their Jeans mass of about $1,000 \, M_\odot$ at $n \approx 10^4 \, \text{cm}^{-3}$ and $T = 200$ K. Once the collapse starts, the major question is whether the cloud will fragment or not. The halo has sufficient angular momentum so that the cloud could form a binary system, or in fact break up into a whole system of fragments. It remains to be tested by detailed numerical simulations whether fragmentation is a frequent or rare occurrence and what the detailed outcome is. If it does occur, the IMF, still unknown, could still be different from that in the Galaxy at present. If numerous small fragments are produced, stars in the $10 \, M_\odot$ range or less would likely be present. If a few fragments formed or if there were a significant number of mergers, the first stars could still have masses in the $100 \, M_\odot$ range.

If fragmentation doesn't occur, as suggested by the earliest simulations of the formation of the first stars, the entire $1,000 \, M_\odot$, or even more mass, has the potential to accrete onto the first protostellar equilibrium core, once it forms. The accretion rate is high, about $10^{-3} \, M_\odot \, \text{yr}^{-1}$. Taking into account angular momentum and disk formation, it has been shown in this case that the equilibrium stellar core reaches the main sequence and starts nuclear burning at about $100 \, M_\odot$. However by this time the surface of the star is quite hot and is radiating a substantial number of UV photons. The formation of an HII region in the infalling material around the star reduces the rate of infall and eventually shuts it off entirely. The resulting mass is in the range $100-200 \, M_\odot$. The IMF still remains to be determined, but it is likely to peak at near $150 \, M_\odot$.

Another channel of first-star formation is possible if additional physics is brought in, namely dark-matter annihilation. The nature of dark matter is unknown, but one of the prime candidates is the neutralino, a particle with mass of order 100 proton masses which is its own antiparticle. Thus if two such particles collide, they annihilate, producing photons, positrons, and electrons, which heat the gas, and neutrinos, which escape. In the primordial halos, the settling of the baryonic matter to the center increases the concentration of dark matter toward the center as well, through a process known as adiabatic contraction. As a result, the dark-matter density in the forming star is thought to be high enough so that the heating from annihilations can be a significant effect. Once the collapsing protostar reaches a density of about $10^{11} \, \text{cm}^{-3}$, the photons and particles are trapped inside the protostar and the additional heating begins to take effect. The equilibrium star remains larger, and with cooler surface temperature, than the analog without dark-matter heating. As the star gains mass its surface temperature remains in the range 5,000–15,000 K, and as a result the feedback effects from the formation of an HII region do not come into play. The mass of the star can increase well beyond $150 \, M_\odot$. The final mass depends on when the supply of dark matter inside the star runs out, an unknown at the present time. Once it does, the star has to contract to the main sequence, which it reaches with surface temperatures above 100,000 K, and the feedback effect shuts off accretion. This final mass can be anywhere in the range $1{,}000\text{--}10^5\,M_\odot$. In this scenario, eventual massive black hole formation is possible. The IMF would result from the differing properties of the various dark-matter halos, assuming that only one star forms per halo. The implications of this scenario on the primordial nucleosynthesis remains to be explored.

9.6 Massive Stars

Although substantial production of very massive stars is quite possible in the early universe, at the present time in our galaxy they are quite rare but very significant. Their feedback into the molecular cloud material self-limits star formation. Their ionizing photons create HII regions, which are hot bubbles in the molecular cloud, causing expansion and eventual loss of the material in the cloud. Because of this effect, reinforced by supernova explosions and stellar winds, most molecular clouds dissipate after only a few percent of their material has been converted into stars. The details of why this number is so low is not well understood theoretically. The HII regions also contribute to the maintenance of turbulence, which on larger scales tends to suppress star formation, while on small scales encouraging it through shock compression. Also the HII region can initiate star formation in the dense shell of swept-up material outside it.

In the past a major difficulty in the formation of massive stars was identified, in the form of feedback of the radiation pressure from the central star on the dust in the infalling envelope. It was considered difficult to form a star even with a mass as low as $6\text{--}7\,M_\odot$. However several different mechanisms have been suggested to suppress this effect. First, the accretion rate onto massive stars is high, around $10^{-3}\,M_\odot\,\text{yr}^{-1}$, and the infall momentum can counter the effect of radiation pressure to some extent. Second, much of the accretion onto the massive star occurs through a disk, which can shield some material from the effects of radiation pressure. Third, various instabilities, found through three-dimensional simulations of massive star formation, can allow dense fingers of material to accrete with radiation escaping between them. Simulations, which involve various degrees of approximation, have produced stars up to about $50\,M_\odot$. However stars more massive than that exist, and it is not clear what determines the upper mass limit to stars.

Another fact about massive stars is that they are predominantly in binary or multiple systems. A relatively large fraction of them are found in spectroscopic binaries with short periods and with the masses comparable or even nearly equal. The most massive object whose mass has been accurately dynamically measured (WR20a) is an equal-mass short-period binary with a total mass of $165\,M_\odot$. (There undoubtedly exist stars with still higher masses; see Sect. 5.1.1). It remains to be explained how the components build up to $80\,M_\odot$ each, in the face of the radiation pressure from the combined masses, and how the binary separation decreases from its initial value, after fragmentation in a circumstellar disk, of perhaps tens of AU down to only a few stellar radii. This example indicates that there are many unsolved problems in massive star formation.

The origin of the WR20a system could have been a single turbulent molecular cloud core, whose total mass, allowing for an (uncertain) efficiency factor of 1/3 resulting mainly from the protostellar outflow, would have to be close to $500 M_\odot$. One might expect a core this massive, once the turbulence decays and collapse sets in, to fragment into many pieces. However, numerical simulations suggest that radiative feedback from the earliest-forming fragments can heat the gas and prevent further fragmentation, provided that the surface density of the molecular cloud core were high enough. A magnetic field could also suppress fragmentation. Also, some of the fragments could later merge. However, the further question arises as to how such a core could have formed in the first place.

The main alternative to formation from a single isolated core is formation in a cluster environment, a reasonable possibility since most massive stars are found in clusters. The central parts of a protocluster cloud form a natural environment for the formation of a massive star, since much of the mass of the cloud will tend to collapse into that region. In fact massive stars are often found near the centers of clusters, but it is still an ongoing debate whether they actually formed there or whether mass segregation occurred, with massive stars preferentially falling toward the center, after the cluster formed. A further question arises regarding how the feedback from the massive star, in the form of outflows and radiation, affects the properties and the mass function of the lower-mass stars forming in the cluster. These questions concerning massive star formation will need to be clarified through further detailed observational and theoretical studies.

9.7 Young Stars and Disks

The end product of star formation in the range 0.2–$2 M_\odot$ is a classical T Tauri star. Below $0.2 M_\odot$ young stars also display T Tauri-like characteristics. Above about $2 M_\odot$ young stars are known as Herbig Ae–Be stars, also with T Tauri-like characteristics. Above a mass in the range 7–$10 M_\odot$ there are no T-Tauri like stars. The objects pass through the pre-main-sequence phase while still hidden from view in the optical spectrum by their infalling circumstellar envelopes. The hallmarks of T Tauri stars are strong Hα in emission, usually other strong emission lines, irregular variability, lithium abundances characteristic of the interstellar medium and those observed in meteorites, as well as infrared and ultraviolet excess. These objects are still accreting mass from a disk, at the relatively slow rate of $10^{-8} M_\odot \text{ yr}^{-1}$ or less, and at the same time they are ejecting mass in a stellar wind as well as in a protostellar outflow. Among the issues associated with the T Tauri stars are the following: how does the rotational period of the star evolve, as influenced by disk accretion, angular momentum loss in the wind, and contraction? How are the winds and outflows generated? Why do the masses of a number of young stars as determined from evolutionary tracks differ noticeably from the masses derived from

9.7 Young Stars and Disks

dynamical considerations, for example in eclipsing binaries? What is the spread in ages of stars in a young cluster? What physical process determines the effective viscosity in the disk that allows material to accrete onto the star? The disks are of relatively low mass compared with the stellar mass, so the known mechanism of gravitational instability is unlikely to operate here. Additionally what causes disks to dissipate rather suddenly, and what causes the large spread in lifetimes (1–10 Myr) of disks around different stars?

In particular, the accretion mechanism, which must transfer angular momentum outward to allow mass flow inward, is not understood. It is usually parameterized by an effective viscosity $\alpha_{\text{visc}} c_s H$, where α_{visc} is a constant, c_s is the sound speed, and H is the pressure scale height of the disk. Observed accretion rates can be reproduced in model disks if $\alpha_{\text{visc}} = 10^{-2} - 10^{-3}$. However the actual form of the effective viscosity is probably different from that assumed, and α_{visc} is probably not a constant as a function of position in the disk. Possible mechanisms divide into those controlled by pure hydrodynamics and those by magnetohydrodynamics. Strong doubts have been raised over whether any of the hydrodynamic processes, including turbulent convection generated in the vertical structure, shear instabilities, or baroclinic instabilities, can do the job; however research is continuing on these complicated processes. The magnetorotational instability (MRI) has been shown to amplify a weak field and to generate turbulence with the right properties to allow accretion. However it also has the problem that an initial magnetic field structure has to be assumed, and it is not clear that real disks have this structure. Also, there are regions in disks, most notably between about 1 and 10 AU from the star, where the ionization is insufficient to couple the field to the matter, and the MRI doesn't work. In a closely related problem, the masses of disks are not well determined observationally, as the method depends on dust opacity, where the properties of the dust are not well known and are likely to vary in time and with position in the disk.

The regions of disks interior to about 1 AU are crucial to the understanding of the workings of young stars. On this size scale lie the transitions from the dust-rich optically thick outer regions to the region where the dust has mainly evaporated and the optical depth is much reduced. Here occurs the complicated physical interaction that determines the nature of the accretion flow onto the star and the generation of the outflow. The physics of radiation transfer and magnetohydrodynamics, as well as the chemistry, need to be studied in more detail in this context, and future observations, which will solve the difficult problem of resolving these inner regions, will provide the empirical tests of the models which are developed.

The FU Orionis-type outbursts still present a major problem in interpretation. The various observed objects show differing times to reach maximum light and differing decline times. It remains to be clarified what the time interval between recurring events is, and how these outbursts affect the dynamics of young-star outflows. The basic mechanism is thought to be a sudden rise of the accretion rate of a disk onto its central star, with a peak value of about $10^{-4} \, M_\odot \, \text{yr}^{-1}$. This event could be induced by a gravitational encounter with a nearby star, a thermal instability in the very inner regions of the disk where hydrogen ionization is occurring, a gravitational instability in the disk, or a combination of gravitational and magnetohydrodynamic effects. The event is probably associated with the phase of evolution when accretion from the molecular cloud core onto the disk is still occurring, toward the end of the protostar phase. Further theoretical calculations are required to clarify the mechanism and provide a more detailed comparison with observations.

9.8 Rate and Efficiency

The overall star formation rate in the Galaxy as a whole is only a few solar masses per year, low compared with the available mass in molecular clouds divided by their free-fall time. Molecular clouds are known to be supported against collapse by a combination of turbulence and magnetic fields; thus the key to maintaining a low star formation rate is the regeneration of turbulence, which, if left to itself, would decay on relatively short time scales. There are several known mechanisms for regenerating turbulence (Sect. 9.1), but the problem is to show that they are adequate for maintaining the overall level of turbulence in molecular clouds for the entire lifetime of the clouds. Qualitatively then the star formation rate depends on a number of factors: the rate at which turbulence-induced shocks generate regions dense enough to go into collapse, the rate at which material in magnetically dominated regions can slip past the magnetic field to the point of collapse, and the rate at which violent events, such as supernovae or expanding HII regions, can induce collapse. The overall star formation rate in the Galaxy must also be controlled to some extent by the rate at which gas clouds pass through spiral arms, whose shock compression can induce molecular cloud formation behind the shocks, as well as the rate at which gas clouds collide, which also can lead to the formation of molecular clouds.

Star formation efficiency has three different aspects. The first applies to the collapse of an individual cloud core and is defined as the fraction of the mass in the core (or region) that ends up in stars. For the individual cores, assuming that the mass function of the cores determines the IMF of stars, this factor is $\sim 1/3$. It is roughly understood how bipolar flows from protostars can produce this result, but the uncertainties are large, and it is not clear why the factor 1/3 should be universal over the full mass range. The dissipation of circumstellar or circumbinary disks by photoevaporation could also play a role. The second aspect applies to cluster formation from a relatively large unit of molecular cloud material. The observed efficiency in young clusters is $\sim 5 - 30\%$. However these measurements only give a snapshot of the fraction of the material in stars at some arbitrary time in the star-forming history of the cluster. The overall efficiency, after star formation in the cluster is complete, is not known. The same mechanisms that limit the efficiency in individual cores no doubt apply to the cluster situation as well; furthermore once massive stars form in the cluster, the heating of the residual gas by the HII regions will sweep the gas away. The third aspect is the global efficiency, applying to all of the molecular clouds in the galaxy, defined as the fraction of the mass of a molecular cloud that is converted into stars during the lifetime of the cloud.

9.8 Rate and Efficiency

This number, probably less than 5%, is uncertain for the most part because the lifetime of the clouds is not well determined. A revised definition of efficiency, namely the efficiency per free-fall time, is a more useful quantity for comparing theory and observations, first, because it is measurable, and second because many numerical simulations run for only 1–2 free-fall times. Again observed values are low, of the order of 1%, for molecular cloud material up to densities of $10^5\,\mathrm{cm}^{-3}$, pointing to the fact that many numerical simulations have produced too high an efficiency and require additional physics to improve the situation. Again, in general the overall efficiency is low because of magnetic and turbulent inhibition of star formation, but also because of the finite lifetimes of molecular clouds, limited by the feedback effects of HII regions and supernovae, generated by star formation itself.

Clearly the full understanding of the star formation rate, including its dependence on density, and the star formation efficiency require considerable future research. However these quantities are useful tools for investigating the evolution of galaxies, when the fine details of the star formation process do not have to be taken into account.

References

1. T. Abel, G. Bryan, M. Norman, Science **295**, 93 (2002)
2. H.A. Abt, Annu. Rev. Astron. Astrophys. **21**, 343 (1983)
3. F.C. Adams, ApJ **542**, 964 (2000)
4. F.C. Adams, W. Benz, in *Complementary Approaches to Double and Multiple Star Research*, ed. by H. McAlister, W. Hartkopf (Astron. Soc. of the Pacific, San Francisco, 1992), p. 185
5. F.C. Adams, J.P. Emerson, G.A. Fuller, ApJ **357**, 606 (1990)
6. F.C. Adams, D. Hollenbach, G. Laughlin, U. Gorti, ApJ **611**, 360 (2004)
7. F.C. Adams, C.J. Lada, F.H. Shu, ApJ **312**, 788 (1987)
8. F.C. Adams, C.J. Lada, F.H. Shu, ApJ **326**, 865 (1988)
9. F.C. Adams, S. Ruden, F.H. Shu, ApJ **347**, 959 (1989)
10. F.C. Adams, F.H. Shu, ApJ **308**, 836 (1986)
11. R.G. Aitkin, *The Binary Stars* (McGraw-Hill, New York, 1935), p. 275
12. R.D. Alexander, C.J. Clarke, J.E. Pringle, MNRAS **369**, 229 (2006)
13. J.F. Alves, C.J. Lada, E.A. Lada, ApJ **515**, 265 (2001)
14. J.F. Alves, C.J. Lada, E.A. Lada, Nature **409**, 159 (2001)
15. J.F. Alves, M. Lombardi, C.J. Lada, Astron. Astrophys. **462**, L17 (2007)
16. V.A. Ambartsumian, *Stellar Evolution and Astrophysics* (Acad. Sci. Armenian SSR, Yerevan, 1947)
17. P. André, T. Montmerle, ApJ **420**, 837 (1994)
18. P. André, D. Ward-Thompson, M. Barsony, ApJ **406**, 122 (1993)
19. P. André, D. Ward-Thompson, M. Barsony, in *Protostars and Planets IV*, ed. by V. Mannings, A.P. Boss, S.S. Russell (University of Arizona Press, Tucson, 2000), p. 59
20. P. André, D. Ward-Thompson, F. Motte, Astron. Astrophys. **314**, 625 (1996)
21. S.M. Andrews, J.P. Williams, ApJ **631**, 1134 (2005)
22. H.G. Arce, D. Shepherd, F. Gueth et al., in *Protostars and Planets V*, ed. by B. Reipurth, D. Jewitt, K. Keil (University of Arizona Press, Tucson, 2007), p. 245
23. P.J. Armitage, M. Livio, J.E. Pringle, MNRAS **324**, 705 (2001)
24. A. Bacmann, P. André, J.-L. Puget, A. Abergel, S. Bontemps, D. Ward-Thompson, Astron. Astrophys. **361**, 555 (2000)
25. S.A. Balbus, J. Hawley, ApJ **376**, 214 (1991)
26. S.A. Balbus, J. Hawley, J. Stone, ApJ **467**, 76 (1996)
27. S.A. Balbus, J.C.B. Papaloizou, ApJ **521**, 650 (1999)
28. J. Ballesteros-Paredes, L. Hartmann, Rev. Mex. Astron. Astrofis. **43**, 123 (2007)
29. J. Ballesteros-Paredes, R. Klessen, M.-M. Mac Low, E. Vázquez-Semadeni, in *Protostars and Planets V*, ed. by B. Reipurth, D. Jewitt, K. Keil (University of Arizona Press, Tucson, 2007), p. 63
30. I. Baraffe, G. Chabrier, F. Allard, P. Hauschildt, Astron. Astrophys. **337**, 403 (1998)

31. I. Baraffe, G. Chabrier, F. Allard, P. Hauschildt, Astron. Astrophys. **382**, 563 (2002)
32. I. Baraffe, G. Chabrier, T.S. Barman, F. Allard, P. Hauschildt, Astron. Astrophys. **402**, 701 (2003)
33. G. Basri, in *Star-Disk Interaction in Young Stars*, ed. by J. Bouvier, I. Appenzeller (Cambridge University Press, Cambridge, 2007), p. 13
34. G. Basri, G.W. Marcy, J.A. Valenti, ApJ **390**, 622 (1992)
35. S. Basu, T. Mouschovias, ApJ **432**, 720 (1994)
36. M.R. Bate, ApJ **508**, L95 (1998)
37. M.R. Bate, MNRAS **392**, 590 (2009)
38. M.R. Bate, MNRAS **392**, 1363 (2009)
39. M.R. Bate, I.A. Bonnell, V. Bromm, MNRAS **336**, 705 (2002)
40. M.R. Bate, I.A. Bonnell, V. Bromm, MNRAS **339**, 577 (2003)
41. M.R. Bate, I.A. Bonnell, N. Price, MNRAS **277**, 362 (1995)
42. M.R. Bate, A. Burkert, MNRAS **288**, 1060 (1997)
43. S.V.W. Beckwith, in *The Origin of Stars and Planetary Systems*, ed. by C.J. Lada, N.D. Kylafis (Kluwer, Dordrecht, 1999), p. 579
44. S.V.W. Beckwith, A.I. Sargent, in *Protostars and Planets III*, ed. by E.H. Levy, J.I. Lunine (University of Arizona Press, Tucson, 1993), p. 521
45. S.V.W. Beckwith, A.I. Sargent, R. Chini, R. Güsten, Astron. J. **99**, 924 (1990)
46. C.A. Beichman, P.C. Myers, J.P. Emerson, S. Harris, R.D. Mathieu, P.J. Benson, R.E. Jennings, ApJ **307**, 337 (1986)
47. K.R. Bell, P.M. Cassen, J.T. Wasson, D.S. Woolum, in *Protostars and Planets IV*, ed. by V. Mannings, A.P. Boss, S.S. Russell (University of Arizona Press, Tucson, 2000), p. 897
48. K.R. Bell, D.N.C. Lin, L.W. Hartmann, S.J. Kenyon, ApJ **444**, 376 (1995)
49. P.J. Benson, P.C. Myers, ApJ Suppl. **71**, 89 (1989)
50. M.J. Berger, J. Oliger, J. Comput. Phys. **53**, 484 (1984)
51. C. Bertout, Annu. Rev. Astron. Astrophys. **27**, 351 (1989)
52. H. Beuther, E.B. Churchwell, C.F. McKee, J.C. Tan, in *Protostars and Planets V*, ed. by B. Reipurth, D. Jewitt, K. Keil (University of Arizona Press, Tucson, 2007), p. 165
53. H. Beuther, P. Schilke, Science **303**, 1167 (2004)
54. J.H. Bieging, M. Cohen, ApJ **289**, L5 (1985)
55. A. Blaauw, in *The Physics of Star Formation and Early Stellar Evolution*, ed. by C.J. Lada, N.D. Kylafis (Kluwer, Dordrecht, 1991), p. 125
56. D.C. Black, P. Bodenheimer, ApJ **199**, 619 (1975)
57. D.C. Black, E.H. Scott, ApJ **263**, 696 (1982)
58. R.D. Blandford, D.G. Payne, MNRAS **199**, 883 (1982)
59. L. Blitz, Y. Fukui, A. Kawamura, A. Leroy, N. Mizuno, E. Rosolowsky, in *Protostars and Planets V*, ed. by B. Reipurth, D. Jewitt, K. Keil (University of Arizona Press, Tucson, 2007), p. 81
60. G.R. Blumenthal, S.M. Faber, R. Flores, J.R. Primack, ApJ **301**, 27 (1986)
61. A.F. Boden, A.I. Sargent, R.L. Akeson, J.M. Carpenter, G. Torres et al., ApJ **635**, 442 (2005)
62. P. Bodenheimer, ApJ **224**, 488 (1978)
63. P. Bodenheimer, in *The Formation and Evolution of Planetary Systems*, ed. by H.A. Weaver, L. Danly (Cambridge University Press, Cambridge, 1989), p. 243
64. P. Bodenheimer, Annu. Rev. Astron. Astrophys. **33**, 199 (1995)
65. P. Bodenheimer, A. Burkert, R.I. Klein, A.P. Boss, in *Protostars and Planets IV*, ed. by V. Mannings, A.P. Boss, S.S. Russell (University of Arizona Press, Tucson, 2000), p. 675
66. P. Bodenheimer, G.P. Laughlin, M. Różyczka, H.W. Yorke, *Numerical Methods in Astrophysics: An Introduction* (Taylor and Francis, New York, 2007)
67. P. Bodenheimer, D.N.C. Lin, Annu. Rev. Earth Planet. Sci. **30**, 113 (2002)
68. P. Bodenheimer, A. Sweigart, ApJ **152**, 515 (1968)
69. P. Bodenheimer, J.E. Tohline, D.C. Black, ApJ **242**, 209 (1980)
70. R.C. Bohlin, B.D. Savage, J.F. Drake, ApJ **224**, 132 (1978)
71. E. Böhm-Vitense, Zeitschr. f. Astrophys. **46**, 108 (1958)

References

72. B.J. Bok, Publ. Astron. Soc. Pacific **89**, 597 (1977)
73. W. Boland, T. de Jong, Astron. Astrophys. **134**, 87 (1984)
74. A.C. Boley, A.C. Mejía, R.H. Durisen, K. Cai, M.K. Pickett, P. D'Alessio, ApJ **651**, 517 (2006)
75. A.Z. Bonanos, K.Z. Stanek, A. Udalski, L. Wyrzykowski, K. Żebruń et al., ApJ **611**, L33 (2004)
76. H. Bondi, MNRAS **112**, 195 (1952)
77. I.A. Bonnell, M.R. Bate, MNRAS **336**, 659 (2002)
78. I.A. Bonnell, M.R. Bate, MNRAS **362**, 915 (2005)
79. I.A. Bonnell, M.R. Bate, S.G. Vine, MNRAS **343**, 413 (2003)
80. I.A. Bonnell, C.L. Dobbs, T.P. Robitaille, J.E. Pringle, MNRAS **365**, 37 (2006)
81. I.A. Bonnell, R.B. Larson, H. Zinnecker, in *Protostars and Planets V*, ed. by B. Reipurth, D. Jewitt, K. Keil (University of Arizona Press, Tucson, 2007), p. 149
82. S. Bontemps, P. André, S. Terebey, S. Cabrit, Astron. Astrophys. **311**, 858 (1996)
83. A.P. Boss, ApJ Suppl. **62**, 519 (1986)
84. A.P. Boss, Comments Astrophys. **12**, 169 (1988)
85. A.P. Boss, ApJ **346**, 336 (1989)
86. A.P. Boss, ApJ **439**, 224 (1995)
87. A.P. Boss, R.T. Fisher, R.I. Klein, C.F. McKee, ApJ **528**, 325 (2000)
88. A.P. Boss, S.I. Ipatov, S.A. Keiser, E.A. Myhill, H.A.T. Vanhala, ApJ **686**, L119 (2008)
89. A.P. Boss, S.A. Keiser, S.I. Ipatov, E.A. Myhill, H.A.T. Vanhala, ApJ **708**, 1268 (2010)
90. J. Bouvier, in *Star-Disk Interaction in Young Stars*, ed. by J. Bouvier, I. Appenzeller (Cambridge University Press, Cambridge, 2007), p. 231
91. C. Briceño, K.L. Luhman, L. Hartmann, J.R. Stauffer, J.D. Kirkpatrick, ApJ **580**, 317 (2002)
92. V. Bromm, R.B. Larson, Annu. Rev. Astron. Astrophys. **42**, 79 (2004)
93. A. Burkert, ApJ **447**, L25 (1995)
94. A. Burkert, M.R. Bate, P. Bodenheimer, MNRAS **289**, 497 (1997)
95. A. Burkert, P. Bodenheimer, ApJ **543**, 822 (2000)
96. A. Burrows, M. Marley, W.B. Hubbard et al., ApJ **491**, 856 (1997)
97. C.J. Burrows, K.R. Stapelfeldt, A.M. Watson, J.E. Krist, G.E. Ballester et al., ApJ **473**, 437 (1996)
98. F. Bürzle, P.C. Clark, F. Stasyszyn et al., MNRAS **412**, 171 (2011) arXiv:1008.3790 (2010)
99. N. Calvet, C. Briceño, J. Hernández et al., Astron. J. **129**, 935 (2005)
100. N. Calvet, P. D'Alessio, L. Hartmann, D. Wilner, A. Walsh, M. Sitko, ApJ **568**, 1008 (2002)
101. N. Calvet, P. D'Alessio, D.M. Watson, R. Franco-Hernandez, E. Furlan et al., ApJ **630**, L185 (2005)
102. N. Calvet, L. Hartmann, S.E. Strom, in *Protostars and Planets IV*, ed. by V. Mannings, A.P. Boss, S. Russell (University of Arizona Press, Tucson, 2000), p. 377
103. D. Calzetti, S.-Y. Wu, S. Hong, R.C. Kennicutt et al., ApJ **714**, 1256 (2010)
104. A.G.W. Cameron, J.W. Truran, Icarus **30**, 447 (1977)
105. R. Cesaroni, R. Neri, L. Olmi, L. Testi, C.M. Walmsley, P. Hofner, Astron. Astrophys. **434**, 1039 (2005)
106. S.-H. Cha, A.P. Whitworth, MNRAS **340**, 91 (2003)
107. G. Chabrier, Publ. Astron. Soc. Pacific **115**, 763 (2003)
108. G. Chabrier, in *The Initial Mass Function 50 years Later*, ed. by E. Corbelli, F. Palle, H. Zinnecker (Springer, Dordrecht, 2005), p. 41
109. G. Chabrier, I. Baraffe, F. Allard, P.H. Hauschildt, ApJ **542**, 464 (2000)
110. S. Chandrasekhar, *An Introduction to the Study of Stellar Structure* (University of Chicago Press, Chicago, 1939)
111. E.I. Chiang, P. Goldreich, ApJ **490**, 368 (1997)
112. E.I. Chiang, P. Goldreich, ApJ **519**, 279 (1999)
113. E.I. Chiang, M.K. Joung, M.J. Creech-Eakman, C. Qi, J.E. Kessler, G.A. Blake, E.F. van Dishoeck, ApJ **547**, 1077 (2001)
114. N. Christlieb, M.S. Bessell, T.C. Beers et al., Nature **419**, 904 (2002)

115. G.E. Ciolek, S. Basu, ApJ **547**, 272 (2001)
116. P.C. Clark, S. Glover, R.S. Klessen, ApJ **672**, 757 (2008)
117. C.J. Clarke, in *Binary Stars as Critical Tools and Tests in Contemporary Astrophysics*, ed. by W.I. Hartkopf, E.F. Guinan, P. Harmanec (Cambridge University Press, Cambridge, UK, 2007), p. 337
118. C.J. Clarke, J.E. Pringle, MNRAS **249**, 584 (1991)
119. D. Clayton, *Principles of Stellar Evolution and Nucleosynthesis* (McGraw-Hill, New York, 1968)
120. L.M. Close, B. Zuckerman, I. Song, T. Barman et al., ApJ **660**, 1492 (2007)
121. R. Courant, K.O. Friedrichs, H. Lewy, Mathematische Annalen **100**, 32 (1928)
122. E. Covino, S. Catalano, A. Frasca et al., Astron. Astrophys. **361**, L49 (2000)
123. E. Covino, A. Frasca, J.M. Alcalá, R. Paladino, M.F. Sterzik, Astron. Astrophys. **427**, 637 (2004)
124. P.A. Crowther, O. Schnurr, R. Hirschi, N. Yusof et al., MNRAS **408**, 731 (2010)
125. R.M. Crutcher, ApJ **520**, 706 (1999)
126. R.M. Crutcher, B. Wandelt, C. Heiles, E. Falgarone, T. Troland, ApJ **725**, 466 (2010)
127. K. Cudworth, G.H. Herbig, Astron. J. **84**, 548 (1979)
128. S.E. Dahm, in *Handbook of Star Forming Regions Vol. I*, ed. by B. Reipurth (Astron. Society of the Pacific, San Francisco, 2008), p. 966
129. S.E. Dahm, T. Simon, Astron. J. **129**, 829 (2005)
130. J.E. Dale, I.A. Bonnell, A.P. Whitworth, MNRAS **375**, 1291 (2007)
131. A. Dalgarno, Proc. Nat. Acad. Sci. **103**, 12269 (2006)
132. F. D'Antona, I. Mazzitelli, Mem. Soc. Astron. Italiana **68**, 807 (1997)
133. L. Deharveng, B. Lefloch, A. Zavagno, J. Caplan, A.P. Whitworth, D. Nadeau, S. Martín, Astron. Astrophys. **408**, L25 (2003)
134. T. de Jong, W. Boland, A. Dalgarno, Astron. Astrophys. **91**, 68 (1980)
135. X. Delfosse, J.-L. Beuzit, L. Marchal, X. Bonfils et al., in *Spectroscopically and Spatially Resolving the Components of Close Binary Stars*, ed. by R.W. Hilditch, H. Hensberge, K. Pavlovski (Astron. Society of the Pacific, San Francisco, 2004), p. 166
136. E.J. Delgado-Donate, C.J. Clarke, M.R. Bate, S.T. Hodgkin, MNRAS **351**, 617 (2004)
137. S.E. Dodson-Robinson, K. Willacy, P. Bodenheimer, N.J. Turner, C.A. Beichman, Icarus **200**, 672 (2009)
138. C. Dominik, J. Blum, J.N. Cuzzi, G. Wurm, in *Protostars and Planets V*, ed. by B. Reipurth, D. Jewitt, K. Keil (University of Arizona Press, Tucson, 2007), p. 783
139. E. Dorfi, Astron. Astrophys. **114**, 151 (1982)
140. B.T. Draine, H.M. Lee, ApJ **285**, 89 (1984)
141. B. Dubrulle, L. Marié, C. Normand, D. Richard, F. Hersant, J.-P. Zahn, Astron. Astrophys. **429**, 1 (2005)
142. G. Duchêne, H. Beust, F. Adjali, Q.M. Konopacky, A.M. Ghez, Astron. Astrophys. **457**, L9 (2006)
143. G. Duchêne, J. Bouvier, S. Bontemps, P. André, F. Motte, Astron. Astrophys. **427**, 651 (2004)
144. G. Duchêne, E. Delgado-Donate, K.E. Haisch Jr., L. Loinard, L.F. Rodríguez, in *Protostars and Planets V*, ed. by B. Reipurth, D. Jewitt, K. Keil (University of Arizona Press, Tucson, 2007), p. 379
145. C.P. Dullemond, D. Hollenbach, I. Kamp, P. D'Alessio, in *Protostars and Planets V*, ed. by B. Reipurth, D. Jewitt, K. Keil (University of Arizona Press, Tucson, 2007), p. 555
146. M. Dunham, N.J. Evans II, S. Terebey, C. Dullemond, C. Young, ApJ **710**, 470 (2010)
147. A. Duquennoy, M. Mayor, Astron. Astrophys. **248**, 485 (1991)
148. R.H. Durisen, R. Gingold, J.E. Tohline, A.P. Boss, ApJ **305**, 281 (1986)
149. A. Dutrey, S. Guilloteau, M. Simon, Astron. Astrophys. **286**, 149 (1994)
150. H.M. Dyck, T. Simon, B. Zuckerman, ApJ **255**, L103 (1982)
151. R. Ebert, S. von Hoerner, S. Temesvary, *Die Entstehung von Sternen durch Kondensation Diffuser Materie* (Springer, Berlin, 1960), pp. 184–324
152. R. Edgar, New Astron. Rev. **48**, 843 (2004)

153. S. Edwards, T. Ray, R. Mundt, in *Protostars and Planets III*, ed. by E.H. Levy, J.I. Lunine (University of Arizona Press, Tucson, 1993), p. 567
154. M.P. Egan, R.F. Shipman, S.D. Price, S.J. Carey, F.O. Clark, M. Cohen, ApJ **494**, L199 (1998)
155. P. Eggleton, L. Kiseleva, ApJ **455**, 640 (1995)
156. B. Elmegreen, ApJ **232**, 729 (1979)
157. B. Elmegreen, in *Star Formation in Stellar Systems*, ed. by G. Tenorio-Tagle, M. Prieto, F. Sánchez (Cambridge University Press, Cambridge, 1992), p. 381
158. B. Elmegreen, ApJ **419**, L29 (1993)
159. B. Elmegreen, in *Origins, ASP Conference Series*, vol. 148, ed. by C.E. Woodward, J.M. Shull, H.A. Thronson (Astron. Soc. of the Pacific, San Francisco, 1998), p. 150
160. B. Elmegreen, in *Proceedings of Fourth Spitzer Science Conference*, ed. by K. Sheth, A. Noriega-Crespo, J. Ingalls, R. Paladini, published online at http://ssc.spitzer.caltech.edu/mtgs/ismevol/ (2009)
161. B. Elmegreen, T. Kimura, M. Tosa, ApJ **451**, 675 (1995)
162. B. Elmegreen, C.J. Lada, ApJ **214**, 725 (1977)
163. M.L. Enoch, N.J. Evans II, A.I. Sargent, J. Glenn, ApJ **692**, 973 (2009)
164. N.J. Evans II, M. Dunham, J. Jorgensen, M. Enoch et al., ApJ Suppl. **181**, 321 (2009)
165. D. Fabrycky, S. Tremaine, ApJ **669**, 1298 (2007)
166. E. Falgarone, T. Troland, R. Crutcher, G. Paubert, Astron. Astrophys. **487**, 247 (2008)
167. C. Fallscheer, H. Beuther, Q. Zhang, E. Keto, T.K. Sridharan, Astron. Astrophys. **504**, 127 (2009)
168. X. Fan, Mem. Soc. Astron. Italiana **77**, 635 (2006)
169. E.D. Feigelson, T. Montmerle, Annu. Rev. Astron. Astrophys. **37**, 363 (1999)
170. F.C. Fekel, ApJ **246**, 879 (1981)
171. R.A. Fiedler, T. Mouschovias, ApJ **415**, 680 (1993)
172. D.F. Figer, Nature **434**, 192 (2005)
173. D.A. Fischer, G.W. Marcy, ApJ **396**, 178 (1992)
174. A. Frebel, W. Aoki, N. Christlieb, H. Ando et al., Nature **434**, 871 (2005)
175. R.S. Freedman, M.S. Marley, K. Lodders, ApJ Suppl. **174**, 504 (2008)
176. D. Forgan, K. Rice, MNRAS **402**, 1349 (2010)
177. K. Freese, C. Ilie, D. Spolyar, M. Valluri, P. Bodenheimer, ApJ **716**, 1397 (2010)
178. D. Froebrich, ApJ Suppl. **156**, 169 (2005)
179. D. Froebrich, S. Schmeja, M.D. Smith, R.S. Klessen, MNRAS **368**, 435 (2006)
180. L. Frommhold, *Collision-Induced Absorption in Gases* (Cambridge University Press, Cambridge, UK, 1994)
181. C.F. Gammie, ApJ **457**, 355 (1996)
182. C.F. Gammie, ApJ **553**, 174 (2001)
183. G. Garay, S. Faúndez, D. Mardones, L. Bronfman, R. Chini, L.-A. Nyman, ApJ **610**, 313 (2004)
184. M.P. Geyer, A. Burkert, MNRAS **323**, 988 (2001)
185. A.E. Glassgold, E.D. Feigelson, T. Montmerle, in *Protostars and Planets IV*, ed. by V. Mannings, A.P. Boss, S.S. Russell (University of Arizona Press, Tucson, 2000), p. 429
186. P. Goldreich, J. Kwan, ApJ **189**, 441 (1974)
187. P. Goldsmith, ApJ **557**, 736 (2001)
188. P. Goldsmith, R. Arquilla, in *Protostars and Planets II*, ed. by D.C. Black, M.S. Matthews (University of Arizona Press, Tucson, 1985), p. 137
189. P. Goldsmith, W.D. Langer, ApJ **222**, 881 (1978)
190. A. Goodman, P. Bastien, F. Ménard, P.C. Myers, ApJ **359**, 363 (1990)
191. A. Goodman, P. Benson, G.A. Fuller, P.C. Myers, ApJ **406**, 528 (1993)
192. S.P. Goodwin, P. Kroupa, A. Goodman, A. Burkert, in *Protostars and Planets V*, ed. by B. Reipurth, D. Jewitt, K. Keil (University of Arizona Press, Tucson, 2007), p. 133
193. J.N. Goswami, H.A.T. Vanhala, in *Protostars and Planets IV*, ed. by V. Mannings, A.P. Boss, S.S. Russell (University of Arizona Press, Tucson, 2000), p. 963
194. T.P. Greene, B.A. Wilking, P. André, E.T. Young, C.J. Lada, ApJ **434**, 614 (1994)

195. E.M. Gregersen, N.J. Evans II, ApJ **538**, 260 (2000)
196. T.H. Greif, V. Bromm, MNRAS **373**, 128 (2006)
197. M. Gritschneder, T. Naab, S. Walch, A. Burkert, F. Heitsch, ApJ **694**, L26 (2009)
198. F. Gueth, S. Guilloteau, Astron. Astrophys. **343**, 571 (1999)
199. K.E. Haisch, E.A. Lada, C.J. Lada, ApJ **553**, L153 (2001)
200. G. Haro, ApJ **115**, 572 (1952)
201. P. Hartigan, S. Edwards, L. Ghandour, ApJ **452**, 736 (1995)
202. L. Hartmann, *Accretion Processes in Star Formation* (Cambridge University Press, Cambridge, UK, 1998)
203. L. Hartmann, ApJ **585**, 398 (2003)
204. L. Hartmann, N. Calvet, E. Gullbring, P. D'Alessio, ApJ **495**, 385 (1998)
205. L. Hartmann, S.J. Kenyon, ApJ **299**, 462 (1985)
206. L. Hartmann, S.J. Kenyon, Annu. Rev. Astron. Astrophys. **34**, 207 (1996)
207. J.F. Hawley, S.A. Balbus, ApJ **376**, 223 (1991)
208. J.F. Hawley, C.F. Gammie, S.A. Balbus, ApJ **440**, 742 (1995)
209. C. Hayashi, Publ. Astron. Soc. Jpn. **13**, 450 (1961)
210. C. Hayashi, Prog. Theor. Phys. Suppl. **70**, 35 (1981)
211. M. Hayashi, N. Ohashi, S. Miyama, ApJ **418**, L71 (1993)
212. A. Heger, S.E. Woosley, ApJ **567**, 532 (2002)
213. F. Heitsch, A. Burkert, L.W. Hartmann, A.D. Slyz, J.E.G. Devriendt, ApJ **633**, L113 (2005)
214. P. Hennebelle, G. Chabrier, ApJ **684**, 395 (2008)
215. P. Hennebelle, G. Chabrier, ApJ **702**, 1428 (2009)
216. L.G. Henyey, R. Lelevier, R.D. Levée, Publ. Astron. Soc. of the Pacific **67**, 154 (1955)
217. G.H. Herbig, ApJ **113**, 697 (1951)
218. G.H. Herbig, ApJ **125**, 612 (1957)
219. G.H. Herbig, Adv. Astron. Astrophys. **1**, 47 (1962)
220. G.H. Herbig, ApJ **214**, 747 (1977)
221. G.H. Herbig, K.R. Bell, *Lick Observatory Bulletin No. 1111* (1988)
222. G.H. Herbig, B. Jones, Astron. J. **86**, 1232 (1981)
223. W. Herbst, J. Eislöffel, R. Mundt, A. Scholz, in *Protostars and Planets V*, ed. by B. Reipurth, D. Jewitt, K. Keil (University of Arizona Press, Tucson, 2007), p. 297
224. M.H. Heyer, C.M. Brunt, ApJ **615**, L45 (2004)
225. R.H. Hildebrand, Quart. J. R. A. S. **24**, 267 (1983)
226. L.A. Hillenbrand, Astron. J. **113**, 1733 (1997)
227. L.A. Hillenbrand, Phys. Script. **130**, 014024 (2008)
228. L.A. Hillenbrand, R.J. White, ApJ **604**, 741 (2004)
229. J.G. Hills, ApJ **235**, 986 (1980)
230. J.R. Hind, MNRAS **24**, 65 (1864)
231. N. Hirano, N. Ohashi, K. Dobashi, H. Shinnaga, M. Hayashi, in *8th Asian-Pacific Regional Meeting*, Vol. II, ed. by S. Ikeuchi, J. Hearnshaw, T. Hanawa (Astron. Soc. Japan, Tokyo, 2002), p. 141
232. D.J. Hollenbach, D. Johnstone, S. Lizano, F.H. Shu, ApJ **428**, 654 (1994)
233. D.J. Hollenbach, H.W. Yorke, D. Johnstone, in *Protostars and Planets IV*, ed. by V. Mannings, A.P. Boss, S.S. Russell (University of Arizona Press, Tucson, 2000), p. 401
234. F. Hoyle, ApJ **118**, 513 (1953)
235. F. Hoyle, R.A. Lyttleton, Proc. Cam. Phil. Soc. **35**, 405 (1939)
236. A.M. Hughes, D.J. Wilner, N. Calvet, P. D'Alessio, M.J. Claussen, M.R. Hogerheijde, ApJ **664**, 536 (2007)
237. A.M. Hughes, S.M. Andrews, C. Espaillat, D.J. Wilner, N. Calvet et al., ApJ **698**, 131 (2009)
238. C.A. Iglesias, F.J. Rogers, ApJ **464**, 943 (1996)
239. A. Isella, J.M. Carpenter, A.I. Sargent, ApJ **701**, 260 (2009)
240. A.-K. Jappsen, R.S. Klessen, R.B. Larson, Y. Li, M.-M. Mac Low, Astron. Astrophys. **435**, 611 (2005)
241. J.H. Jeans, *Astronomy and Cosmogony* (The University Press, Cambridge, UK, 1928)

242. C.M. Johns-Krull, J.A. Valenti, C. Koresko, ApJ **516**, 900 (1999)
243. D. Johnstone, M. Fich, G.F. Mitchell, G. Moriarty-Schieven, ApJ **559**, 307 (2001)
244. J.K. Jorgensen, E.F. van Dishoeck, R. Visser, T.L. Bourke, D.J. Wilner, D. Lommen, M.R. Hogerheijde, P.C. Myers, Astron. Astrophys. **507**, 861 (2009)
245. A.H. Joy, ApJ **102**, 168 (1945)
246. N. Kaifu, T. Hasegawa, M. Morimoto, J. Inatani et al., Astron. Astrophys. **134**, 7 (1984)
247. R.C. Kennicutt, Jr., ApJ **498**, 541 (1998)
248. S.J. Kenyon, N. Calvet, L. Hartmann, ApJ **414**, 676 (1993)
249. S.J. Kenyon, L.W. Hartmann, K.M. Strom, S.E. Strom, Astron. J. **99**, 869 (1990)
250. E. Keto, ApJ **580**, 980 (2002)
251. R. Kippenhahn, A. Weigert, *Stellar Structure and Evolution* (Springer, Berlin, 1990)
252. S. Kitsionas, A.P. Whitworth, MNRAS **378**, 507 (2007)
253. S. Kitsionas, A.P. Whitworth, R.S. Klessen, in *Exoplanets: Detection, Formation, and Dynamics, IAU Symposium No. 249*, ed. by Y.-S Sun, S. Ferraz-Mello, J.-L. Zhou (Cambridge University Press, Cambridge, 2008), p. 271
254. H.H. Klahr, P. Bodenheimer, ApJ **582**, 869 (2003)
255. R.I. Klein, P. Colella, C.F. McKee, in *Evolution of the Interstellar Medium*, ed. by L. Blitz (Astron. Soc. of the Pacific, San Francisco, 1990), p. 117
256. R.I. Klein, C.F. McKee, P. Colella, ApJ **420**, 213 (1994)
257. R.I. Klein, R.W. Whitaker, M.T. Sandford II, in *Protostars and Planets II*, ed. by D.C. Black, M.S. Matthews (University of Arizona Press, Tucson, 1985), p. 340
258. R.S. Klessen, A. Burkert, M.R. Bate, ApJ **501**, L205 (1998)
259. R.S. Klessen, F. Heitsch, M.-M. Mac Low, ApJ **535**, 887 (2000)
260. D. Koerner, A. Sargent, S. Beckwith, ApJ **408**, L93 (1993)
261. R. Köhler, Journ. of Phys.: Conference Series **131**, 012028 (2008)
262. R. Köhler, W. Brandner, in *The Formation of Binary Stars*, ed. by H. Zinnecker, R.D. Mathieu (Astron. Soc. of the Pacific, San Francisco, 2001), p. 147
263. R. Köhler, C. Leinert, Astron. Astrophys. **331**, 977 (1998)
264. R. Köhler, M.G. Petr-Gotzens, M.J. McCaughrean, J. Bouvier, G. Duchêne, A. Quirrenbach, H. Zinnecker, Astron. Astrophys. **458**, 461 (2006)
265. A. Königl, ApJ **370**, L39 (1991)
266. C.D. Koresko, ApJ **531**, L147 (2000)
267. Y. Kozai, Astron. J. **67**, 591 (1962)
268. J.E. Krist, K.R. Stapelfeldt, D.A. Golimowski et al., Astron. J. **130**, 2778 (2005)
269. P. Kroupa, MNRAS **277**, 1507 (1995)
270. P. Kroupa, MNRAS **322**, 231 (2001)
271. P. Kroupa, Science **295**, 82 (2002)
272. P. Kroupa, S. Aarseth, J. Hurley, MNRAS **321**, 699 (2001)
273. M. Krumholz, in *Massive Star Formation: Observations Confront Theory*, ed. by H. Beuther, H. Linz, Th. Henning (Astron. Soc. of the Pacific, San Francisco, 2008), p. 200
274. M. Krumholz, I. Bonnell, in *Structure Formation in Astrophysics*, ed. by G. Chabrier (Cambridge University Press, Cambridge, UK, 2009), p 288
275. M. Krumholz, A.J. Cunningham, R.I. Klein, C.F. McKee, ApJ **713**, 1120 (2010)
276. M. Krumholz, R.I. Klein, C.F. McKee, ApJ **656**, 959 (2007)
277. M. Krumholz, R.I. Klein, C.F. McKee, S.S.R. Offner, A.J. Cunningham, Science **323**, 754 (2009)
278. M. Krumholz, C.F. McKee, ApJ **630**, 250 (2005)
279. M. Krumholz, C.F. McKee, R.I. Klein, ApJ **611**, 399 (2004)
280. M. Krumholz, C.F. McKee, R.I. Klein, ApJ **618**, L33 (2005)
281. M. Krumholz, C.F. McKee, R.I. Klein, Nature **438**, 332 (2005)
282. M. Krumholz, C.F. McKee, J. Tumlinson, ApJ **699**, 850 (2009)
283. M. Krumholz, J.C. Tan, ApJ **654**, 304 (2007)
284. R. Kudritzki, ApJ **577**, 389 (2002)

285. C.J. Lada, in *Star Forming Regions* (IAU Symposium 115), ed. by M. Peimbert, J. Jugaku (Reidel, Dordrecht, 1987), p. 1
286. C.J. Lada, in *The Origin of Stars and Planetary Systems*, ed. by C.J. Lada, N.D. Kylafis: (Kluwer, Dordrecht, 1999), p. 143
287. C.J. Lada, ApJ **640**, L63 (2006)
288. C.J. Lada, E.A. Lada, in *The Formation and Evolution of Star Clusters, ASP Conference Series No. 13*, ed. by K.A. Janes (Astron. Soc. of the Pacific, San Francisco, 1991), p. 3
289. C.J. Lada, E.A. Lada, Annu. Rev. Astron. Astrophys. **41**, 57 (2003)
290. C.J. Lada, M. Margulis, D. Dearborn, ApJ **285**, 141 (1984)
291. C.J. Lada, B. Wilking, ApJ **287**, 610 (1984)
292. E.A. Lada, ApJ **393**, L25 (1992)
293. E.A. Lada, J. Bally, A. Stark, ApJ **368**, 432 (1991)
294. E.A. Lada, K.M. Strom, P.C. Myers, in *Protostars and Planets III*, ed. by E.H. Levy, J.I. Lunine (University of Arizona Press, Tucson, 1993), p. 245
295. E.F. Ladd, F.C. Adams, S. Casey, J.A. Davidson, G.A. Fuller, D.A. Harper, P.C. Myers, R. Padman, ApJ **366**, 203 (1991)
296. A.-M. Lagrange, M. Bonnefoy, G. Chauvin, D. Apai et al., Science **329**, 57 (2010)
297. M.H. Lamm, R. Mundt, C.A.L. Bailer-Jones, W. Herbst, Astron. Astrophys. **430**, 1005 (2005)
298. L.D. Landau, E.M. Lifshitz, *Fluid Mechanics* (Pergamon, London, 1959)
299. R.B. Larson, MNRAS **145**, 271 (1969)
300. R.B. Larson, Fund. Cosmic Phys. **1**, 1 (1973)
301. R.B. Larson, MNRAS **194**, 809 (1981)
302. R.B. Larson, MNRAS **214**, 379 (1985)
303. R.B. Larson, in *The Formation of Binary Stars*, ed. by H. Zinnecker, R.D. Mathieu (Astron. Soc. of the Pacific, San Francisco, 2001), p. 93
304. R.B. Larson, MNRAS **359**, 211 (2005)
305. J.C. Lattanzio, J.J. Monaghan, H. Pongracic, M.P. Schwarz, MNRAS **215**, 125 (1985)
306. G.P. Laughlin, P. Bodenheimer, ApJ **436**, 335 (1994)
307. G.P. Laughlin, M. Różyczka, ApJ **456**, 279 (1996)
308. N.R. Lebovitz, ApJ **190**, 121 (1974)
309. P. Ledoux, ApJ **94**, 537 (1941)
310. C.W. Lee, P.C. Myers, M. Tafalla, ApJ **526**, 788 (1999)
311. T. Lee, D.A. Papanastassiou, G.J. Wasserburg, ApJ **211**, L107 (1977)
312. D. Leisawitz, ApJ **359**, 319 (1990)
313. D. Leisawitz, F.N. Bash, P. Thaddeus, ApJ Suppl. **70**, 731 (1989)
314. G. Lesur, G.I. Ogilvie, MNRAS **404**, L64 (2010)
315. G. Lesur, J.C.B. Papaloizou, Astron. Astrophys. **513**, 60 (2010)
316. C.D. Levermore, G.C. Pomraning, ApJ **248**, 321 (1981)
317. P.S. Li, M.L. Norman, M.-M. Mac Low, F. Heitsch, ApJ **605**, 800 (2004)
318. Z.-Y. Li, C.F. McKee, ApJ **464**, 373 (1996)
319. Z.-Y. Li, F. Nakamura, ApJ **640**, L187 (2006)
320. G.H.R.A. Lima, S.H.P. Alencar, N. Calvet, L. Hartmann, J. Muzerolle, Astron. Astrophys. **522**, A104 (2010)
321. D.N.C. Lin, J.C.B. Papaloizou, MNRAS **191**, 37 (1980)
322. D.N.C. Lin, J.C.B. Papaloizou, in *Protostars and Planets II*, ed. by D.C. Black, M.S. Matthews (University of Arizona Press, Tucson, 1985), p. 981
323. D.N.C. Lin, J.E. Pringle, MNRAS **225**, 607 (1987)
324. M. Liu, Science **305**, 1442 (2004)
325. S. Lizano, F.H. Shu, ApJ **342**, 834 (1989)
326. C. Low, D. Lynden-Bell, MNRAS **176**, 367 (1976)
327. L.B. Lucy, Astron. Astrophys. **457**, 629 (2006)
328. H.-G. Ludwig, B. Freytag, M. Steffen, Astron. Astrophys. **346**, 111 (1999)
329. K.L. Luhman, ApJ **617**, 1216 (2004)
330. K.L. Luhman, ApJ Suppl. **173**, 104 (2007)

331. K.L. Luhman, J.R. Stauffer, A.A. Muench, G.H. Rieke, E.A. Lada, J. Bouvier, C.J. Lada, ApJ **593**, 1093 (2003)
332. D. Lynden-Bell, J.E. Pringle, MNRAS **168**, 603 (1974)
333. R.A. Lyttleton, *The Stability of Rotating Liquid Masses* (Cambridge University Press, Cambridge, UK, 1953)
334. M.-M. Mac Low, R.S. Klessen, A. Burkert, M. Smith, Phys. Rev. Lett. **80**, 2754 (1998)
335. M.-M. Mac Low, R.S. Klessen, Rev. Mod. Phys. **76**, 125 (2004)
336. M.N. Machida, S. Inutsuka, T. Matsumoto, ApJ **676**, 1088 (2008)
337. R.J. Maddalena, M. Morris, J. Moscowitz, P. Thaddeus, ApJ **303**, 375 (1986)
338. R.K. Mann, J.P. Williams, ApJ **694**, L36 (2009)
339. R.K. Mann, J.P. Williams, ApJ **699**, L55 (2009)
340. D. Mardones, P.C. Myers, M. Tafalla et al., ApJ **489**, 719 (1997)
341. B.D. Mason, D.R. Gies, W.I. Hartkopf, W.G. Bagnuolo, Jr., T. ten Brummelaar, H.A. McAlister, Astron. J. **115**, 821 (1998)
342. R.D. Mathieu, Annu. Rev. Astron. Astrophys. **32**, 465 (1994)
343. R.D. Mathieu, I. Baraffe, M. Simon, K.G. Stassun, R. White, in *Protostars and Planets V*, ed. by B. Reipurth, D. Jewitt, K. Keil (University of Arizona Press, Tucson, 2007), p. 411
344. R.D. Mathieu, A. Ghez, E. Jensen, M. Simon, in *Protostars and Planets IV*, ed. by V. Mannings, A.P. Boss, S.S. Russell (University of Arizona Press, Tucson, 2000), p. 703
345. S. Matt, R.E. Pudritz, ApJ **632**, L135 (2005)
346. C.D. Matzner, C.F. McKee, ApJ **545**, 364 (2000)
347. M. Mayor, D. Queloz, Nature **378**, 355 (1995)
348. T. Mazeh, M. Simon, L. Prato, B. Markus, S. Zucker, ApJ **599**, 1344 (2003)
349. M.J. McCaughrean, in *The Formation of Binary Stars*, ed. by H. Zinnecker, R.D. Mathieu (Astron. Soc. of the Pacific, San Francisco, 2001), p. 169
350. M.J. McCaughrean, C.R. O'Dell, Astron. J. **111**, 1977 (1996)
351. C.F. McKee, L.L. Cowie, ApJ **195**, 715 (1975)
352. C.F. McKee, E.C. Ostriker, Annu. Rev. Astron. Astrophys. **45**, 565 (2007)
353. C.F. McKee, J.C. Tan, Nature **416**, 59 (2002)
354. C.F. McKee, J.C. Tan, ApJ **585**, 850 (2003)
355. C.F. McKee, J.C. Tan, ApJ **681**, 771 (2008)
356. C.F. McKee, J.P. Williams, ApJ **476**, 144 (1997)
357. C.F. McKee, E.G. Zweibel, A.A. Goodman, C. Heiles, in *Protostars and Planets III*, ed. by E.H. Levy, J.I. Lunine (University of Arizona Press, Tucson, 1993), p. 327
358. S.T. Megeath, T.L. Wilson, M.R. Corbin, ApJ **622**, L141 (2005)
359. F. Ménard, C. Bertout, in *The Origin of Stars and Planetary Systems*, ed. by C.J. Lada, N.D. Kylafis (Kluwer, Dordrecht, 1999), p. 341
360. F. Meru, M.R. Bate, MNRAS **406**, 2279 (2010)
361. L. Mestel, R.B. Paris, Astron. Astrophys. **136**, 98 (1984)
362. L. Mestel, L. Spitzer Jr., MNRAS **116**, 503 (1956)
363. M.R. Meyer, D.E. Backman, A.J. Weinberger, M.C. Wyatt, in *Protostars and Planets V*, ed. by B. Reipurth, D. Jewitt, K. Keil (University of Arizona Press, Tucson, 2007), p. 573
364. D. Mihalas, *Stellar Atmospheres*, 2nd Ed. (Freeman, San Francisco, 1978)
365. S. Mohanty, F.H. Shu, ApJ **687**, 1323 (2008)
366. S. Molinari, F. Faustini, L. Testi, S. Pezzuto, R. Cesaroni, J. Brand, Astron. Astrophys. **487**, 1119 (2008)
367. S. Molinari, L. Testi, J. Brand, R. Cesaroni, F. Palla, ApJ **505**, L39 (1998)
368. J.J. Monaghan, Rep. Prog. Phys. **68**, 1703 (2005)
369. E. Moraux, J. Bouvier, J.R. Stauffer, J.-C. Cuillandre, Astron. Astrophys. **400**, 891 (2003)
370. F. Motte, P. André, R. Neri, Astron. Astrophys. **336**, 150 (1998)
371. T. Ch. Mouschovias, E. Paleologou, ApJ **230**, 204 (1979)
372. T. Ch. Mouschovias, E. Paleologou, ApJ **237**, 877 (1980)
373. T. Ch. Mouschovias, K. Tassis, M.W. Kunz, ApJ **646**, 1043 (2006)
374. A.A. Muench, E.A. Lada, C.J. Lada, J. Alves, ApJ **573**, 366 (2002)

375. N. Murray, ApJ **729**, 133 (2011)
376. J. Muzerolle, K.L. Luhman, C. Briceño, L. Hartmann, N. Calvet, ApJ **625**, 906 (2005)
377. P.C. Myers, in *Star Forming Regions*, ed. by M. Peimbert, J. Jugaku (Reidel, Dordrecht, 1987), p. 33
378. P.C. Myers, F.C. Adams, H. Chen, E. Schaff, ApJ **492**, 703 (1998)
379. P.C. Myers, R. Bachiller, P. Caselli, G.A. Fuller, D. Mardones, M. Tafalla, D.J. Wilner, ApJ, **449**, L65 (1995)
380. P.C. Myers, P.J. Benson, ApJ **266**, 309 (1983)
381. P.C. Myers, N.J. Evans II, N. Ohashi, in *Protostars and Planets IV*, ed. by V. Mannings, A.P. Boss, S.S. Russell (University of Arizona Press, Tucson, 2000), p. 217
382. P.C. Myers, E.F. Ladd, ApJ **413**, L47 (1993)
383. J.R. Najita, S.E. Strom, J. Muzerolle, MNRAS **378**, 369 (2007)
384. F. Nakamura, Z.-Y. Li, ApJ **631**, 411 (2005)
385. F. Nakamura, Z.-Y. Li, ApJ **662**, 395 (2007)
386. T. Nakano, Fund. Cosmic Phys. **9**, 139 (1984)
387. T. Nakano, ApJ **494**, 587 (1998)
388. A. Natta, V.P. Grinin, V. Mannings, in *Protostars and Planets IV*, ed. by V. Mannings, A.P. Boss, S.S. Russell (University of Arizona Press, Tucson, 2000), p. 559
389. A. Natta, L. Testi, S. Randich, Astron. Astrophys. **452**, 245 (2006)
390. J.F. Navarro, C.S. Frenk, S.D.M. White, ApJ **462**, 563 (1996)
391. A.F. Nelson, MNRAS **373**, 1039 (2006)
392. J. Nittmann, S.A.E.G. Falle, P.H. Gaskell, MNRAS **201**, 833 (1982)
393. M. Norman, J. Wilson, ApJ **224**, 497 (1978)
394. C.R. O'Dell, Z. Wen, ApJ **436**, 194 (1994)
395. S. Ogino, K. Tomisaka, F. Nakamura, Publ. Astron. Soc. Jpn. **51**, 637 (1999)
396. T. Ohkubo, H. Umeda, K. Maeda et al., ApJ **645**, 1352 (2006)
397. B.R. Oppenheimer, D. Brenner, S. Hinkley, N. Zimmerman, A. Sivaramakrishnan et al., ApJ **679**, 1574 (2008)
398. B.W. O'Shea, M.L. Norman, ApJ **654**, 66 (2007)
399. V. Ossenkopf, Th. Henning, Astron. Astrophys. **291**, 943 (1994)
400. J.P. Ostriker, P. Bodenheimer, ApJ **180**, 171 (1973)
401. N. Ouellette, S.J. Desch, J.J. Hester, ApJ **662**, 1268 (2007)
402. R. Ouyed, R.E. Pudritz, J.M. Stone, Nature **385**, 409 (1997)
403. D.L. Padgett, L. Cieza, K.R. Stapelfeldt, N.J. Evans II, D. Koerner et al., ApJ **645**, 1283 (2006)
404. P. Padoan, A. Nordlund, ApJ **576**, 870 (2002)
405. P. Padoan, A. Nordlund, A. Kritsuk, M. Norman, P.S. Li, ApJ **661**, 972 (2007)
406. F. Palla, S.W. Stahler, ApJ **392**, 667 (1992)
407. F. Palla, S.W. Stahler, ApJ **525**, 772 (1999)
408. F. Palla, S.W. Stahler, ApJ **553**, 299 (2001)
409. J. Patience, A.M. Ghez, I.N. Reid, K. Matthews, Astron. J. **123**, 1570 (2002)
410. J. Patience, A.M. Ghez, I.N. Reid, A.J. Weinberger, K. Matthews, Astron. J. **115**, 1972 (1998)
411. D.W. Peaceman, H.H. Rachford, J. Soc. Indust. Appl. Math. **3**, 28 (1955)
412. B.E. Penprase, W.L.W. Sargent, I.T. Martinez, J.X. Prochaska, D.J. Beeler, in *First Stars III*, ed. by B.W. O'Shea, A. Heger, T. Abel (American Institute of Physics, College Park, MD, 2008), p. 499
413. M.V. Penston, MNRAS **144**, 425 (1969)
414. M. Pérault, A. Omont, G. Simon, P. Seguin, D. Ojha et al., Astron. Astrophys. **315**, L165 (1996)
415. M.R. Petersen, G.R. Stewart, K. Julien, ApJ **658**, 1252 (2007)
416. M. Pinsonneault, Annu. Rev. Astron. Astrophys. **35**, 557 (1997)
417. R.A. Piontek, E.C. Ostriker, ApJ **601**, 905 (2004)
418. R. Plume, D.T. Jaffe, N.J. Evans II, J. Martín-Pintado, J. Gómez-González, ApJ **476**, 730 (1997)

419. T. Preibisch, A. Brown, T. Bridges, E. Guenther, H. Zinnecker, Astron. J. **124**, 404 (2002)
420. T. Preibisch, V. Ossenkopf, H.W. Yorke, Th. Henning, Astron. Astrophys. **279**, 577 (1993)
421. T. Preibisch, G. Weigelt, H. Zinnecker, in *The Formation of Binary Stars*, ed. by H. Zinnecker, R.D. Mathieu (Astron. Soc. of the Pacific, San Francisco, 2001), p. 69
422. D.J. Price, J. Comput. Phys., in press, arXiv:1012.1885 (2010)
423. D.J. Price, M.R. Bate, MNRAS **385**, 1820 (2008)
424. D.J. Price, M.R. Bate, MNRAS **398**, 33 (2009)
425. J.E. Pringle, Annu. Rev. Astron. Astrophys. **19**, 137 (1981)
426. J.E. Pringle, MNRAS **239**, 361 (1989)
427. R.E. Pudritz, R. Ouyed, C. Fendt, A. Brandenburg, in *Protostars and Planets V*, ed. by B. Reipurth, D. Jewitt, K. Keil (University of Arizona Press, Tucson, 2007), p. 277
428. J.M. Rathborne, J.M. Jackson, R. Simon, ApJ **641**, 389 (2006)
429. G. Rauw, M. De Becker, Y. Nazé, P.A. Crowther, E. Gosset, H. Sana, K.A. van der Hucht, J.-M. Vreux, P.M. Williams, Astron. Astrophys. **420**, L9 (2004)
430. B. Reipurth (ed.), *Handbook of Star Forming Regions* (Astron. Soc. of the Pacific, San Francisco, 2008)
431. B. Reipurth, C. Clarke, Astron. J. **122**, 432 (2001)
432. T.W. Rettig, J. Haywood, T. Simon, S.D. Brittain, E. Gibb, ApJ **616**, L163 (2004)
433. D. Richard, J.-P. Zahn, Astron. Astrophys. **347**, 734 (1999)
434. J.S. Richer, D.S. Shepherd, S. Cabrit, R. Bachiller, E. Churchwell, in *Protostars and Planets IV*, ed. by V. Mannings, A.P. Boss, S.S. Russell (University of Arizona Press, Tucson, 2000), p. 867
435. R.D. Richtmyer, *Difference Methods for Initial Value Problems* (Interscience, New York, 1957)
436. E. Ripamonti, F. Haardt, A. Ferrara, M. Colpi, MNRAS **334**, 401 (2002)
437. A. Roberge, A.J. Weinberger, E.M. Malumuth, ApJ **622**, 1171 (2005)
438. T.P. Robitaille, B.A. Whitney, R. Indebetouw, K. Wood, ApJ Suppl. **169**, 328 (2007)
439. C. Roddier, F. Roddier, M. Northcott, J. Graves, K. Jim, ApJ **463**, 326 (1996)
440. L.F. Rodríguez, P. D'Alessio, D.J. Wilner, P.T.P. Ho, J.M. Torrelles et al., Nature **395**, 355 (1998)
441. D. Ryu, J. Goodman, ApJ **388**, 438 (1992)
442. S.I. Sadavoy, J. Di Francesco, S. Bontemps, S. Megeath, L.M. Rebull et al., ApJ **710**, 1247 (2010)
443. E. Salpeter, ApJ **121**, 161 (1955)
444. A.I. Sargent, S.V.W. Beckwith, ApJ **323**, 294 (1987)
445. F. Sato, T. Hasegawa, J.B. Whiteoak, R. Miyawaki, ApJ **535**, 857 (2000)
446. D. Saumon, G. Chabrier, H. van Horn, ApJ Suppl. **99**, 713 (1995)
447. J.M. Scalo, Fund. Cosmic Phys. **11**, 1 (1986)
448. D. Schaerer, Astron. Astrophys. **382**, 28 (2002)
449. M. Schmidt, ApJ **129**, 243 (1959)
450. O. Schnurr, J. Casoli, A.-N. Chené, A.F.J. Moffat, N. St-Louis, MNRAS **389**, L38 (2008)
451. M. Schwarzschild, *Structure and Evolution of the Stars* (Princeton University Press, Princeton, 1958)
452. D. Semenov, Th. Henning, Ch. Helling, M. Ilgner, E. Sedlmayr, Astron. Astrophys. **410**, 611 (2003)
453. N.I. Shakura, R.A. Sunyaev, Astron. Astrophys. **24**, 337 (1973)
454. H. Shang, Z.-Y. Li, N. Hirano, in *Protostars and Planets V*, ed. by B. Reipurth, D. Jewitt, K. Keil (University of Arizona press, Tucson, 2007), p. 261
455. D.S. Shepherd, E. Churchwell, ApJ **472**, 225 (1996)
456. F.H. Shu, ApJ **214**, 488 (1977)
457. F.H. Shu, *The Physics of Astrophysics, Vol. I: Radiation* (University Science Books, Mill Valley, CA, 1991)
458. F.H. Shu, *The Physics of Astrophysics, Vol. II: Gas Dynamics* (University Science Books, Mill Valley, CA, 1992)

459. F.H. Shu, F.C. Adams, S. Lizano, Annu. Rev. Astron. Astrophys. **25**, 23 (1987)
460. F.H. Shu, Z.-Y. Li, A. Allen, in *Star Formation in the Interstellar Medium*, ed. by D. Johnstone et al. (Astron. Soc. of the Pacific, San Francisco, 2004), p. 37
461. F.H. Shu, S. Lizano, F.C. Adams, S. Ruden, in *Formation and Evolution of Low Mass Stars*, ed. by A. Dupree, M. Lago (Kluwer, Dordrecht, 1988), p. 123
462. F.H. Shu, J. Najita, E. Ostriker, F. Wilkin, S.P. Ruden, S. Lizano, ApJ **429**, 781 (1994)
463. F.H. Shu, H. Shang, A.E. Glassgold, T. Lee, Science **277**, 1475 (1997)
464. L. Siess, E. Dufour, M. Forestini, Astron. Astrophys. **358**, 593 (2000)
465. M. Simon, L. Prato, ApJ **450**, 824 (1995)
466. R. Simon, J.M. Jackson, J.M. Rathborne, E.T. Chambers, ApJ **639**, 227 (2006)
467. C.L. Slesnick, L.A. Hillenbrand, J.M. Carpenter, ApJ **610**, 1045 (2004)
468. B.A. Smith, R. Terrile, Science **226**, 1421 (1984)
469. M.D. Smith, Irish Astron. J. **27**, 25 (2000)
470. N. Smith, B. Whitney, P. Conti, C. de Pree, J. Jackson, MNRAS **399**, 952 (2009)
471. C. Sneden, A. McWilliam, G.W. Preston, J.J. Cowan, D.L. Burris, B.J. Armosky, ApJ **467**, 819 (1996)
472. R.L. Snell, R.B. Loren, R.L. Plambeck, ApJ **239**, L17 (1980)
473. D.R. Soderblom, B.F. Jones, S. Balachandran, J.R. Stauffer, D.K. Duncan et al., Astron. J. **106**, 1059 (1993)
474. J. Solf, K.-H. Böhm, ApJ **375**, 618 (1991)
475. P.M. Solomon, A.R. Rivolo, J. Barrett, A. Yahil, ApJ **319**, 730 (1987)
476. L. Spitzer, Jr., *Physical Processes in the Interstellar Medium* (Wiley, New York, 1978)
477. D. Spolyar, P. Bodenheimer, K. Freese, P. Gondolo, ApJ **705**, 1031 (2009)
478. D. Spolyar, K. Freese, P. Gondolo, Phys. Rev. Lett. **100**, 051101 (2008)
479. T.K. Sridharan, H. Beuther, M. Saito, F. Wyrowski, P. Schilke, ApJ **634**, L57 (2005)
480. A. Stacy, T.H. Greif, V. Bromm, MNRAS **403**, 45 (2010)
481. S. Stahler, F. Palla, *The Formation of Stars* (Wiley-VCH, Weinheim, 2004)
482. S. Stahler, F. Shu, R. Taam, ApJ **241**, 637 (1980)
483. S. Stahler, F. Shu, R. Taam, ApJ **242**, 226 (1980)
484. S. Stahler, F. Shu, R. Taam, ApJ **248**, 727 (1981)
485. D. Stamatellos, M.J. Griffin, J.M. Kirk, S. Molinari, B. Sibthorpe et al., MNRAS **409**, 12 (2010)
486. K.G. Stassun, R.D. Mathieu, J.A. Valenti, Nature **440**, 311 (2006)
487. K.G. Stassun, R.D. Mathieu, J.A. Valenti, ApJ **664**, 1154 (2007)
488. A.T. Steffen, R.D. Mathieu, M.G. Lattanzi, D.W. Latham, T. Mazeh et al., Astron. J. **122**, 997 (2001)
489. R.F. Stein, A. Nordlund, ApJ **499**, 914 (1998)
490. M.F. Sterzik, A.A. Tokovinin, Astron. Astrophys. **384**, 1030 (2002)
491. J.T. Stocke, P. Hartigan, S.E. Strom, K.M. Strom, E.R. Anderson, L.W. Hartmann, S.J. Kenyon, ApJ Suppl. **68**, 229 (1988)
492. J.M. Stone, C.F. Gammie, S.A. Balbus, J.F. Hawley, in *Protostars and Planets IV*, ed. by V. Mannings, A.P. Boss, S.S. Russell (University of Arizona Press, Tucson, 2000), p. 589
493. J.M. Stone, M.L. Norman, ApJ **390**, L17 (1992)
494. J.M. Stone, E. Ostriker, C. Gammie, ApJ **508**, L99 (1998)
495. H. Susa, ApJ **659**, 908 (2007)
496. G. Suttner, H.W. Yorke, ApJ **551**, 461 (2001)
497. J.C. Tan, ApJ **536**, 173 (2000)
498. J.C. Tan, C.F. McKee, ApJ **603**, 383 (2004)
499. J.-L. Tassoul, *Theory of Rotating Stars* (Princeton University Press, Princeton, NJ, 1978)
500. G. Tenorio-Tagle, Astron. Astrophys. **71**, 59 (1979)
501. G. Tenorio-Tagle, M. Różyczka, Astron. Astrophys. **155**, 120 (1986)
502. S. Terebey, F.H. Shu, P. Cassen, ApJ **286**, 529 (1984)
503. A.G.G.M. Tielens, *The Physics and Chemistry of the Interstellar Medium* (Cambridge University Press, Cambridge, UK, 2005)

References

504. J.J. Tobin, L. Hartmann, N. Calvet, P. D'Alessio, ApJ **679**, 1364 (2008)
505. J.E. Tohline, Annu. Rev. Astron. Astrophys. **40**, 349 (2002)
506. R.C. Tolman, ApJ **90**, 568 (1939)
507. K. Tomisaka, S. Ikeuchi, T. Nakamura, ApJ **362**, 202 (1990)
508. A. Toomre, ApJ **139**, 1217 (1964)
509. D.E. Trilling, G. Bryden, C.A. Beichman, G.H. Rieke, K.Y.L. Su et al., ApJ **674**, 1086 (2008)
510. T.H. Troland, R.M. Crutcher, ApJ **680**, 457 (2008)
511. J.K. Truelove, R.I. Klein, C.F. McKee, J.H. Holliman, II, L.H. Howell, J.A. Greenough, ApJ **489**, L179 (1997)
512. J.K. Truelove, R.I. Klein, C.F. McKee, J.H. Holliman, II, L.H. Howell, J.A. Greenough, D.T. Woods, ApJ **495**, 821 (1998)
513. J. Tsai, E. Bertschinger, Bull. A. A. S. **21**, 1089 (1989)
514. W.M. Tscharnuter, K.-H. Winkler, Comput Phys. Commun. **18**, 171 (1979)
515. T. Tsuji, Publ. Astron. Soc. Jpn. **18**, 127 (1966)
516. T. Tsuribe, S. Inutsuka, ApJ **523**, L155 (1999)
517. J. Tumlinson, A. Venkatesan, J.M. Shull, ApJ **612**, 602 (2004)
518. M.J. Turk, T. Abel, B. O'Shea, Science **325**, 601 (2009)
519. N.J. Turner, P. Bodenheimer, M. Różyczka, ApJ **524**, 129 (1999)
520. N.J. Turner, E. Quataert, H.W. Yorke, ApJ **662**, 1052 (2007)
521. N.J. Turner, T. Sano, N. Dziourkevitch, ApJ **659**, 729 (2007)
522. Y. Uchida, K. Shibata, Publ. Astron. Soc. Jpn. **37**, 515 (1985)
523. S. Udry, M. Mayor, J.-L. Halbwachs, F. Arenou, in *Microlensing 2000: A New Era of Microlensing Astrophysics*, ed. by J. Menzies, P.D. Sackett (Astron. Soc. of the Pacific, San Francisco, 2001), p. 91
524. R.K. Ulrich, ApJ **210**, 377 (1976)
525. T. Umebayashi, T. Nakano, Publ. Astron. Soc. Jpn. **33**, 617 (1981)
526. A. Urban, H. Martel, N.J. Evans II, ApJ **710**, 1343 (2010)
527. T. van Albada, Bull. Astron. Inst. Neth. **19**, 479 (1968)
528. H.A.T. Vanhala, A.P. Boss, ApJ **538**, 911 (2000)
529. H.A.T. Vanhala, A.P. Boss, ApJ **575**, 1144 (2002)
530. J. von Neumann, R.D. Richtmyer, J. Appl. Phys. **21**, 232 (1950)
531. E.I. Vorobyov, S. Basu, ApJ **633**, L137 (2005)
532. F.J. Vrba, S.E. Strom, K.M. Strom, Astron. J. **81**, 958 (1976)
533. A. Wachmann, Beob. Zirk. **21**, 12 (1939)
534. C.K. Walker, J. Carlstrom, J. Bieging, ApJ **402**, 655 (1993)
535. C.K. Walker, C.J. Lada, E.T. Young, P.R. Maloney, B.A. Wilking, ApJ **309**, L47 (1986)
536. M. Walmsley, Rev. Mex. Astron. Astrofis. Ser. de Conf. **1**, 137 (1995)
537. D. Ward-Thompson, P. André, R. Crutcher, D. Johnstone, T. Onishi, C. Wilson, in *Protostars and Planets V*, ed. by B. Reipurth, D. Jewitt, K. Keil (University of Arizona Press, Tucson, 2007), p. 33
538. D. Ward-Thompson, F. Motte, P. André, MNRAS **305**, 143 (1999)
539. D. Ward-Thompson, P. Scott, R. Hills, P. André, MNRAS **268**, 276 (1994)
540. S.J. Weidenschilling, Astrophys. Sp. Sci. **51**, 153 (1977)
541. S.J. Weidenschilling, J.N. Cuzzi, in *Protostars and Planets III*, ed. by E.H. Levy, J.I. Lunine (University of Arizona Press, Tucson, 1993), p. 1031
542. C. Weidner, P. Kroupa, MNRAS **348**, 187 (2004)
543. G. Weigelt, R. Albrecht, C. Barbieri, J.C. Blades, A. Boksenberg et al., ApJ **378**, L21 (1991)
544. R.J. White, A.M. Ghez, ApJ **556**, 265 (2001)
545. R.J. White, A.M. Ghez, I.N. Reid, G. Schultz, ApJ **520**, 811 (1999)
546. S.C. Whitehouse, M.R. Bate, MNRAS **367**, 32 (2006)
547. A.P. Whitworth, MNRAS **186**, 59 (1979)
548. A.P. Whitworth, A.S. Bhattal, S.J. Chapman, M.J. Disney, J.A. Turner, MNRAS **268**, 291 (1994)
549. B.A. Wilking, C.J. Lada, E.T. Young, ApJ **340**, 823 (1989)

550. H. Williams, J.E. Tohline, ApJ **334**, 449 (1988)
551. J.P. Williams, E.A. Bergin, P. Caselli, P.C. Myers, R. Plume, ApJ **503**, 689 (1998)
552. J.P. Williams, C.F. McKee, ApJ **476**, 166 (1997)
553. D.J. Wilner, P.T.P. Ho, J.H. Kastner, L.F. Rodríguez, ApJ **534**, L101 (2000)
554. D.J. Wilner, P.C. Myers, D. Mardones, M. Tafalla, ApJ **544**, L69 (2000)
555. K.-H. Winkler, M. Newman, ApJ **236**, 201 (1980)
556. S.C. Wolff, S.E. Strom, D. Dror, L. Lanz, K. Venn, Astron. J. **132**, 749 (2006)
557. M.G. Wolfire, J.P. Cassinelli, ApJ **319**, 850 (1987)
558. S.J. Wolk, B.D. Spitzbart, T.L. Bourke, R.A. Gutermuth, M. Vigil, F. Comerón, Astron. J. **135**, 693 (2008)
559. A. Wolszczan, D.A. Frail, Nature **355**, 145 (1992)
560. D. Wood, E. Churchwell, ApJ Suppl. **69**, 831 (1989)
561. A. Wootten, ApJ **337**, 858 (1989)
562. Y. Wu, Y. Wei, M. Zhao, Y. Shi, W. Yu, S. Qin, M. Huang, Astron. Astrophys. **426**, 503 (2004)
563. G. Wuchterl, R.S. Klessen, ApJ **560**, L185 (2001)
564. H.W. Yorke, P. Bodenheimer, ApJ **525**, 330 (1999)
565. H.W. Yorke, P. Bodenheimer, G.P. Laughlin, ApJ **443**, 199 (1995)
566. H.W. Yorke, C. Sonnhalter, ApJ **569**, 846 (2002)
567. N. Yoshida, K. Omukai, L. Hernquist, Science **321**, 669 (2008)
568. N. Yoshida, K. Omukai, L. Hernquist, T. Abel, ApJ **652**, 6 (2006)
569. C.H. Young, N.J. Evans II, ApJ **627**, 293 (2005)
570. A. Zavagno, L. Deharveng, F. Comerón, J. Brand, F. Massi, J. Caplan, D. Russeil, Astron. Astrophys. **446**, 171 (2006)
571. S. Zhou, ApJ **442**, 685 (1995)
572. S. Zhou, N.J. Evans II, C. Kömpe, C.M. Walmsley, ApJ **404**, 232 (1993)
573. S. Zhou, N.J. Evans II, Y. Wang, R. Peng, K.Y. Lo, ApJ **433**, 131 (1994)
574. Z. Zhu, L. Hartmann, C.F. Gammie, L.G. Book, J.B. Simon, E. Engelhard, ApJ **713**, 1134 (2010)
575. H. Zinnecker, H.W. Yorke, Annu. Rev. Astron. Astrophys. **45**, 481 (2007)
576. B. Zuckerman, N.J. Evans II, ApJ **192**, L149 (1974)

Index

AB Aur, 20, 25, 144
Absorption line, 26, 149, 174, 285, 296, 303, 305
Accretion luminosity, 75, 107, 108, 110, 114, 124, 135, 149, 175, 261, 263, 264, 315
Accretion rate, 10, 31, 114, 115, 118, 119, 122, 124, 131, 140, 143, 149, 150, 152, 154, 167, 175, 179, 181, 194, 261, 262, 266, 267, 271, 272, 274, 276, 278, 315, 318–321
Accretion time scale, 108, 110, 122, 123, 263
Adaptive mesh refinement, 88
Adiabatic collapse, 93, 102, 124, 236, 254, 258
Adiabatic contraction, 270, 272, 274
Adiabatic gradient, 282, 291
Alfvén velocity, 53, 60–62, 69, 90, 164, 177, 178, 311
Alfvén waves, 13, 44, 53, 60–62, 90, 130, 314
Aluminum 26, 77, 79, 87
AM Aur, 143
Ambipolar diffusion, 12, 37, 53, 64–68, 70, 77, 86, 184, 312, 314
Ammonia, 13, 14, 30, 43, 49, 91
Angular momentum, 11, 12, 28, 30, 32, 49, 50, 59–62, 66, 84, 89, 127–132, 135, 138, 151, 165, 170, 177, 179, 180, 200, 204, 223, 225, 229, 231–233, 241, 246, 249, 250, 255, 256, 276, 279, 280, 297, 313, 314, 318
Angular momentum transport, 11, 12, 28, 31, 32, 59, 84, 127, 128, 133, 137, 138, 140–142, 151, 154, 160, 164–166, 177, 221, 233, 242, 248, 260, 298, 314, 317, 321
Arecibo, 55

Associations, 1, 3, 6, 83, 185
AU Mic, 22, 144

B335, 122
Barnard 68, 14, 15, 98
Baroclinic instability, 161, 166, 321
Beta Pictoris, 20, 26, 144, 179
Binary frequency, 222, 223, 227
Birth line, 31, 279, 290, 291, 299
Black body, 26–28, 58, 112, 116, 118, 132, 146, 156, 157, 289, 290
Black holes, 265, 268, 274, 275, 277, 317, 319
Bok globules, 14
Bolometric temperature, 112, 113, 118, 119, 123
Bondi-Hoyle accretion, 203
Bonnor-Ebert sphere, 48, 79, 84, 94, 96–98, 101, 122–124, 176, 313
Brown dwarf, 8, 10, 23, 209, 215, 218, 223, 228, 245, 283, 286, 292, 294, 301, 305, 308, 309, 316
Brunt-Väisälä frequency, 164

Carbon monoxide, 11, 13, 14, 16, 19, 24, 38–40, 49, 58, 144, 170–172, 179, 254, 303
Circumbinary disk, 16, 18, 84, 122, 228–230, 241, 242, 246, 248, 303, 313, 314, 317, 322
Class 0, 28, 29, 31, 108, 118, 119, 122, 123, 128, 143, 170–175, 180, 190, 315
Class I, 26–31, 81, 93, 108, 114, 116–118, 122, 123, 128, 143, 170, 172–175, 223, 315, 316

Class II, 26–29, 31, 128, 143, 146, 149, 150, 170, 171, 177, 295, 315
Class III, 27, 29, 171, 295
Cloud-cloud collisions, 83, 85
Clusters, 6–9, 23, 24, 27, 32, 33, 38, 42, 45, 46, 53, 66, 71, 73, 74, 76, 81, 82, 115, 148, 169, 183–189, 196, 199, 207–211, 214–216, 221, 223, 224, 226, 231, 232, 238, 241–244, 247, 248, 256, 276, 298–302, 306–308, 312, 313, 316, 320, 322
Collision-induced emission, 254, 258, 259
Competitive accretion, 208–211, 216, 217
Compton cooling, 253
Continuity equation, 90, 132, 133, 152, 201
Convection, 3, 24, 125, 164, 252, 271, 272, 281–283, 290, 292, 299, 305–310, 321
Cooling curve, 294
Cooling time, 57, 78–80, 138, 139, 160, 166, 180, 231, 244, 254, 256, 276, 316
Cosmic rays, 57, 58, 63, 64, 164
Courant-Friedrichs-Lewy (CFL) condition, 88, 134, 154
Critical density, 39, 256

Dark matter, 251, 255, 256, 265, 268–271, 273–278, 318
Dark star, 270, 273, 277, 278
Dead zone, 164
Debris disks, 23, 144, 179
Degeneracy, 283, 285, 292, 308, 309
Deuterium, 111, 274, 281, 294, 307
Disk dispersal, 128, 167, 179, 321
Disk lifetimes, 11, 128, 143, 148, 149
Disk luminosity, 155, 156
Disk masses, 23, 143, 145–148, 160
Disk sizes, 15, 20, 143, 144, 158, 181
Disk surface densities, 147, 149, 152, 154, 157, 159, 160, 162, 180, 250
Disk temperatures, 147–149, 157–160, 162, 163, 169, 250
Disk wind, 178
Dissociation, 3, 57, 93, 103–106, 123, 125, 253, 254, 258, 259, 262, 265, 271, 276
Doppler shift, 130
Dust destruction front, 111, 181, 194
Dust emission, 20, 40, 45, 46, 58, 146, 167, 172
Dust grains, 3, 13, 21, 27, 29, 40, 58, 80, 102, 115, 142, 145, 158, 159, 167, 179, 251, 254

Dust opacity, 23, 27, 102, 103, 115–117, 123, 147, 157, 164, 167, 179, 180, 230, 286, 321
Dust photosphere, 110, 111, 116, 117, 123

Eagle Nebula, 86
Eccentricity, 226, 227, 231, 246–248
Eclipsing binary, 19, 24, 229, 308
Eddington factor, 202
Eddington limit, 195, 261–264, 278
Effective temperature, 110, 153–155, 180, 271, 276, 282, 292, 297, 299, 305–309, 318
Electron scattering, 195, 261, 278, 285, 309
Emission coefficient, 287
Emission line, 15, 18, 21, 30, 38, 39, 41, 43, 49, 120, 121, 174, 295, 296, 308, 320
Energy equation, 103, 132–134, 139, 152, 153, 201, 281
Equation of state, 3, 95–97, 103–105, 107, 125, 133, 153, 252, 271, 283, 284, 293

Fission, 221, 230–234, 248, 249, 316
Flared disk, 156–158
Flux-limited diffusion, 132, 133, 197, 200, 202
Forbidden lines, 170, 174, 175
Fragmentation, 6, 12, 52, 73, 82, 93, 124, 127, 137, 161, 193, 197, 200, 203–205, 207–211, 216, 218, 221, 230, 231, 233, 236, 237, 239–242, 244, 247, 248, 256, 257, 268, 276, 312, 314, 316, 318, 320
Free-fall time, 4, 6, 25, 30, 34, 58, 61, 62, 65–67, 72, 76, 78, 90, 91, 99, 100, 103, 107, 134, 193, 196, 198, 203, 208, 209, 219, 220, 236, 240, 254, 256, 279, 323
FU Orionis objects, 17, 18, 22, 31, 128, 179, 280, 315, 321
Funnel flow, 177, 178

GG Tau, 143, 227–229, 303
GM Aur, 158, 167
Gravitational instability, 11, 17, 32, 114, 127, 128, 136–140, 142, 150, 154, 160, 163, 179, 181, 203, 204, 210, 217, 221, 250, 257, 316, 321
Gravitational radius, 169, 266

Index

Hα, 5, 18, 20, 81, 170, 174, 175, 295–297, 299, 303
Hayashi track, 114, 115, 125, 290, 292, 294, 299, 305, 307, 309
Henyey track, 290
Herbig Ae-Be stars, 20, 295, 320
Herbig–Haro objects, 11, 15–17, 19, 21, 170, 171
Herschel satellite, 149
Hertzsprung–Russell diagram, 18, 19, 24, 31, 32, 111–113, 115, 116, 124, 149, 227–229, 273, 278, 279, 282, 291, 292, 300, 303, 307, 313
HH1, 17, 21, 170
HH2, 17, 21, 170
HH211, 171, 172, 174
HH30, 15, 17, 144
HII regions, 1, 5, 33, 70, 71, 77, 80–84, 87, 183, 184, 187, 188, 190–193, 211, 265–267, 276, 312, 318, 319, 322, 323
HL Tauri, 143
Homologous contraction, 283, 284, 309
Hubble Space Telescope, 20, 23, 25–27, 42, 144, 229, 303
Hyades, 307
Hydrostatic equilibrium, 4, 10, 12, 25, 30, 44, 95, 105–109, 112, 123–125, 136, 152, 232, 260, 269, 271, 279, 280, 282, 307, 311

IC 348, 301, 302
Infall, 10, 11, 13, 28, 31, 119–123, 128, 136, 143
Infrared dark cloud, 189, 190
Infrared excess, 17, 27, 29, 31, 128, 145, 148, 167, 168, 174, 179, 295, 296, 299, 308, 320
Infrared source, 13, 16, 26, 30, 63, 67, 82, 304, 312
Initial mass function (IMF), 8–10, 14, 45, 46, 75, 185, 186, 199, 207–211, 216, 217, 220, 226, 245, 249, 251, 268, 277, 280, 301, 302, 305, 308, 313, 317–319, 322
Inside-out collapse, 102, 122, 131
Interstellar medium, 1, 2
Ionization, 1–3, 6, 7, 12, 17, 28, 38, 57, 61, 63–67, 80, 82, 84, 103–106, 125, 164, 169, 183, 191–193, 211, 247, 251, 260, 264, 265, 267, 271, 277, 283, 314, 315, 321
IRAS 16293-2422, 15, 122, 222

IRAS satellite, 14, 20, 145, 190, 191
Iron 60, 77, 79, 80, 87
Isothermal collapse, 89, 93, 99–101, 108, 131, 221, 230, 234, 235, 248

Jacobi ellipsoid, 232
Jeans condition, 237
Jeans length, 48, 73, 78, 79, 89, 90, 96, 100, 123, 237, 250, 312
Jeans mass, 48, 51, 56, 57, 61, 69, 71, 73, 75, 77, 79, 80, 86, 90, 93, 122, 124, 193, 198, 199, 208, 217–219, 230, 238, 239, 241, 242, 250, 252, 256, 257, 276, 302, 318
Jets, 11, 15–17, 19, 31, 113, 170–172, 175, 177–179, 188

Kelvin-Helmholtz instability, 78
Kelvin-Helmholtz time, 25, 108, 183, 210, 219, 263, 276, 290, 294, 298
Kolmogorov law, 68
Kozai mechanism, 247, 317

L1527, 122
L1544, 99
L1551 IRS5, 14–16, 19, 20, 143, 171, 222
Large Magellanic Cloud, 8, 10, 33, 38, 185, 186
Larson's Laws, 43
Linewidth-size relation, 42–44, 54, 208, 242
Lithium, 17, 18, 294, 295, 303, 305–308, 320
Local thermodynamic equilibrium, 287, 288
Luminosity problem, 114, 115, 142, 316
Lyman band, 40, 265

Maclaurin spheroid, 232
Magnetic braking, 37, 50, 53, 59–63, 128, 130, 314
Magnetic diffusivity, 51
Magnetic fields, 3, 7, 12, 13, 16, 19, 24, 28, 30, 32, 33, 37, 44, 50–52, 54–56, 59, 64–70, 75, 77, 87, 89, 91, 93, 122, 128, 132, 145, 149, 155, 163, 165, 175–179, 184, 193, 206, 208–211, 216–218, 221, 233, 247, 251, 279, 280, 297, 298, 311, 314, 320–322
Magnetic flux, 12, 51, 52, 56, 63, 64, 66, 314
Magnetic induction equation, 51, 163
Magnetic Reynolds number, 52, 163
Magnetohydrodynamics, 12, 17, 59, 64, 176, 217, 321

Magnetorotational instability, 68, 128, 161, 163–165, 321
Magnetosphere, 31, 155, 295
Mass absorption coefficient, 287
Mass ratios, 224, 226, 246, 248, 249, 317
Mass spectrum, 8, 44, 45, 71
Mass-luminosity relation, 313
Mass-to-flux ratio, 12, 52–54, 56, 62, 64, 66, 73, 75, 86
Mean free path–particle, 68
Mean free path–photon, 133, 287
Minimum mass solar nebula, 159
Mixing length, 282, 283, 292, 303, 305
Molecular cloud cores, 12–14, 24, 28, 37, 39, 41, 46, 56, 59, 63, 66, 75, 79, 80, 94, 97–99, 109, 114, 115, 117, 123, 128, 129, 131, 176, 223, 225, 230, 233–235, 237, 238, 241, 246, 249, 297, 312, 313, 316, 320
Molecular hydrogen, 38, 40, 48, 55–58, 89, 91, 93, 94, 103–105, 172, 252–254, 256, 257, 260, 269, 276, 318
Molecular opacity, 110, 292
Momentum equation, 90, 132, 133, 139, 201
Monolithic collapse, 207, 210, 217, 233
Multiplicity fraction, 10, 222, 224, 245

Navier-Stokes equations, 139, 140
Neutralino, 269, 318
Neutrinos, 271, 318
NGC 2264, 42, 66, 298–300

Optical depth, 38, 93, 97, 99, 102, 105, 109–112, 116, 120, 132, 146, 152, 153, 180, 181, 206, 254, 258, 259, 282, 288
Orion Nebula cluster, 23, 40, 46, 145, 223, 298, 299, 301, 302, 305, 312
Outflows, 6, 11, 12, 14, 16, 18, 19, 21, 31, 70, 71, 74–76, 93, 118, 119, 122, 170, 172–179, 187, 188, 190–193, 198, 199, 206, 207, 210, 211, 218, 246, 265, 279, 313, 314, 316, 320–322

Pauli exclusion principle, 294
Photoevaporation, 32, 169, 170, 265, 267, 322
Photoionization, 285
Photosphere, 272, 279, 282, 283, 290, 295, 299, 311
Pipe Nebula, 46, 313
Planck mean opacity, 58, 116, 202

Planet formation, 12, 19, 22, 32, 127, 128, 145, 148, 159, 160, 168, 179, 231, 279, 299
Pleiades, 306, 307
Poisson equation, 90, 133, 201
Polytrope, 232–234, 262, 268, 271, 272, 290, 309
Population II, 251, 277
Population III, 251
Present-day mass function, 186, 212, 216
Propagating star formation, 82, 83

R136, 185, 186
Radiation energy, 133, 201
Radiation pressure, 8, 21, 104, 145, 173, 179, 183, 186, 192, 194–197, 200, 205, 206, 210, 211, 261, 265, 266, 272, 278, 288, 308, 319
Radiative diffusion, 102, 103, 108, 110, 152, 180, 281, 289
Radiative flux, 23, 26, 102, 109, 117, 132, 146, 152, 156, 180, 195, 197, 202, 286, 289
Radiative transfer, 3, 11, 75, 112, 120, 122, 130, 131, 135, 141, 142, 161, 195–197, 200, 202, 203, 206, 237, 245–247, 252, 255, 281, 285, 287, 316
Rayleigh criterion, 166
Rayleigh number, 164
Rayleigh-Taylor instability, 78, 80, 81, 196, 206
RCW79, 80, 83
Recombination, 63, 64, 191, 192
Reynolds number, 68, 166
Rho Ophiuchi, 10, 28, 33, 45, 223
Rosseland mean opacity, 102, 103, 108, 116, 132, 153, 197, 202, 261, 272, 281, 282, 285, 286, 289
Rotation, 11, 12, 15, 18, 24, 30, 31, 37, 48–50, 59, 60, 62, 89, 93, 94, 112, 117, 119, 122, 129, 132, 170, 175, 177, 179–181, 191, 196, 198–200, 204, 205, 230, 241, 249, 256, 260, 265, 266, 268, 279, 297, 298, 314

Saha equation, 105
Scale height, 24, 153, 154, 160, 282, 303, 321
Schwarzschild criterion, 166, 272, 281
Sharpless 104, 80
Shocks, 1, 3, 7, 10, 30, 38, 66, 67, 69, 70, 72, 77–83, 86, 87, 91, 94, 105–107, 109,

111, 115, 123, 131, 133, 135, 137, 145, 149, 160, 169, 171, 172, 190, 198, 200, 206, 208, 217, 236, 252, 259, 311, 312, 319
Singular isothermal sphere, 97, 99, 100, 102, 122, 128, 131
Sink particle, 89, 200, 201, 203, 208, 218, 257
Sink zone, 135, 197, 200
Smoothed particle hydrodynamics, 87, 88, 199, 208, 240, 242
Snow line, 160
Sobolev approximation, 121
Specific intensity, 286
Spectral energy distribution (SED), 15, 16, 23, 26, 28, 29, 108, 111, 115–118, 122, 123, 131, 141, 142, 144, 145, 149, 156–158, 190, 191, 296
Spectroscopic binary, 19, 186, 223, 224, 226, 228, 229, 302, 319
Spitzer Space Telescope, 23, 28, 76, 81, 83, 145, 158
Starless cores, 13, 67, 98, 312
Stellar core, 30, 31, 94, 106, 116, 123, 176, 233, 258–260, 271, 276, 290
Stellar evolution, 1, 2, 12, 19, 24, 280, 290, 291, 293, 305
Subcritical cloud, 52, 55, 61, 65, 67, 77
Supercritical cloud, 52, 55, 61, 64, 66, 67, 73, 86
Supernova, 1–3, 7, 67, 70, 77–81, 87, 183, 251, 255, 264, 268, 274, 276, 277, 312, 317, 319, 322, 323

T Tau, 19, 23, 157, 229, 304
T Tauri stars, 14, 17–20, 23, 26, 28, 31, 50, 116, 129, 141, 157, 167, 170, 171, 175, 177, 178, 180, 223, 233, 236, 249, 280, 295–297, 299, 300, 303, 305, 314, 315, 320
Taurus-Aurigae region, 10, 14, 28, 33, 99, 145, 223, 299, 301, 302, 312, 313
Thermal equilibrium, 271, 272, 281
Thermal instability, 321
Three-body capture, 221, 231
Tidal capture, 221, 231
Toomre Q, 136, 137, 139, 140, 160, 231, 244, 248
Transition disk, 167, 170
Trapezium cluster, 23, 27, 169
Triple systems, 222, 229, 242, 243, 246, 247, 304, 317
Turbulence, 3, 5, 7, 8, 11, 13, 24, 32, 37, 42–44, 51, 59, 65–70, 72, 73, 76, 77, 84, 86, 87, 93, 94, 115, 119, 154, 163, 164, 170, 189, 193, 196, 203, 210, 217, 218, 242, 252, 302, 311–314, 319, 322
TW Hydrae, 144, 148–150, 167, 223

Ultraviolet excess, 145, 149, 168, 174, 175, 296, 308, 320

Virial theorem, 43, 44, 108, 123, 124, 280, 292, 309
Viscosity, 11, 32, 33, 128, 135, 139, 140, 143, 145, 150–152, 154, 160, 164, 166, 169, 170, 179, 180, 232, 321
Viscosity, artificial, 133, 134
Viscous time, 154, 161, 168
Visual binary, 225
Volume filling factor, 37
Vortices, 166

Werner band, 265
Winds, 6, 7, 32, 80, 87, 113, 128, 168, 170, 171, 183, 187, 265, 280, 298, 319, 320
Wolf-Rayet stars, 80, 186

X rays, 19, 169, 171, 175, 297, 300, 308
X-wind, 176–178

Zeeman effect, 54

ASTRONOMY AND ASTROPHYSICS LIBRARY

Series Editors: G. Börner · A. Burkert · W. B. Burton · A. Coustenis
M. A. Dopita · B. Leibundgut · A. Maeder
P. Schneider · V. Trimble

The Stars By E. L. Schatzman and F. Praderie
Modern Astrometry 2nd Edition
By J. Kovalevsky
The Physics and Dynamics of Planetary Nebulae By G. A. Gurzadyan
Galaxies and Cosmology By F. Combes, P. Boissé, A. Mazure and A. Blanchard
Observational Astrophysics 2nd Edition
By P. Léna, F. Lebrun and F. Mignard
Physics of Planetary Rings Celestial Mechanics of Continuous Media
By A. M. Fridman and N. N. Gorkavyi
Tools of Radio Astronomy 4th Edition, Corr. 2nd printing
By K. Rohlfs and T. L. Wilson
Tools of Radio Astronomy Problems and Solutions 1st Edition, Corr. 2nd printing
By T. L. Wilson and S. Huttemeister
Astrophysical Formulae 3rd Edition
(2 volumes)
Volume I: Radiation, Gas Processes and High Energy Astrophysics
Volume II: Space, Time, Matter and Cosmology
By K. R. Lang
Galaxy Formation 2nd Edition
By M. S. Longair
Astrophysical Concepts 4th Edition
By M. Harwit
Astrometry of Fundamental Catalogues
The Evolution from Optical to Radio Reference Frames
By H. G. Walter and O. J. Sovers
Compact Stars. Nuclear Physics, Particle Physics and General Relativity 2nd Edition
By N. K. Glendenning
The Sun from Space By K. R. Lang
Stellar Physics (2 volumes)
Volume 1: Fundamental Concepts and Stellar Equilibrium
By G. S. Bisnovatyi-Kogan

Stellar Physics (2 volumes)
Volume 2: Stellar Evolution and Stability
By G. S. Bisnovatyi-Kogan
Theory of Orbits (2 volumes)
Volume 1: Integrable Systems and Non-perturbative Methods
Volume 2: Perturbative and Geometrical Methods
By D. Boccaletti and G. Pucacco
Black Hole Gravitohydromagnetics
By B. Punsly
Stellar Structure and Evolution
By R. Kippenhahn and A. Weigert
Gravitational Lenses By P. Schneider, J. Ehlers and E. E. Falco
Reflecting Telescope Optics (2 volumes)
Volume I: Basic Design Theory and its Historical Development. 2nd Edition
Volume II: Manufacture, Testing, Alignment, Modern Techniques
By R. N. Wilson
Interplanetary Dust
By E. Grün, B. Å. S. Gustafson, S. Dermott and H. Fechtig (Eds.)
The Universe in Gamma Rays
By V. Schönfelder
Astrophysics. A New Approach 2nd Edition
By W. Kundt
Cosmic Ray Astrophysics
By R. Schlickeiser
Astrophysics of the Diffuse Universe
By M. A. Dopita and R. S. Sutherland
The Sun An Introduction. 2nd Edition
By M. Stix
Order and Chaos in Dynamical Astronomy
By G. J. Contopoulos
Astronomical Image and Data Analysis
2nd Edition By J.-L. Starck and F. Murtagh
The Early Universe Facts and Fiction
4th Edition By G. Börner

# ASTRONOMY AND ASTROPHYSICS LIBRARY

Series Editors: G. Börner · A. Burkert · W. B. Burton · A. Coustenis
M. A. Dopita · B. Leibundgut · A. Maeder
P. Schneider · V. Trimble

The Design and Construction of Large Optical Telescopes By P.Y. Bely

The Solar System 4th Edition
By T. Encrenaz, J.-P. Bibring, M. Blanc, M. A. Barucci, F. Roques, Ph. Zarka

General Relativity, Astrophysics, and Cosmology By A. K. Raychaudhuri, S. Banerji, and A. Banerjee

Stellar Interiors Physical Principles, Structure, and Evolution 2nd Edition
By C. J. Hansen, S.D. Kawaler, and V. Trimble

Asymptotic Giant Branch Stars
By H. J. Habing and H. Olofsson

The Interstellar Medium
By J. Lequeux

Methods of Celestial Mechanics (2 volumes)
Volume I: Physical, Mathematical, and Numerical Principles
Volume II: Application to Planetary System, Geodynamics and Satellite Geodesy
By G. Beutler

Solar-Type Activity in Main-Sequence Stars
By R. E. Gershberg

Relativistic Astrophysics and Cosmology
A Primer By P. Hoyng

Magneto-Fluid Dynamics
Fundamentals and Case Studies
By P. Lorrain

Compact Objects in Astrophysics
White Dwarfs, Neutron Stars and Black Holes
By Max Camenzind

Special and General Relativity
With Applications to White Dwarfs, Neutron Stars and Black Holes
By Norman K. Glendenning

Stellar Physics
2: Stellar Evolution and Stability
By G. S. Bisnovatyi-Kogan

Principles of Stellar Interferometry
By A. Glindemann

Principles of Star Formation
By Peter H. Bodenheimer